CAMBRIDGE LIBRARY COLLECTION

Books of enduring scholarly value

Life Sciences

Until the nineteenth century, the various subjects now known as the life sciences were regarded either as arcane studies which had little impact on ordinary daily life, or as a genteel hobby for the leisured classes. The increasing academic rigour and systematisation brought to the study of botany, zoology and other disciplines, and their adoption in university curricula, are reflected in the books reissued in this series.

Arctic Zoology

In the 'Advertisement' to this 1784 two-volume work, Thomas Pennant (1726–98), zoologist and traveller, explains that his original intention was to record the zoology of North America 'when the empire of Great Britain was entire'. After the War of Independence, he changed his focus to the zoology (and people, archaeology and geology) of the Arctic regions of America, Europe and Siberia. The content of the volumes, one of the earliest works of systematic zoology published in Britain, is based on the writings of earlier zoologists, information obtained by Pennant from his scientific correspondents all over Europe and America, and his studies in private museums and collections. It is embellished with engravings of animals, birds, landscapes and artefacts. Volume 2 deals with land and water birds, including some, such as the passenger pigeon, which are now extinct. Other works by Thomas Pennant are also reissued in the Cambridge Library Collection.

Cambridge University Press has long been a pioneer in the reissuing of out-of-print titles from its own backlist, producing digital reprints of books that are still sought after by scholars and students but could not be reprinted economically using traditional technology. The Cambridge Library Collection extends this activity to a wider range of books which are still of importance to researchers and professionals, either for the source material they contain, or as landmarks in the history of their academic discipline.

Drawing from the world-renowned collections in the Cambridge University Library and other partner libraries, and guided by the advice of experts in each subject area, Cambridge University Press is using state-of-the-art scanning machines in its own Printing House to capture the content of each book selected for inclusion. The files are processed to give a consistently clear, crisp image, and the books finished to the high quality standard for which the Press is recognised around the world. The latest print-on-demand technology ensures that the books will remain available indefinitely, and that orders for single or multiple copies can quickly be supplied.

The Cambridge Library Collection brings back to life books of enduring scholarly value (including out-of-copyright works originally issued by other publishers) across a wide range of disciplines in the humanities and social sciences and in science and technology.

Arctic Zoology

VOLUME 2:
CLASS II. BIRDS

THOMAS PENNANT

CAMBRIDGE
UNIVERSITY PRESS

CAMBRIDGE
UNIVERSITY PRESS

University Printing House, Cambridge, CB2 8BS, United Kingdom

Cambridge University Press is part of the University of Cambridge.
It furthers the University's mission by disseminating knowledge in the pursuit of
education, learning and research at the highest international levels of excellence.

www.cambridge.org
Information on this title: www.cambridge.org/9781108073660

© in this compilation Cambridge University Press 2014

This edition first published 1785
This digitally printed version 2014

ISBN 978-1-108-07366-0 Paperback

This book reproduces the text of the original edition. The content and language reflect
the beliefs, practices and terminology of their time, and have not been updated.

ARCTIC ZOOLOGY.

V O L.   II.

C L A S S II.   B I R D S.

PIED DUCK, No 488.

L O N D O N.

PRINTED BY HENRY HUGHS.

M.DCC.LXXXV.

CLASS II. BIRDS.

CLASS II. BIRDS.

DIV. I. LAND BIRDS.
II. WATER BIRDS.

DIV. I. ORDER I. RAPACIOUS.

Genus.
I. **V**ULTURE.
II. FALCON.
III. OWL.

II. PIES.

IV. SHRIKE.
V. PARROT.
VI. CROW.
   * Roller.
VII. ORIOLE.
VIII. GRAKLE.
IX. CUCKOO.
   Wryneck.
X. WOODPECKER.
XI. KINGFISHER.
XII. NUTHATCH.
XIII. TODY.
   Hoopoe.

* The *Genera* which have not the number prefixed, are not found in *America.*

XIV. CREEPER.

Genus.

XIV. CREEPER.

XV. HONEY SUCKER.

III. GALLINACEOUS.

XVI. TURKEY.

XVII. GROUS.

XVIII. PARTRIDGE.

XIX. BUSTARD.

IV. COLUMBINE.

XX. PIGEON.

V. PASSERRINE.

XXI. STARE.

XXII. THRUSH.

XXIII. CHATTERER.

XXIV. GROSBEAK.

XXV. BUNTING.

XXVI. TANAGRE.

XXVII. FINCH.

XXVIII. FLYCATCHER.

XXIX. LARK.

Wagtail.

XXX. WARBLERS.

XXXI. TITMOUSE.

XXXII. SWALLOW.

XXXIII. GOATSUCKER.

DIV. II. WATER BIRDS,

VI. CLOVEN-FOOTED.

XXXIV. SPOONBILL.

XXXV. HERON.

XXXVI. IBIS.

Genus.

XXXVI. Ibis.

XXXVII. Curlew.

XXXVIII. Snipe.

XXXIX. Sandpiper.

XL. Plover.

XLI. Oyster-Catcher.

XLII. Rail.

XLIII. Gallinule.

VII. PINNATED FEET.

XLIV. Phalarope.

XLV. Coot.

XLVI. Grebe.

VIII. WEB-FOOTED.

XLVII. Avoset.

XLVIII. Flammant.

XLIX. Albatross.

L. Auk.

LI. Guillemot.

LII. Diver.

LIII. Skimmer.

LIV. Tern.

LV. Gull.

LVI. Petrel.

LVII. Merganser.

LVIII. Duck.

LIX. Pelecan.

CLASS

CLASS II. BIRDS.

DIV. I. LAND BIRDS.

ORDER I. RAPACIOUS.

I. VULTURE, *Gen. Birds* I.

Urubu, Aura Tzopilotl, *Mexic. Margrave*, 207, 208.—*Wil. Orn.* 68.—*Raii Syn.*   86. CARRION.
    Av. 180.

Carrion Crow, *Sloane Jam.* ii. 294.—*Brown Jam.* 471.

Corvus Sylvaticus, *Barrere*, 129.

Gallinazo, *Ulloa voy.* i. 60. 201.

Turkey Buzzard, *Joffelyn.*—*Lawfon*, 138.—*Catefby*, i. 6.—*Bancroft*, 152.—*Du*
    Pratz, ii. 77.

Vultur Aura, *Lin. Syft.* 122.—*De Buffon*, i. 175.—*Pl. Enl.* N° 187.

Le Vautour du Brafil, *Briffon*, i. 468.—*Latham*, i. 9. N° 5.—LEV. MUS.

WEIGHT four pounds and an half. Head fmall, covered with   DESCRIPTION.
a naked wrinkled red fkin, befet with black briftles. This
gives it fome refemblance to a Turkey; from which it derives one of
the names. The noftrils are very large, and pervious: the whole
plumage is dufky, dafhed with purple and green: legs of a dirty flefh-
color: claws black.

Thefe birds are common from *Nova Scotia* to *Terra del Fuego*; but   PLACE.
fwarm in the hotter parts of *America*; and are found in the iflands,
where they are faid to be far inferior in fize to thofe of *North Ame-*
rica.

In the warm climates they keep in vaft flocks. Perch at night on   MANNERS.
rocks or trees; fitting with difhevelled wings to purify their bodies,

2   which

which are moft offenfively fetid.   Towards morning they take flight, foaring at a vaft height, with the gentle motion of a kite; expecting notice of their banquet by the tainted effluvia of carrion, excrements, or any filth.   They have moft fagacious noftrils, and fmell their prey at a vaft diftance; to which they refort from all quarters, wheeling about, and making a gradual defcent till they reach the ground. They do not confine themfelves to dead animals, but feed on Snakes, and fometimes on Lambs.   They are very tame, and, while they are at their meals, will fuffer a very near approach.

In the torrid zone, particularly about *Carthagena*, they haunt inhabited places, and are feen in numbers fitting on the roofs of the houfes, or walking along the ftreets with a fluggifh pace.   In thofe parts they are ufeful, as the IBIS in *Egypt*, devouring the noifome fubjects, which would otherwife, by the intolerable ftench, render the climate ftill more unwholefome than it is.

When thefe birds find no food in the cities, they are driven by hunger among the cattle of the neighboring paftures.   If they fee a beaft with a fore on the back, they inftantly alight on it, and attack the part affected.   The poor animal attempts in vain to free itfelf from the devourers, rolling on the ground with hideous cries: but in vain; for the Vultures never quit hold, till they have effected its deftruction.   Sometimes an Eagle prefides at the banquet, and keeps thefe cowardly birds at a diftance, until it has finifhed its repaft.

USES.   Mifchievous as they are in a few inftances, yet, by the wife and beneficent difpenfations of Providence, they make in the hot climates full recompence, by leffening the number of thofe deftructive animals the Alligators, which would otherwife become intolerable by their multitudes.   During the feafon in which thefe reptiles lay their eggs in the fand, the Vultures will fit hid in the leaves of the trees, watching the coming of the female Alligator to depofit its eggs, who then covers them with fand, to fecure them, as fhe imagines, from all danger: but no fooner does fhe retire into the water,

than

than the birds dart on the fpot, and with claws, wings, and beak, tear away the fand, and devour the whole contents of the depofitory.

No birds of this genus are found in northern regions of *Europe* or *Afia*, at left in thofe latitudes which might give them a pretence of appearing here. I cannot find them in our quarter of the globe higher than the *Grifon Alps* *, or *Silefia* †; or at fartheft *Kalifh*, in *Great Poland* ‡. Certainly the Count *De Buffon* was misinformed as to the habitation of the fpecies, which he afcribes to *Norway* ‖. In the *Ruffian* dominions, the Bearded Vulture of Mr. *Edwards*, iii. tab. 106. breeds on the high rocks of the great *Altaic* chain, and beyond lake *Baikal* §; which may give it in *Europe* a latitude of 52. 20. in *Afia* of 55.

---

* *Wil. Orn.* 67.   † *Schwenckfeldt av. Silefia*, 375.   ‡ *Rzaczynfki, Hift. Nat. Polon.* 298.   ‖ *Hift D'Oif.* i. 164. *Pl. Enl.* 449.

§ Dr. PALLAS's *Catalogue of the Birds of the Ruffian empire*, which he favored me with in MS. my fureft clue to the *Arctic* birds.

II. FALCON,

II. FALCON, *Gen. Birds* II.

36. A. Sea Eagle.

Br. Zool. i. Nº 44.
Falco Offifragus, *Lin. Syft.* 124.—*Latham*, i. 30.—*Pl. Enl.* 12. 415.
Grey Eagle, *Lawfon*, 137.
Land Oern, *Leems*, 230.
L'Orfraie, *De Buffon*, i. 112. pl. 3.—Lev. Mus.

VARIES a little from the *Britifh* fpecies, and is much fuperior in fize. The length three feet three inches; of wing, twenty-five inches.

Feathers on head, neck, and back, brown, edged with dirty white: chin white: breaft and belly brown, fpotted with white: coverts of wings brown, clouded; primaries black: tail dufky; the middle mottled with white: legs feathered half down.

Place. Very common in the northern parts of *America*, and endures its fevereft winters, even as high as *Newfoundland*. Thefe birds prey on fea fowls, as well as land, and on young Seals, which they feize floating, and carry out of the water.

Eagles, and all forts of birds of prey, abound in *America*, where fuch quantity of game is found. Multitudes are always feen below the falls of *Niagara*, invited by the carcafes of Deer, Bears, and other animals, which are fo frequently hurried down in attempting to crofs the river above this ftupendous cataract.

This fpecies is very frequent in *Kamtfchatka*; and is found during fummer even on the Arctic coaft: is very common in *Ruffia* and *Sibiria*; nor is it more rare about the *Cafpian* fea, where they breed on the loftieft trees.

Br. Zool. i. N° 43.
Falco Fulvus, *Lin. Syſt.* 125.—*Latham*, i. 32, N° 6.
White-tailed Eagle, *Edw.* i. 1.—Lev. Mus.
L'Aigle commun, *De Buffon*, i. 86,—*Pl. Enl.* 409.—Lev. Mus.

THE whole plumage is of a duſky-brown: the breaſt marked with triangular ſpots of white; in which it varies from the *Britiſh* kind: the tail white, tipt with black; but in young birds duſky, blotched with white: legs covered to the toes with ſoft ruſt-colored feathers: vent feathers of the ſame color.

Inhabits *Hudſon's Bay*, and northern *Europe* as far as *Drontheim* *. Is found on the higheſt rocks of the *Uralian* chain, where it is not covered with wood †; but is moſt frequent on the *Sibirian*, where it makes its neſt on the loftieſt rocks. It is rather inferior in ſize to the former; but is a generous, ſpirited, and docile bird. The independent *Tartars* train it for the chace of Hares, Foxes, Antelopes, and even Wolves. The uſe is of conſiderable antiquity; for *Marco Polo*, the great traveller of 1269, obſerved and admired the diverſion of the great *Cham* of *Tartary*; who had ſeveral Eagles, which were applied to the ſame purpoſes as they are at preſent ‡. I muſt add, that the *Tartars* eſteem the feathers of the tail as the beſt they have for pluming their arrows.

The *Kalmucs* uſe, beſides this ſpecies of Eagle, that which the *French* call *Jean le Blanc* ‖, and alſo the *Lanner*; all which breed among them: but people of rank, who are curious in their Falcons, procure from the *Baſchkirians* the Gyrfalcon and the Peregrine, which inhabit the lofty mountains of the country §.

* Eſpecially in the winter, *Leems*, 233. † *Dr. Pallas.* ‡ *M. Polo, in Purchas*, iii. 85. in *Bergeron*. 74. ‖ *De Buffon.*
§ *Extracts*, iii. 303. A name by which I quote an abridgement of the travels of Pallas, Gmelin, Lepechin, and others, publiſhed by the Societe Typographique, at *Berne*, under the title of Histoire des Decouvertes, faites par divers ſavans voyageurs dans pluſieurs contreès de la *Ruſſie* et de la *Perſe*, 4 vols. 8vo.

F. With

F. With a dufky and blue bill; yellow cere: head, neck, and breaft, of a deep afh-color: each cheek marked with a broad black bar paffing from the corner of the mouth beyond the eyes: back, belly, wings, and tail, black: legs yellow, feathered below the knees.

Is about the fize of the laft. Communicated to me by the late *Taylor White*, Efq; who informed me that it came from *North America*. Is defcribed by Mr. *Latham*, i. 35, N° 10; and feems to be the fpecies engraven by *M. Robert*, among the birds in the menagery of *Louis* XIV.

Falco Leucocephalus, *Lin. Syft*. 124.
Bald Eagle, *Lawfon*, 137.—*Catefby*, i. 1*. *Brickell*, 173.—*Latham*, i. 29.—LEV. MUS.
Le Pygargue a tête blanche, *De Buffon*, i. 99.—*Pl. Enl*. 411.—LEV. MUS.

BILL, cere, and feet, pale yellow: head, neck, and tail, of a pure white: body and wings of a chocolate-color. It does not acquire its white head till the fecond year.

This Eagle is leffer than the foregoing fpecies, but of great fpirit: preys on Fawns, Pigs, Lambs, and fifh: is the terror of the Ofprey, whofe motions it watches. The moment the latter has feized a fifh, the former purfues till the Ofprey drops its prey; which, with amazing dexterity, it catches before it falls to the ground, be the diftance ever fo great. This is matter of great amufement to the inhabitants of *North America*, who often watch their aerial contefts. This fpecies frequently attends the fportfman, and fnatches up the game he has fhot, before he can reach it.

Thefe birds build in vaft decayed cypreffes †, or pines, impending over the fea, or fome great river, in company with Ofpreys, Herons, and other birds: and their nefts are fo numerous, as to refemble a rookery. The nefts are very large, and very fetid by reafon of

---

* Le Pygargue a tête blanche, *De Buffon*, i. 99. *Pl. Enl*. 411.   † *Catefby*.

the

the reliques of their prey. *Lawſon* ſays, they breed very often, laying again under their callow young; whoſe warmth hatches the eggs.
In *Bering*'s iſle they make their neſts on the cliffs, near ſix feet wide,
and one thick; and lay two eggs in the beginning of *July*.

THIS moſt beautiful and ſcarce-ſpecies is entirely white, except
the tips of the wings, which are black. We know nothing of
this bird, but what is collected from *Du Pratz**. The natives of
Louiſiana ſet a high value on the feathers, and give a large price for
thoſe of the wings; with them they adorn the Calumet, or pipe
of peace. Different nations make uſe of the wings, or feathers of
different birds; but, according to *Hennepin*, always decorate it with
the moſt beautiful.

The Calumet is an inſtrument of the firſt importance among the
Americans. It is nothing more than a pipe, whoſe bowl is generally
made of a ſoft red marble †: the tube of a very long reed, orna-
mented with the wings and feathers of birds. No affair of conſe-
quence is tranſacted without the Calumet. It ever appears in meet-
ings of commerce, or exchanges; in congreſſes for determining of
peace or war; and even in the very fury of a battle. The accept-
ance of the Calumet is a mark of concurrence with the terms pro-
poſed; as the refuſal is a certain mark of rejection. Even in the
rage of a conflict this pipe is ſometimes offered; and if accepted, the
weapons of deſtruction inſtantly drop from their hands, and a truce
enſues. It ſeems the ſacrament of the Savages; for no compact is
ever violated, which is confirmed by a whiff from this holy reed.
The *Dance of the Calumet* is a ſolemn rite which always confirms a
peace, or precedes a war. It is divided into three-parts: the firſt,
appears an act of devotion, danced in meaſured time: the ſecond,
is a true repreſentation of the Pyrrhic dance ‡: the third, is
attended with ſongs expreſſive of the victories they had obtained,
the nations they had conquered, and the captives they had made.

90 WHITE
EAGLE.

CALUMET.

* *Du Pratz*, ii. 75. *Latham*, i. 36. † *Du Pratz*, i. 298. *Kalm*, iii. 230.
‡ *Strabo*, lib. x. p. 736. edit. *Amſtel.* 1707.

From

From the winged ornaments of the *Calumet*, and its conciliating ufes, writers compare it to the *Caduceus* of *Mercury*, which was carried by the *Caduceatores*, or meffengers of peace, with terms to the hoftile ftates. It is fingular, that the moft remote nations, and the moft oppofite in their other cuftoms and manners, fhould in fome things have, as it were, a certain confent of thought. The *Greeks* and the *Americans* had the fame idea, in the invention of the *Caduceus* of the one, and the *Calumet* of the other. Some authors imagine, that among the *Greeks* the wings were meant as a fymbol of eloquence. I rather think that the twifted Serpents expreffed that infinuating faculty; and that the emblem was originally taken from the fatal effect the rhetoric of *Satan* had on our great mother, when he affumed the form of that reptile, which the higheft authority reprefents as *more fubtile than any beaft of the field*. On this the heathen mythology formed their tale of *Jupiter* taking the figure of a Serpent, to infinuate himfelf into the good graces of *Olympias*; who, like *Eve*, fell a victim to his perfuafive tongue. As to the wings, it is moft probable that they were to fhew the flight of difcord; which the reconciled parties gave, with all the horrors of war, to the air, and fport of the winds.

The *Oole*, or Eagle, is a facred bird among the *Americans*. In cafe of ficknefs, they invoke this bird to defcend from heaven (which in its exalted flight it approaches nearer than any other) and bring down refrefhing things; as it can dart down on its rapid wing quick as a flafh of lightning *.

* *Adair's Hift. Am. Indians*, 179

Fishing Hawk, *Catesby*, i. 2.—*Lawson*, 137.—*Brickell*, 173.
Ofprey, *Joffelyn's Rarities*, ii.—*Br. Zool.* i. N° 46.—*Latham*, i. 45.
Le Balbuzard, *De Buffon*, i. 103. pl. 2.
Falco Haliætus. Blafot. Fifk-orn, *Faun. Suec.* N° 63.
Fifk Gjoe, *Leems*, 234.—*Pl. Enl.* 414.—Lev. Mus.

91. Osprey.

F. With blue cere, and feet : head, and lower part of the body, white : upper part brown : two middle feathers of the tail plain brown ; the reft barred with white and brown.

Manners.

This, in all refpects, refembles the *European* kind. Notwithftanding it is fo perfecuted by the *Bald Eagle*, yet it always keeps near its haunts. It is a fpecies of vaft quicknefs of fight ; and will fee a fifh near the furface from a great diftance * : defcend with prodigious rapidity, and carry the prey with an exulting fcream high into the air. The Eagle hears the note, and inftantly attacks the Ofprey ; who drops the fifh, which the former catches before it can reach the ground, or water. It fometimes happens that the Ofprey perifhes in taking its prey ; for if it chances to fix its talons in an over-grown fifh, it is drawn under water before it can difengage itfelf, and is drowned.

Place.

It is very frequent in *Kamtfchatka* ; and in fummer, even under the *Arctic* zone of *Europe* and *Afia*. Is very common in *Sibiria*, and fpreads far north ; probably common to the north of *America*, and *Afia*. Is rare in *Ruffia*. It is likewife very frequent as low on the *Wolga* as the tract between *Syfran* and *Saratoff*, where they are faid to be the fupport of the *Ern Eagle*, as they are of the White-tailed Eagle in *America*, each living by the labors of the Ofprey. The *Tartars* have a fuperftition, that a wound from its claws is mortal, either to man or fifh, and confequently dread its attack †.

* That agreeable traveller, the reverend Dr. *Burnaby*, adds, that it is often feen refting on the wing for fome minutes, without any vifible change of place, before it defcends. *Travels in America*, 2d ed. p. 48.
† *Extracts*, i. 479.

Falco

92. ROUGH-
LEGGED.

Falco Lagopus *Brunnicb*, N° 15.—*Leems Lapm.* 236.
Rough-legged Falcon, *Br. Zool.* ii. *App.* 529.—*Latham*, i. 75.—LEV. MUS.

F. With a yellow cere, and feet: head, neck, and breaft, of a
  yellowifh white, marked with a few oblong brown fpots: belly
of a deep brown: thighs white, ftriped with brown: fcapulars
blotched with yellowifh white and brown: coverts of the wings
edged with ruft-color; primaries black: tail, little longer than the
wings; the part next to the rump white; the end marked with a
black bar; the tips white: legs feathered to the toes: feet yellow.

SIZE.   Length two feet two inches.

PLACE.   Inhabits *England*, *Norway*, *Lapmark*, and *North America*. Was fhot
in *Connecticut*.

93. ST. JOHN'S.   *Latham*, i. 77, N° 58.

F. With a fhort dufky bill: head of a deep brown: hind part of
  the neck, back, fcapulars, and coverts of the tail, marked with
bars of black, and dull white, pointing obliquely: coverts of the
wings deep brown; the greater fpotted on their inner fides with
white; the primaries dufky, the lower part white, barred with deep
afh-color and black: the under fide of the body brown, marked
fparingly with white and yellowifh fpots: tail fhorter than the extremity of the wings; the end white; beneath that is a bar of black,
fucceeded by two or three black and cinereous bands; the reft of
the tail marked with broad bars of white, and narrower of afh-color:
the legs are cloathed with feathers to the toes, which are yellow, and
very fhort.

SIZE.   Length, one foot nine inches.

PLACE.   Inhabits *Hudfon's Bay* and *Newfoundland*.   BL. MUS.

Latham,

M. Griffith del

P. Mazell Sculp.

St. John's Falcon N.º 93.   Chocolate Colored Falcon N.º 94.

Latham, i. 54. N° 34. A; 76. N° 57.

94. CHOCOLATE-
COLORED.

F. With a fhort and black bill, and yellow cere. The whole plumage of a deep bay or chocolate-color, in parts tinged with ferruginous: primaries black; the lower exterior fides of a pure white, forming a confpicuous fpot or fpeculum: the wings reach to the end of the tail: the exterior fides of the five outmoft feathers of the tail, dufky; their inner fides blotched with black and white; the two middle, black and cinereous: the legs and toes feathered; the laft remarkably fhort. LENGTH one foot ten inches.

Inhabits *Hudfon's Bay* and *Newfoundland.* Preys much on Ducks. Sits on a rock and watches their rifing, when it inftantly ftrikes at them.

PLACE.

Latham, i. 79. N° 60.

95. NEWFOUND-
LAND.

F. With a yellow cere: deep yellow irides: hind part of the head ferruginous: crown, back, fcapulars, and coverts of wings, brown, edged with a paler color: belly ruft-colored, blotched with deeper fhades: thighs of a mottled afh, marked with round dufky fpots, and on the lower parts with four large dark blotches: the tail croffed by four bars of deeper and lighter brown: legs yellow, ftrong, and feathered half way down. LENGTH twenty inches. The defcription borrowed from Mr. *Latham.*

Inhabits *Newfoundland.*

PLACE.

D d

Belon,

96. SACRE.   *Belon, Hift. des Oif.* 108.—*Buffon,* i. 246.
              Speckled Patridge Hawk of *Hudfon's Bay, Phil. Tranf.* lxii. 383.—*Latham,* i. 78.
                 Nᵒˢ 58, 59.

F. With a dufky bill; upper mandible toothed: irides yellow: cere and legs bluifh.   Head, and upper part of the body, of a dufky brown: hind part of the head mottled with white: whole under fide of the body, from chin to vent, white; the middle of each feather marked with a dufky fpot: wings reach almoft to the end of the tail: coverts, fcapulars, and primaries, of a deep brown, elegantly barred tranfverfely with white: tail brown, marked on each fide with oval tranfverfe fpots of red: feathers on the thighs very long, brown fpotted with white: the fore part of the legs covered with feathers almoft to the feet.   LENGTH two feet.   Weight two pounds and an half.

PLACE.   Inhabits *Hudfon's Bay* and *Newfoundland:* found alfo in *Tartary,* and is a fpecies celebrated there for the fport of falconry.   It is a hardy fpecies; for it never quits the rigorous climate of *Hudfon's Bay.*   Preys on the white Grous, which it will feize even while the fowler is driving them into his nets.   It breeds in *April* and *May,* in defert places.   The young fly in the middle of *June.*   The females are faid to lay only two eggs.

97. PEREGRINE.   *Br. Zool.* i. Nᵒ 48.—*Latham,* i. 68, Nᵒ 49; 73. Nᵒ 52.
                  Spotted Hawk, or Falcon; and Black Falcon, *Edw.* i. 3, 4.
                  Le Faucon, *De Buffon,* i. 249. pl. 16.—LEV. MUS.

F. With a fhort ftrong bill, toothed on the upper mandible, of a bluifh color: cere yellow: irides hazel: forehead whitifh: crown, and hind part of the head, dufky: the back, fcapulars, and coverts of wings, elegantly barred with deep blue and black: the

9   **primaries**

primaries dufky, with tranfverfe oval white fpots: the throat, chin, and breaft, of a pure white, the laft marked with a few dufky lines pointing down : the belly white, croffed with numerous dufky bars, pointed in the middle : legs yellow : toes very long.

The *American* fpecies is larger than the *European.* They are fubject to vary. The black Falcon, and the fpotted Falcon of Mr. *Edwards,* are of this kind; each preferve a fpecific mark, in the black ftroke which drops from beneath the eyes, down towards the neck. The differences in the marks in the tail may poffibly proceed from the different ages of the birds; for few kinds differ fo much in the feveral periods of life as the Rapacious.

Inhabits different parts of *North America,* from *Hudfon's Bay* as low as *Carolina.* In *Afia,* is found on the higheft parts of the *Uralian* and *Sibirian* chain. Wanders in fummer to the very Arctic circle. Is common in *Kamtfchatka.*

PLACE.

Gentil Falcon, *Br. Zool.* i. N° 50.
F. Gentilis. Falk. *Faun. Suec.* N° 58.—*Latham,* i. 64.—Lev. Mus.

98. GENTIL.

F. With a dufky bill : yellow cere, irides, and legs : head and upper fide of the neck ferruginous, ftreaked with black : under fide, from chin to tail, white, marked with dufky heart-fhaped fpots : back, coverts of wings, and fcapulars, brown, edged with ruft-color : primaries dufky, barred on the exterior fide with black : wings reach only half the length of the tail : tail long, barred with four or five broad bands of black cinereous; each of the firft bounded by a narrow line of dirty white.

In fize fuperior to the *European* kind, being two feet two inches long. Shot in the province of *New York.* Is found in northern *Europe,* as far as *Finmark* *.

SIZE.
PLACE.

* *Leems,* 337. *Strom.* 224.

D d 2

Br. Zool.

Br. Zool. i. Nº 52.

F. Palumbarius, *Faun. Suec.* Nº 67.—*De Buffon*, i. 230.—*Latham*, i. 58.—LEV. MUS.

F. With a bluiſh bill, black at the tip : yellowiſh green cere : yellow legs : head brown ; hind part mottled with white : over each eye extends a long whitiſh line : hind part of the neck, back, and wings, of a deep brown color : breaſt and belly white, croſſed with numerous undulated lines of brown : tail of a cinereous brown, croſſed by four or five bars of black : wings ſhorter than the tail.

That which I ſaw in the *Leverian Muſeum*, was ſuperior in ſize to
the *European*. Mr. *Lawſon* ſays, they abound in *Carolina :* are ſpirited birds, but leſſer than thoſe of *Muſcovy*. Is common in that
country, and *Sibiria*. Dr. *Pallas* adds, that there is a large white
variety on the *Uralian* mountains, mottled with brown and yellow.
Theſe are yet more frequent in the eaſt part of *Sibiria* ; and in
Kamtſchatka they are entirely white. Theſe are the beſt of all Hawks
for falconry. They extend to the river *Amur* ; and are uſed by the
emperor of *China* in his ſporting progreſſes *, attended by his grand
falconer, and a thouſand of the ſubordinate. Every bird has a ſilver
plate faſtened to its foot, with the name of the falconer who had the
charge of it ; that in caſe it ſhould be loſt, it might be brought to
the proper perſon : but if he could not be found, the bird is de-
livered to another officer, called the *Guardian of loſt birds* ; who
keeps it till it is demanded by the falconer to whom it belonged.
That this great officer may the more readily be found, among the
army of hunters, who attend the emperor, he erects a ſtandard in the
moſt conſpicuous place †.

The emperor often carries a Hawk on his hand, to let fly at any
game which preſents itſelf ; which are uſually Pheaſants, Partridges,
Quails, or Cranes. *Marco Polo* ſaw this diverſion about the year

* *Bell.* ii. 87.   † *Bergeron*, 75, 76.

1269 ;

1269 * ; a proof of its antiquity in thefe parts, when it formed fo
regular and princely an eftablifhment in the ftate of this great eaftern
monarch; the origin of which might have been in fome long preceding age. The cuftom of carrying a Falcon extended to many
countries, and was efteemed a diftinction of a man of rank. The
Welfh had a faying, that you may know a gentleman by his *Hawk,
Horfe, and Grehound.* In fact, a perfon of rank feldom went without
one on his hand. *Harold*, afterwards king of *England*, is painted
going on a moft important embaffy, with a Hawk on his hand, and
a Dog under his arm †. *Henry* VI. is reprefented at his nuptials, attended by a nobleman and his Falcon ‡. Even the ladies were not
without them, in earlier times; for in an antient fculpture in the
church of *Milton Abbas*, in *Dorfetfhire*, appears the confort of King
Athelftan with a Falcon on her royal fift § tearing a bird : and, perhaps
to indulge his queen in her paffion for the diverfion, he demanded
of my countrymen (befides an immenfe tribute) fome of their moft
excellent Hounds, and of their beft Hawks : which proves the high
efteem in which our Dogs and Falcons were held in thofe early
days ‖.

American Buzzard, *Latham*, i. 50.—LEV. MUS.   100. RED-TAILED.

F. With a dufky bill, and yellow cere : head, lower part of the
neck, and chin, brown, mixed with white : breaft and belly
white, varied with long ftripes of brown, pointing downwards : fe-
moral feathers very long, white, and marked with long dentated
ftripes of pale brown : upper part of the neck, and back, of a very
deep brown : coverts and tertials brown, barred or edged with white :
primaries dufky, barred with cinereous : tail of a pale ruft-color,
marked near the end with a dufky narrow bar : legs yellow. SIZE
of the *Gófhawk*.

Inhabits *North America*. Sent from *Carolina* to Sir *Afhton Lever*.   PLACE

* *Bergeron*, 75. 76.   † *Monumens de la Monarchie Françoife*, i. 372.   ‡ *Mr.
Walpole's Anecdotes of Painting*, i. 33.   § *Hutchins's Dorfetfhire*, ii. 443.
‖ *Malmfbury*, lib. ii. c. 6.

3   F. With

101. Leverian. **F.** With a dufky bill, greatly hooked: head ftriped with brown and white: upper part of the body and wings of a deep brown; each feather elegantly marked at the end with a large white fpot: the whole under fide of the body white: the outmoft feathers of the tail marked with nine white, and the fame number of dufky bars; middle feathers with dufky and cinereous: the wings extend beyond the end of the tail: legs ftrong and yellow.

Place.   Size of a Buzzard. Sent to Sir *Afhton Lever* from *Carolina*.

102. Red-shouldered.   Barred-breafted F. *Latham*, i. 56, Nº 26.—Lev. Mus.

F. With a flender dufky bill; yellow cere; and legs, head, and neck, of a yellowifh white, ftreaked downwards with dufky lines: back of a deep brown, edged with ruft-color: leffer coverts of wings ferruginous, fpotted with black; primaries and fecondaries black, fpotted on each fide moft diftinctly with white: breaft and belly of a light tawny; the firft ftreaked downwards with black; the laft traverfed with deeper tawny: tail fhort and dufky, croffed by feven narrow bands of white; the two neareft to the ends more remote than the others: legs weak.   Length twenty-two inches.

Place.   Inhabits *Long Ifland*.   This is a new fpecies, preferved in Mrs. Blackburne's *Mufeum*.

Afh-

Aſh-colored Buzzard, *Edw*. ii. 53.—*Latham*, i. 55, N° 35. 48; N° 28.—*De Buffon*, i. 223.

Falco Buteo. Quidfogel, *Faun. Suec.* N° 60.—*Br. Zool.* i. 54.—LEV. Mus.

F. With a duſky bill, and bluiſh yellow cere: head, and hind part of the neck, of a cinereous brown, ſtreaked with yellow: back brown; lower part and rump barred with ruſt-color, ſometimes with white: the coverts of the wings brown; the greater and ſcapulars ſpotted with white; the three firſt quil-feathers black, white at their baſes; the interior webs of the reſt blotched with black and white: the throat and breaſt yellowiſh, marked thinly with oblong brown ſpots: belly white, varied with great ſpots of brown: feathers of the thighs long, white, croſſed with ſagittal bars of yellow: tail marked with about nine bands of black and light cinereous; the tip white: legs ſhort, ſtrong, and yellow. LENGTH two feet two inches.

The *American* varies in ſize, and ſometimes ſlightly in color; but in both has ſo much the habit of the *Engliſh* Buzzard, as not to merit ſeparation. It is called in *New York*, the great Hen Hawk, from its feeding on poultry. It continues there the whole year. Lays in *May* five eggs: the young fly about the middle of *June*. It is alſo an inhabitant of *Hudſon's Bay* and *Newfoundland*; and in *Europe* as high as *Sondmor*, in *Norway*; where, from its attacking the Eagle, it is called *Orne-Falk*. Migrates, before winter, from *Sweden*. Is ſcarce in *Ruſſia*; and very few are found in *Sibiria*. Is found in winter as low as *Woroneſch* *.

PLACE.

* In *Ruſſia*, *lat.* 52 north.

F. With

104. PLAIN.

F. With the bill black: head dusky: nape spotted with white: back, and coverts of wings, and tail, of an uniform deep brown: under side of the neck, breast, and belly and thighs, deep brown, slightly spotted with white: primaries dusky: inner webs marked with great oval spots of white, mottled with brown: middle feathers of the tail plain brown; inner webs of the rest mottled with white; exterior webs and ends slightly edged with the same: legs strong: yellow? Wing reaches near the length of the tail. LENGTH, from bill to tail, two foot one.

PLACE.   Inhabits *Hudson's Bay*

105. MARSH.

Marsh Hawk, *Edw.* iv. 291.—*Latham*, i. 90.—LEV. MUS.—BL. MUS.

F. With a bluish bill; orange cere, orbits, and legs: irides hazel: a black line extends from the corner of the bill beyond the eyes; above that is another of white, which encircles the cheeks, and meets in front of the neck: head, throat, and upper part of the breast, varied with black and rust-color: back, and coverts of the wings, brown: rump white: breast and belly, and thighs, of a bright ferruginous: tail dusky brown, crossed by four black bands: legs strong, thick, and short; which are specific distinctions from the next. LENGTH two feet.

PLACE.   Inhabits *Pensylvania*: frequents, during the summer, marshy places; where it feeds on the small birds, Frogs, Snakes, and Lizards. At approach of winter quits the country.

Br. Zool. i. N° 59.—Edw. iii. 107.—Latham, i. N°ˡˢ 75, 75 A, and N° 34, is a   106. RINGTAIL.
    ruſt-colored variety.
Falco Pygargus, F. Hudſonius, Lin. Syſt. 128.—Muller, N° 72.—BL. MUS.

F.   With a duſky bill and yellow cere : a white line over each eye :
    head, upper part of the neck, and back, duſky brown : coverts
and primaries of the ſame color ; the inner ſides of the laſt white :
breaſt, belly, and thighs, whitiſh, marked with ferruginous ſpots :
vent and rump white, encircling the root of the tail : the middle
feathers of the tail duſky ; the next of a bluiſh aſh-color ; the outmoſt white, all marked tranſverſely with orange bars : legs long, and
very ſlender.

This ſpecies is ſuperior in ſize to the Britiſh Ringtail ; but having   SIZE.
moſt of the characters of that bird, we doubt not but that it is the
ſame.   Like the European kind, ſkims along the ground in ſearch of
prey, which is Mice, Frogs, and ſmall birds.   Builds its neſt indifferently on the ground, or on the lower parts of trees. It is ſubject
to vary to a deep ruſt-color ; plain, except on the rump and tail.

Inhabits Hudſon's Bay. Weight, in Hudſon's Bay, ſeventeen ounces SIZE AND PLACE.
and a half.   Length twenty-one inches. Extent three feet ſeven.   Is
very common in the open and temperate parts of Ruſſia and Sibiria ;
and extends as far as lake Baikal*. It is not found far in the north of
Europe.   Linnæus omits it among the birds of his country ; but Mr.
Brunnick deſcribes one, which had been ſhot in lat. 58, on the little
iſland of Chriſtianſoe †.

F.   With a black bill ; yellow cere : head of a deep brown : back   107. WINTER.
    the ſame, tinged with ruſt : hind part of the neck ſtreaked with
white : the coverts of the wings duſky, edged with dull white ; thoſe
on the ridge with orange ; ends of the primaries duſky ; the other
parts barred with brown and white : breaſt and belly white, marked
with heart-ſhaped ſpots : thighs ſulphur-colored, ſpeckled with

* DR. PALLAS.   † In the Baltic, a little north-eaſt of Bornholm.

E e   duſky :

dufky: vent feathers white: tip of the tail white; then fucceeds a broad dufky bar; the remaining part barred with brown, tawny, and black: legs long, and very flender.

SIZE.

Is of an elegant form, and about the fize of the RINGTAIL.

PLACE.

Inhabits the province of *New York*: appears at approach of winter, and·retires in the fpring.   BL. MUS.

Mr. *Latham*'s northern Falcon, N° 62, feems to differ from this only in age, or fex.

---

108. SWALLOW-TAILED.

Hirundo maxima Peruviana, avis prædatoris calcaribus inftructa, *Feuillee voy.* Peru, tom. ii. 33.

Herring, or Swallow-tailed Hawk, *Lawfon*, 138.—*Brickell*, 175.—*Catefby*, i. 4.

Le Melan de la Caroline, *Briffon*, i. 418.—*De Buffon*, i. 221.

Falco Furcatus, *Lin. Syft.* 129.—*Latham*, i. 60.—LEV. MUS.

F. With a black bill, lefs hooked than ufual with rapacious birds; bafe of the bill hid in feathers, and briftly: the eyes large; irides red: head, neck, breaft, and belly, of a fnowy whitenefs: back, coverts of wings, and fcapulars, black, gloffed with purple and green: inner webs of the primaries and fecondaries white towards their bafe; the tertials white: tail of the fame color with the back; and moft extremely forked; the outmoft feather above eight inches longer than the middlemoft: the legs yellow.

PLACE.

This moft elegant fpecies inhabits only the fouthern parts of *North America*; and that only during fummer. Like Swallows, they feed chiefly flying; for they are much on wing, and prey on various forts of infects. They alfo feed on Lizards and Serpents; and will kill the largeft of the regions it frequents with the utmoft eafe. They quit *North America* before winter. We are not acquainted with their retreat. It probably is in *Peru*: at left we have the proof of one being taken in the South-fea, off the coaft which lies between *Yla* and *Arica*, in about the latitude 23 fouth, on *September* 11th, by the reverend the Father *Louis Feuillee* *.

Journal des Obferv. &c. vol. ii. 33.

F. With

Swallow-tailed Falcon N.º 108.

P. Mazell Sculp.

F. With dufky bill: head, cheeks, neck, breaft, and belly, white, **109. BUZZARDET.**
marked with large brown fpots, more fparingly difperfed over
the breaft and belly: leffer coverts brown; the others colored like
the head: primaries dufky: thighs white, with fmall fagittal fpots
of brown: tail dufky, barred and tipt with white: legs yellow.
LENGTH fifteen inches. It has much the habit of the Buzzard; but
the legs in proportion are rather longer.

In the LEVERIAN Mufeum. Except in the almoft uniform color of
the tail, Mr. *Latham*'s fpecies, p. 97, N° 83, agrees with this.

Little Hawk, *Catefby*, i. 5.—*Latham*, i. 110, N° 94. **110. LITTLE.**
Emerillon de Cayenne, *Buffon*, i. 291.—*Pl. Enl.* N° 444.
Falco Sparverius, *Lin. Syft.* 128.—LEV. MUS.—BL. MUS.

M A L E.

F. With bluifh bill, and yellow cere: crown of fine light grey,
with a red fpot in the middle; on the hind part a femicircle
formed of round black fpots: cheeks white, bounded on each fide
with a large black fpot: throat white: breaft of a pale yellow,
fpotted with black: back of a brilliant bay, croffed by broad black
bars. coverts of the wings of a beautiful grey, thinly fpotted with
black; primaries black, fpotted on their inner webs with white: tail
long; the middle feathers barred near the end with a black band,
and tipt with white; the two exterior feathers white, croffed with
three or four black bars: legs yellow

Length eleven inches and a half. Weight only three ounces and **SIZE.**
an half. This varies in color from the female, in the fame manner
as the *European* Keftrils.

Thefe birds inhabit *America*, from *Nova Scotia* to the *Antilles*; are **PLACE.**
active and fpirited. They prey on fmall birds, Mice, Lizards, and
infects. The FEMALE is the following.

Emerillon de St. Domingue, *De Buffon*, i. 291.—*Pl. Enl.* N° 465.—*Latham*, i. 111,. N° 95.—Lɛv. Mus.—ʙʟ. Mus.

F. With a fhort and very crooked bill: crown of a deep flaty blue, obfcurely fpotted with red: hind part of the neck, back, and tail, of a bright ferruginous color and black, elegantly difpofed in narrow tranfverfe bars: coverts of the wings of the fame colors; primaries black: under fide of the neck, breaft, and belly, of a dirty white, marked with large ferruginous fpots: thighs and vent feathers white: legs long, flender, and orange-colored: tail long, croffed with eleven black, and the fame number of bright ferruginous bars.

The *New York* Merlin of Mr. *Latham*, i. 107, N° 94, bears fo great a refemblance to this, that I do not venture to feparate them.

ɪɪɪ. Pɪɢɛon.   Pigeon Hawk, *Catefby*, i. 3.—*Phil. Tranf.* lxii. 382.—*Latham*, i. 101. Falco Columbarius, *Lin. Syfl.* 128.—Lɛv Mus.—Bʟ. Mus.

F. With a dufky bill, and yellow cere: crown, back, and coverts. of the wings and rump, of a bluifh grey, with the middle of each feather ftreaked with black: the hind part of the head fpotted with reddifh white: cheeks and under fide of the body white, with large oblong fpots of black: primaries and fecondaries dufky; their infides marked with great oval fpots of white: tail long; black tipt with white, and croffed with four bars of bluifh grey: legs yellow.

Sɪzɛ.   Its length is from ten to twelve inches. The weight fix ounces.

Pʟᴀᴄɛ.   It inhabits *America*, from *Hudfon's Bay* as low as *South Carolina*. In the laft it attains to a larger fize. In *Hudfon's Bay* it appears in *May* on the banks of *Severn* river, breeds, and retires fouth in autumn. It feeds on fmall birds; and on the approach of any perfon,

flies.

flies in circles, and makes a great ſhrieking. It forms its neſt in a rock, or ſome hollow tree, with ſticks and graſs; and lines it with feathers: and lays from two to four eggs, white, ſpotted with red. In *Carolina* it preys on Pigeons, and young of the wild Turkies.

F. With a duſky bill: yellow cere and irides: head duſky, ſtreaked with ruſt-color: back and coverts of wings brown, edged with ruſt; the primaries duſky aſh-color, barred with black, and the inner webs marked tranſverſely with oval ferruginous ſpots: tail long, of a deep cinereous, with four broad bars of black: breaſt and belly dirty white, marked with oblong ſtreaks of brown: legs yellow.   *112. DUBIOUS.*

Length about ten inches. Weight ſix ounces. In the marks and colors of the tail it much reſembles the Sparrow Hawk: in the ſpots on the breaſt it agrees with the *Engliſh* Merlin.   SIZE.

Inhabits *New York* and *Carolina*. I have my doubts whether this is any more than a variety of the preceding, eſpecially as the *Engliſh* SPARROW HAWK varies with the ſame colors.   PLACE.

F. With a bluiſh bill; upper mandible armed with a ſharp proceſs; yellow cere: head, back, and coverts of the wings, and tail, a duſky brown, ſlightly edged with ferruginous: hind part of the neck ſpotted with white: primaries duſky; inner webs marked with oval ſpots of a pale ruſt-color: tail *ſhort*, tipped with white, and barred with four broad duſky ſtrokes, and the ſame number of narrow ones of white: the hind part of the head ſpotted with white: from the chin to the tail whitiſh, ſtreaked downwards with diſtinct lines of black: legs deep yellow.   *113. DUSKY.*

Inferior in ſize to the laſt. Inhabits the province of *New York*. SIZE, AND PLACE. BL. MUS.

GOLDEN

A. GOLDEN EAGLE, *Br. Zool.* i. N° 42.
Orn. *Faun. Suec.* N° 54.—LEV: MUS.

F. With a bluish bill: plumage dusky and rust-color: tail dusky brown, blotched at the base with ash-color: legs feathered to the toes. Weight about twelve pounds.

PLACE. Inhabits *Sweden*; perhaps *Norway*. Found about the southern part of the *Urallian* mountains, and the mountains which border *Sibiria* on the south. Grows scarcer towards the east.

B. CINEREOUS EAGLE, *Br. Zool.* i. N° 45.—*Latham*, i. 33.
Vultur Albiulla, *Lin. Syst.* 123.

F. With pale yellow bill, irides, cere, and feet: plumage light cinereous: body and coverts of the wings clouded with darker: primaries dusky: tail white.

SIZE, AND PLACE. In size equal to the Black Eagle. Inhabits *Europe*, as high as *Iceland* and *Lapmark* *. Is common in *Greenland*; but does not extend to *America*: at least, if it does, it varies into the White-headed Eagle, to which it has great affinity, in particular in its feeding much on fish: the *Danes* therefore call it *Fiske-orn* †. Is common in the south of *Russia*, and about the *Volga*, as far as trees will grow. Is very scarce in *Sibiria*; but has been observed in the eastern parts about *Nertschink*. It seems to be the species called by the *Tungusi*, *Elo*;

* *Leems,* 331.　　† *Brunnick,* N° 12.

I

which

which breeds on the banks of the *Kharioufowa*, a river which falls into the *Penfhina* fea *.

It inhabits *Greenland* the whole year, fitting on the rocks with flagging wing, and flies flowly. It makes its neft on the lofty cliffs, with twigs, lining the middle with moffes and feathers. Lays two eggs. Sits in the latter end of *May*, or beginning of *June*.

Thefe birds prey on young Seals, which they feize as they are floating on the water ; but oft-times, by fixing their talons in an old one, they are overmatched, and drawn down to the bottom, fcreaming horribly. They feed alfo on fifh, efpecially the Lumpfifh, and a fort of Trout †; on Ptarmigans, Auks, and Eider Ducks. They fit on the top of rocks, attentive to the motion of the diving birds ; and, with quick eyes, obferve their courfe by the bubbles which rife to the furface of the water, and catch the fowls as they rife for breath.

The *Greenlanders* ufe their fkins for cloathing, next to their bodies. They eat the flefh, and keep the bill and feet for amulets. They kill them with the bow, or take them in nets, placed in the fnow, properly baited ; or tempt them by the fat of Seals, which the Eagles eat to an excefs ; which occafions fuch a torpidity as to make them an eafy prey.

C. CRYING EAGLE, Planga et Clanga, *Ariftot. Hift. An.* lib. ix.
  Morphnos, Clanga, Anataria, *Wil. Orn.* 63.—*Raii Syn. av.* 7, N° 7.
  Spotted Eagle, *Latham*, i. 38.
  Le Petit Aigle, *De Buffon*, i. 91.—BR. MUS.

F. With a dufky bill and yellow cere : color of the plumage a fer ruginous brown ; the coverts of the wings, and fcapulars, elegantly varied with oval white fpots; on the greater coverts very large : primaries dufky; the ends of the greater white : breaft and belly of a deeper color than the reft of the plumage, ftreaked downwards with dull yellow : tail dark brown, tipt with dirty white : legs feathered to the feet, which are yellow. LENGTH two feet.

* *Hift. Kamtfchatka,* 501.   † Salmo Carpio, *Faun. Groenl.* 170, N° 124.

Is

Is found in many parts of *Europe*, but not in *Scandinavia :* is frequent in *Ruſſia* and *Sibiria*, and extends even to *Kamtſchatka*. Is leſs generous and ſpirited than other Eagles ; and is perpetually making a plaintive noiſe, from which it was ſtyled by the antients *Planga & Clanga* ; and *Anataria*, from its preying on Ducks, which *Pliny* * deſcribes with great elegance. The *Arabs* uſed to train it for the chace ; but its quarry was Cranes, and other birds : the more generous Eagle being flown at Antelopes, and various quadrupeds. This ſpecies was even itſelf an object of diverſion ; and made the game of even ſo ſmall a Falcon as the Sparrow Hawk : which would purſue it with great eagerneſs, ſoar above, then fall on the Eagle, and, faſtening with its talons, keep beating it about the head with its wings, till they both fell together to the ground. This Sir *John Chardin* has ſeen practiſed about *Tauris*.

D. ICELAND FALCON, *Gent. Mag.* 1771, p. 297, fig. good.
Falco Iſlandus Fuſcus, *Brunnick*, 2, Nº 9.
Le Gerfault d' Iſland, *Briſſon*, i. 373, tab. xxxi.—*Pl. Enl.* 210.
Falco Gyrfalco, *Lin. Syſt.* 130 —*Faun. Suec.* Nº 64.—*Latham*, i. 82, Nº 68 ; and 71, Nº 50 B. parag. 2d.—LEV. MUS.

F. With a ſtrong bill, much hooked, and the upper mandible ſharply angulated on the lower edges ; cere bluiſh : head of a very pale ruſt-color, ſtreaked downwards with duſky lines : neck, breaſt, and belly, white, marked with cordated ſpots : thighs white, croſſed with ſhort bars of deep brown : back and coverts of wings duſky, ſpotted and edged with white : the exterior webs of the primaries duſky, mottled with reddiſh white ; the inner barred with white : the feathers of the tail croſſed with fourteen or more narrow bars of duſky and white ; the duſky bars regularly oppoſing thoſe of white : the wings, when cloſed, reach almoſt to the end of the train : legs ſtrong and yellow. The LENGTH of the wing, from the pinion to the tip, ſixteen inches.

* Lib. x. c. 3.

This

This species is an inhabitant of *Iceland*, is the most esteemed of any for the sport of falconry, and is, with the two following, reserved for the kings of *Denmark*; who sends his falconer, with two attendants, annually into the island to purchase them. They are caught by the natives; a certain number of whom in every district are licensed for that purpose. They bring all they take, about *Midsummer*, to *Bessested*, to meet the royal falconer; and each brings ten or twelve, capped, and perched on a cross pole, which they carry on horseback, and rest on the stirrup. The falconer examines the birds, rejects those which are not for his purpose, and gives the seller a written certificate of the qualities of each, which entitles him to receive from the king's receiver-general seventeen rixdollars for F, or the purest white Falcon; ten for E, or those which are left white; and seven for this species *. This brings into the island between two and three thousand rixdollars annually †.

They are taken in the following manner:—Two posts are fastened in the ground, not remote from their haunts. To one is tied a Ptarmigan, a Pigeon, a Cock or Hen, fastened to a cord that it may have means of fluttering, and so attract the attention of the Falcon. On the other post is placed a net, distended on a hoop, about six feet in diameter. Through this post is introduced a string, above a hundred yards long, which is fastened to the net, in order to pull it down; and another is fastened to the upper part of the hoop, and goes through the post to which the bait is tied. As soon as the Falcon sees the fowl flutter on the ground, he takes a few circles in the air, to see if there is any danger, then darts on its prey with such violence as to strike off the head, as nicely as if it was done with a razor. He then usually rises again, and takes another circle, to explore the place a second time: after which it makes another stoop; when, at the instant of its descending, the man pulls the dead bird under the net; and, by means of the other cord, covers the Falcon with the net, at the moment it has seized the prey; the person lying

MANNER OF TAKING.

* *Brunnick*, p. 2.   † *Olaffen*, i. 32.

F f

concealed

concealed behind fome ftones, or elfe lies flat on his belly, to elude the fight of the Falcon *.

As foon as one is caught, it is taken gently out of the net, for fear of breaking any of the feathers of the wings or tail; and a cap is placed over its eyes. If any of the tail-feathers are injured, the falconers have the art of grafting others †; which fometimes has occafioned a needlefs multiplication of fpecies.

The *Iceland* Falcons are in the higheft efteem. They will laft ten or twelve years; whereas thofe of *Norway*, and other countries, feldom are fit for fport after two or three years ufe. Yet the *Norwegian* Hawks were in old times in great repute in this kingdom, and even thought bribes worthy of a king. *Geoffry Le Pierre*, chief jufticiary, gave two good *Norway* Hawks to King *John*, that *Walter Le Madina* might have leave to export a hundred weight of cheefe. *John*, the fon of *Ordgar*, gave a *Norway* Hawk to have the king's requeft to the king of *Norway*, to let him have his brother's chattels; and *Ralf Havoc* fined to King *Stephen* in two Girfals (Gyrfalcons) and two *Norway* Hawks, that he might have the fame acquittance that his father had ‡.

ANTIQUITY OF FALCONRY. I cannot fix the precife time of the origin of falconry; the paffage in *Ariftotle*, and the epigram in *Martial*, do by no means fix it to the periods in which they wrote. The philofopher ‖ informs us, that " there was a diftrict in *Thrace*, in which the boys ufed to
" affemble at a certain time of the year, for the fake of bird-catch-
" ing. That the fpot was much frequented by Hawks, which
" were wont to appear on hearing themfelves called: and would
" drive the little birds into the bufhes, where they were caught
" by the children; and that the Hawks would even fometimes take
" the birds and fling them to thefe young fportfmen; who (after
" finifhing their diverfion) gratefully beftowed on their affiftants
" part of their prey." This tale may have fome truth at the bottom;

---

* *Horrebow*, 59, 60.　　　† *Brunnick*, p. 3.　*Horrebow*, 58.　　　‡ *Madox*,
Antiq. Exch. 469. 497.　　‖ *Arift. de Mirabil. Aufcult.*

it being notorious that Larks, and even Partridges, will, by the terror of a Hawk paffing over them, lie fo ftill as to fuffer themfelves to be taken by any paffenger. Here feems to have been no training of thefe *Thracian* Hawks, but a mere cafual concurrence of Hawks and fmall birds, which afforded now and then an amufement to the youth of the country. The thought expreffed on the antient gem, of little *Genii* engaged in the chace of Deer, affifted by an Eagle, may have originated from this ftory.

The Poet only defcribes another kind of bird-catching, in the following epigram on the fate of a Hawk:

> Prædo fuit volucrum, famulus nunc Aucupis, idem
> Decipit, et captas non fibi, mœret, aves †.

By the word *decipit*, it is plain that the Hawk was not trained, but was merely ufed as a ftale, either to entice fmall birds under a net, or to the limed twigs: the laft is a method ftill in ufe in *Italy*. The *Italians* call it *Uccellare con la Civetta*; for inftead of a Hawk, they place a fmall fpecies of Owl on a pole, in the middle of a field; and furround it, at various diftances, with lime-twigs. The fmall birds, from their ftrange propenfity to approach rapacious fowls, fly around, perch on the rods, and are taken in great numbers ‡. A Hawk would ferve the purpofe full as well. *Pliny* mentions the ufe of bird-lime ‖; and *Longus*, in his elegant romance of *Daphnis* and *Chloe*, employs the latter to catch little birds for his beloved §.

I cannot find any certainty of Hawks being trained for diverfion before the time of King *Ethelbert*, the *Saxon* monarch; who died in the year 760 ¶. He wrote into *Germany* for a brace of Falcons, which would fly at Cranes and bring them to the ground *, as there were very few fuch in *Kent*. This fhews how erroneous the opinion was, of thofe who place it in the reign of the emperor *Frederic Bar-*

---

† Lib. xiv. ep. 216.   ‡ *Olina*, 65.   ‖ *Hift. Nat.* lib. xvi. c. 44.
§ *Fr.* ed. octavo, 82.   ¶ *Saxon Chr.* 60.
* Quoted by Mr. *Whitaker* in *Hift. Manchefter*, from *Max. Bibliotheca Patrum*, xiii. p. 85. ep. 40.

baroffa,

baroſſa *, who was drowned in 1189. By the application of *Ethelbert* to *Boniface*, archbiſhop of *Mentz*, for the brace of Falcons, it is evi-dent, that the diverſion was in perfection in *Germany* before the year 752, the time in which that prelate was martyred by the Pagans. It ſeems to me highly probable, that falconry was invented in *Scythia*, and paſſed from thence into the northern parts of *Europe*. *Tartary* is even at preſent celebrated for its fine breed of Falcons ; and the ſport is in ſuch general eſteem, that, according to *Olearius, there was no but but what had its Eagle or Falcon* †. The boundleſs plains of that country are as finely adapted to the diverſion, as the wooded or mountanous nature of moſt part of *Europe* is ill calculated for that rapid amuſement.

The antiquity of falconry in *Tartary* is evinced by the exhibition of the ſport on the very antient tombs ‡ found in that country ; in which are figured horſemen at full ſpeed, with Hawks on their hands : others again, in the ſame attitude, diſcharging their arrows at their game, in the very manner of the antient *Scythians*.

From *Germany*, falconry got footing in *England* ; and became ſo favored a diverſion, that even ſanguinary laws were enacted for the preſervation of rapacious fowls. *Edward* III. made it death for the ſtealing of a Hawk : and to take its eggs, even in a perſon's own ground, was puniſhable with a fine at the king's pleaſure, and im-priſonment for a year and a day. In the reign of *James* I. the amuſe-ment was carried to ſuch an extravagant pitch, that Sir *Thomas Mon-ſon* is ſaid to have given a thouſand pounds for a caſt of Hawks.

E. **DUSKY.** Falco Fuſcus; *Faun. Groen.* 56, N° 34. b.
　　　Grey Falcon, *Crantz,* i. 78.—*Egede,* 64.

F. With duſky irides : lead-colored cere and feet : brown crown, marked with irregular oblong white ſpots : forehead whitiſh : cheeks blackiſh : hind part of the head and throat white : breaſt and

　• *Spelman's Gloſſ.*　　† *Olearius's travels,* 177.　　‡ *Strahlenberg,* tab. A. B.

belly

belly of a yellowifh white, ftriped downwards with dufky ftreaks : the back dufky, tinged with blue, the ends of the feathers lighteft, and fprinkled over with a few white fpots, efpecially towards the rump : wings of the fame colors, variegated beneath with white and black : the upper part of the tail dufky, croffed very faintly with paler bars ; the under fide whitifh.

SIZE.

Leffer than the Collared Falcon.

PLACE.

Inhabits all parts of *Greenland*, from the remoteft hills to thofe which impend over the fea. They are even feen on the iflands of ice remote from fhore. They retire in the breeding-feafon to the fartheft part of the country, and return in autumn with their young. They breed in the fame manner as the Cinereous Eagle, but in more diftant places; and lay from three to five eggs. The tail of the young is black, with great brown fpots on the exterior webs.

They prey on Ptarmigans, Auks, and all the fmall birds of the country : have frequent difputes with the Raven, but feldom come off victors ; for the Raven will, on being attacked, fling itfelf on its back ; and, either by defending itfelf with its claws, or by calling, with its croaking, numbers of others to its help, oblige the Falcon to retire. The *Greenlanders* ufe the fkin, among many others, for their inner garments ; the wings for brufhes; the feet for amulets : but feldom eat the flefh, unlefs compelled by hunger.

It is alfo a native of *Iceland*.

F. GYRFALCON, *Br. Zool.* N° 47, tab. xix.—*Latham*, i. 71, N° 50 A, and N° 50 B, 1ft paragr. and 83, N° 69.
Falco Iflandus, *Faun. Groenl.* 58, N° 35.—*Brunnick*, Nrls 7, 8.—*Crantz*, i. 78.— *Egede*, 64.—*Horrebow*, 58.—LEV. MUS.

F. With a yellow cere : bluifh bill, greatly hooked : eye dark blue : the throat of a pure white : the whole body, wings, and tail, of the fame color, moft elegantly marked with dufky bars, lines, or fpots,

leaving

leaving the white the far prevaling color. There are inſtances, but rare, of its being found entirely white. In ſome, the whole tail is croſſed by remote bars of black or brown; in others, they appear only very faintly on the middle feathers : the feathers of the thighs are very long, and unſpotted : the legs ſtrong, and of a light blue.

SIZE.
PLACE.

Its weight forty-five ounces Troy : length near two feet : extent four feet two. Of the ſame manners and haunts with the former. Is very frequent in *Iceland*; is found in *Lapmark* *, and *Norway* †; and rarely in the *Orknies*, and *North Britain*. In *Aſia*, it dwells in the higheſt points of the *Urallian* and other *Sibirian* mountains, and dares the coldeſt climates throughout the year. It is kept, in the latitude of *Peterſburg*, uninjured in the open air during the ſevereſt winters, when the *Peregrine Falcon*, N° 97, loſes its claws by the froſt.

VERY HARDY.

Mr. *Hutchins* ‡ has often obſerved it about *Albany* fort, where it appears in *May*, and retires before winter. It feeds on the white, and other Grous. This ſpecies ought to be added to the *American* claſs.

This ſpecies is pre-eminent in courage as well as beauty, and is the terror of other Hawks. It was flown at all kinds of fowl, how great ſoever they were; but its chief game uſed to be Herons and Cranes.

G. COLLARED. Falco Ruſticolus, *Lin. Syſt.* 125.—*Faun. Suec.* N° 56.—*Faun. Groenl.* N° 34.—*Latham*, i. 56.

F. With a lead-colored bill, tipt with black : head broad and flat, ſtreaked lengthways with black and white; on the cheeks the white predominates : the throat, under ſide of the neck, and

---

* *Leems*, 235.   † *Strom*.

‡ At the time this ſheet was printing, I had the good fortune to meet with Mr. *Hutchins*, ſurgeon, a gentleman many years reſident in *Hudſon's Bay*; who, with the utmoſt liberality, communicated to me his MS. obſervations, in a large folio volume: in every page of which his extenſive knowlege appears. The benefit which this work will, from the preſent page, receive, is here once for all gratefully acknowleged.

breaſt,

breaſt, are of a pure white; that on the neck almoſt ſurrounds it, forming a ſpecies of collar: the belly is of the ſame color, marked with a few duſky cordated ſpots: the back is waved with aſh-color and white; the tip of each feather white: the coverts of the wings of the ſame colors, but more obſcure: the exterior webs of the primaries duſky: the tail rounded, croſſed with twelve or thirteen whitiſh and duſky bars: the legs yellow. Size of a Hen.

Is rarely found in the remoteſt parts of *Greenland*. Inhabits alſo *Sweden*; and extends eaſtward as far as *Simbirſk, lat.* 54½, in the government of *Caſan**.

PLACE.

H. KITE, *Br. Zool.* i. N° 53 —*Latham,* i. 61, N° 43.
Falco Milvus Glada, *Faun. Suec.* N° 57.
Le Milan Royal, *De Buffon,* i. 197.—*Pl. Enl.* 422.—LEV. MUS.

F. With yellow bill and cere: white head, ſtreaked with black: body ferruginous, with a few duſky ſpots: tail much forked and ferruginous.

Weight forty-four ounces. Length twenty-ſeven inches: extent five feet one.

SIZE.

Inhabits the north of *Europe,* as high as *Jarlſberg,* in the very ſouth of *Norway* †; but does not extend farther. This ſpecies, the *Sea Eagle, Lanner, Buzzard,* and *Keſtril,* quit *Sweden,* in flocks, at approach of winter, and return in ſpring ‡. Of theſe, the *Buzzard* and *Keſtril* winter at *Woroneſch,* in *Ruſſia,* in *lat.* 52 ‖; and, together with the *Lanner* and *Kite,* about *Aſtrakan* §, in *lat.* 46. 30; but the far greater part of the Kites are ſuppoſed to retire into *Egypt,* being ſeen in *September* paſſing by *Conſtantinople* ¶, in their way from the north; and again in *April* returning to *Europe* **, to ſhun the great

PLACE.

* *Extracts,* i. 315. † *Hammer, Faun. Norway.* ‡ *Amœn. Acad.* iv.
‖ *Extracts,* i. 100. § *Vol.* ii. 142. ¶ *Forſkahl, Deſcr. Arab.* 7. ** *Will. Orn.* 75.

heats

heats of the eaft. They are obferved in vaft numbers about *Cairo*, where they are extremely tame, and feed even on dates, I fuppofe for want of other food *. They alfo breed there; fo that, contrary to the nature of other rapacious birds, they encreafe and multiply twice in the year; once in the mild winters of *Egypt*, and a fecond time in the fummers of the north. It makes its appearance in *Greece* in the fpring; and in the early ages, fays *Ariftophanes* †, "it governed " that country: and men fell on their knees when they were firft " bleffed with the fight of it, becaufe it pronounced the flight of " winter, and told them to begin to fheer their vernal fleeces. The " CRANE likewife, by its autumnal departure, warns the mariner to " hang up his rudder, and take his reft, and every prudent man to " provide their winter garments : and the SWALLOW again informed " them when they were to put on thofe of fummer. Thus, adds the " chorus of birds, are we to you as AMMON, DODONA, APOLLO :" meaning, in thofe early days, that man confulted only thefe natural calendars, and needed no other than what they took from the flight of birds ‡, or the flowering of plants.

They inhabit *England* in all feafons. I have feen their young taken, the laft week in *May*, or firft in *June*, in the great woods belonging to Sir *Jofeph Banks*, in *Lincolnfhire*; and have often obferved them in various places in the depth of winter.

I. HONEY BUZZARD, *Br. Zool.* i. N° 56.—*Latham*, i. 52, N° 33.
Falco Apivorus Slaghok, *Faun. Suec.* N° 65;.—LEV. MUS.

F. With an afh-colored head; dark brown above; below white, fpotted or barred with rufty brown : tail brown, barred with two dufky bars, remote from each other: legs ftrong and yellow : bill and cere black. LENGTH twenty-three inches. WEIGHT thirty ounces.

* *Belon Obf.* xxxvi. p. 107. b.　　　　† *Aves.*
‡ See this fubject moft ingenioufly handled in Mr. STILLINGFLEET'S Effays, in the *Calendar of Flora.*

Inhabits

Inhabits as far north as the diſtrict of *Sondmor*, in *Norway* * Is PLACE. found in plenty in the open parts of *Ruſſia* and *Sibiria*, near woods, and preys much upon Lizards.

K. LANNER, *Br. Zool.* i. N° 51.—*Latham*, i. 86.
   Falco Lannarius, *Faun. Suec.* N° 62.—*De Buffon*, i. 243.

F. With a white line over each eye: cere and legs bluiſh: breaſt white, tinged with yellow, and marked with brown ſpots: primaries and tail duſky; the firſt marked with oval ruſt-colored ſpots on the inner webs; the laſt, on both.

Inhabits *Iceland*, the *Feroe* iſles, and *Sweden*; the *Tartarian* deſerts PLACE. and the *Baraba*. Breeds on very low trees. None in the north or eaſt of *Sibiria*. Much eſteemed for falconry.

L. MOOR BUZZARD, *Br. Zool.* N° 57.—*Latham*, i. 53.
   Falco Æruginoſus, *Faun. Suec.* N° 66.
   Hons-tjuf, Le Buſard, *De Buffon*, i. 218. pl. x.—*Pl. Enl.* 424.

F. Entirely of a chocolate brown, tinged with ruſt: on the hind part of the head a light clay-colored ſpot: ſlender long yellow legs: cere black.

Weight twenty ounces. Length twenty-one inches. SIZE.

Found in the *Tranſbaltic* countries, as far north as *Sondmor* †. PLACE. Common in the ſouth of *Ruſſia*: not in *Sibiria*. It continues the whole year in *Sweden*.

* *Strom.* 235. † The ſame.

G g KESTRIL,

M. Kestril, *Br. Zool.* i. N° 60.—*Latham*, i. 94.
Falco Tinnunculus, Kirko-Falk, *Faun. Suec.* N° 61.—*Muller*, N° 65.
La Cresserelle, *De Buffon*, i. 280. pl. xviii.—*Pl. Enl.* 401, 471.

Male. F. With the crown and tail of a fine light grey, the last
marked with a black bar near the end : back and wings
of a purplish red, spotted with black. *Female.* Head reddish ; crown
streaked with black : back, tail, and coverts of wings, dull rust-color, barred with black : legs yellow. Weight of *Male* six ounces
and a half : of *Female* eleven.

Place.   Frequent in the deserts of *Tartary* and *Sibiria*, in the open countries, where small trees are found for it to breed in. Migrates into
Sweden, at the time in which the White Wagtail returns, and the Saf-
fron, Snowdrop, and bulbous Violet, blossom. Each of these birds
quit the country about the same day, in *September* *. Not found
farther north ?

N. Sparrow Hawk, *Br. Zool.* i. N° 62.—*Latham*, i. 99.
Sparshok, *Faun. Suec.* N° 68.—*Muller*, N° 71.—*Strom.* 235.
L'Epervier, *De Buffon*, i. 225. pl. xi.—*Pl. Enl.* 412, 467.

F. With head, back, and coverts of wings and tail, (in some)
of a deep bluish grey ; in others, of a deep brown, edged with
rust-color : breast and belly of a whitish yellow, with waved bars
of deep brown or dull orange : tail cinereous, with five broad
black bars ; the tips white.
Weight of the male five ounces : female nine.

Place.   Found as high as *Sondmor*, and in the *Feroe* islands, in the south of
Russia ; but none in *Sibiria*.

* *Calendar of Flora*, and *Migr. av.* in *Aman. Acad.* v. 397. 382.——Is found as
far south as as the *Holy Land*. *Hasselquist*, *Itin.* 291.

HOBBY,

O. Hobby, *Br. Zool.* i. N° 61.—*Latham*, i. 103.
Falco Subluteo, *Faun. Succ.* N° 59.

F. With crown, back, and coverts of a bluiſh black : from the crown a black ſtroke points down the cheeks, which are white : breaſt white, with oblong black ſpots : thighs and vent pale orange : inner webs of primaries marked with oval reddiſh ſpots : two middle feathers of the tail plain dove-color ; the inner webs of the others marked like the primaries : legs yellow. WEIGHT of the *male* ſeven ounces.

Schonen, the moſt ſouthern province of *Sweden* *, and, I believe, PLACE. does not extend farther north. This ſpecies winters about *Woroneſch* and *Aſtrakan* †; and frequents the ſame places in *Sibiria* with the KESTRIL.

* *Faun. Succ.* † *Extracts,* ii. 142.

III. OWL. *Gen. Birds.* III.

* EARED OWLS.

114. EAGLE.

Great Horned Owl, *Edw.* 60.—*Latham*, i. 119,
Great Grey Owl, *Joffelyn*, 96.—*Lawfon*, 145.
Jacurutu, *Margrave*, 199.
Stria Bubo Uf, *Faun. Succ.* N° 69.

O. With a dufky bill: yellow irides: horns fhorter than the *European* Eagle Owl; thofe, with the head, black, marked with tawny: circle round the eyes cinereous, edged with black: on the throat a large cruciform mark of a pure white, reaching to the beginning of the breaft: upper part of the breaft dufky and tawny; the lower part thickly barred with black afh-color, mixed with yellow: coverts of wings, fcapulars, and back, elegantly painted with zigzag lines, cinereous, black, and orange; the fcapulars alfo marked with a few great white fpots: primaries broadly barred with black and ferruginous: tail of a deep brown, croffed with brown dufky bars, and marked with numerous tranfverfe cinereous lines: legs and feet covered with foft light brown feathers to the very claws, which are very ftrong and hooked.

SIZE.

This fpecies is inferior in fize to the EAGLE OWL, *Br. Zool.* i. N° 64; but feems only a variety.

PLACE.

It is common to *South* and *North America*, as high as *Hudfon's Bay.* Makes, during night, a moft hideous noife in the woods, not unlike the hollowing of a man; fo that paffengers, beguiled by it, often lofe their way.

The favages have their birds of ill omen, as well as the *Romans.* They have a moft fuperftitious terror of the Owl; which they carry fo far as to be highly difpleafed at any one who mimics its hooting *.

This fpecies is common in *Kamtfchatka*, and even extends to the *Arctic* regions; in the firft of which it very often inclines to white. It is found as low as *Aftrakan*.

Br. Zool. i. Nº 65.
Strix Otus, *Faun. Suec.* Nº 71.—*Latham*, i. 121.

115. LONG-EARED.

O. With very long ears, of fix feathers each, yellow and black: irides yellow: back and coverts of wings deep brown, grey, and yellowifh ruft-color: primaries barred with dufky and ferruginous: breaft and belly pale yellow, with flender brown ftrokes, pointing downwards: tail barred with cinereous and dufky; the bars of the middle feathers bound above and below with white: feet feathered to the claws. LENGTH fourteen inches: EXTENT of the *Englifh* fpecimens three feet four †. Weight ten ounces.

PLACE. Obferved by Mr. *Hutchins* about *Severn* fettlement in *Hudfon's Bay*, where it lives in the woods, far from the fea: at night fallies in fearch of prey. Approaches the tents of the inhabitants, and is very clamorous. Builds its neft in trees, and lays four white eggs in *April*. Never migrates.

Inhabits *Sweden*, and the northern and fouthern parts of the *Ruffian* dominions, and the eaftern parts of *Sibiria*. Is found as far fouth as *Aftrakan*, and even in the hot climate of *Egypt* ‡.

Short-eared Owl, *Br. Zool.* i. Nº 66.—*Phil. Tranf.* lxii. 384.—*Latham*, i. 124.
La Chouette ou la grande Chevêche, *De Buffon*, i. 372. tab. xxvii.—*Pl. Enl.* 438.
—BL. MUS.—LEV. MUS.

116. SHORT-EARED.

O. With a leffer head in proportion than the former: bill dufky: irides yellow: head, back, and coverts of the wings, pale

* *Colden's Six Indian Nations*, i. 17.
† If no miftake is made in Mr *Hutchins's* MS. the extent is lefs by far than that of the *Englifh* kind.
‡ *Haffelquift, Itin.* 233.

brown,

brown, edged with dull yellow: breaſt and belly yellowiſh white, marked with a few duſky ſtreaks pointing downwards : thighs, legs, and toes, covered with plain yellow feathers : primaries duſky, barred with red : tail of a deep brown, marked on each ſide of the middle feathers with a large yellow circle, with a brown ſpot in the middle. In ſome, the feathers are yellowiſh, obliquely barred with black. The horns, or ears, conſiſt of only a ſingle feather, which it can raiſe or depreſs at pleaſure. The wings reach beyond the end of the tail. Length fourteen inches. Weight fourteen ounces.

SIZE.
PLACE.

Found in plenty in the woods near *Chateau Bay*, on the *Labrador* coaſt. It is alſo an inhabitant of the *Falkland Iſlands*; ſo probably is common to *North* and *South America.* In *Hudſon's Bay* it is called the *Mouſe Hawk.* It never flies, like other Owls, in ſearch of prey; but ſits quiet on a ſtump of a tree, watching, like a Cat, the appearance of Mice. It breeds near the coaſt; makes its neſt with dry graſs upon the ground; and migrates ſouthwards in autumn. Father *Feuillée* ſpeaks of an Owl he found in *Peru* that has ſome reſemblance to this, particularly in the Hawk-like ſhape of the bill. He ſays it burrows under ground to a great depth, like a Rabbet; for which reaſon he names it *Ulula Cunicularia**. It is very common in the northern and woody parts of *Sibiria.* Comes boldly to the night fires, and aſſaults men, when it is often killed with ſticks.

In *Europe* it is found in *Great Britain,* and reaches to the *Orkney* iſles. Does not perch, but ſits on the ground, on which it lays its eggs amidſt the heath. Appears and diſappears in *Lincolnſhire* with the Woodcock. Perhaps migrates to *Sweden* or *Norway,* where it is alſo found, and even as high as *Iceland* †. Flies and preys by day, in dark and cloudy weather. Friendly to the farmer, by being an excellent mouſer. Does not fly far; but if diſturbed, ſoon alights, and ſits looking about; at which times its horns are very conſpicuous. This circumſtance hitherto unattended to; ſo that it has been ranked among the Earleſs Owls.

* *Voy. Peru,* ii. 562.
† See Strix Funerea, *Faun. Suec.* N° 75. *Pontop. Atlas Danica,* tab. 25. *Olaſſen's Iceland,* ii. tab. 46.

Little Owl, *Catefby*, i. 7.—*Latham*, i. 123.   117. RED.
Strix Afio, *Lin. Syft*. 132.—BL. MUS.—LEV. MUS.

O.   With yellow irides: horns, head, back, and wings, of a plea-
fant tawny red, ftreaked with black: the fcapulars marked with
large white fpots: primaries barred with black, red, and white:
breaft pale tawny, marked with oblong black fpots: tail red, barred
with dufky: feet covered with feathers to the claws.   LENGTH ten
inches and a half.

Inhabits *New York*, and as low as the *Carolinas*.   Lives in the   PLACE.
woods near the coaft.

Latham, i. 126.—BL. MUS.—LEV. MUS.   118. MOTTLED.

O.   With the face white, fpotted with brown: head, wings, and
upper part of the body, mottled with afh-color and pale red:
the fcapulars marked with great white fpots; as are the coverts of the
wings: the primaries with black and pale ferruginous: breaft and
belly whitifh, varied with dufky ragged ftripes, pointing downwards:
toes feathered to the claws.   LENGTH eleven inches.

Inhabits the province of *New York*. Breeds in *May*, and continues   PLACE.
in the country the whole year.

** WITHOUT EARS.

O.   With gloffy black bill, and claws much incurvated: bafe of 119. WAPACUTHU
the bill befet with ftrong briftles: irides bright yellow: fpace
between the eyes, cheeks, and throat, white: the ends of the feathers
on the head black; fcapulars, and all the coverts of the wings, white,
elegantly barred with dufky reddifh marks, pointing downwards:

primaries,

7

primaries, fecondaries, and tail feathers, irregularly fpotted and barred
with pale red and black : back and coverts of the tail white, mixed
with a few dufky fpots : breaft and belly dirty white, croffed with
innumerable reddifh lines : vent white : legs feathered to the toes,
which are covered with hairs. WEIGHT five pounds : length two
feet : extent four.

PLACE.

Inhabits the woods about *Hudfon's Bay :* makes its neft on the
mofs, on the dry ground. The young are hatched in *May,* and fly
in *June* ; and are white for a long time after. Feeds on Mice and
fmall birds. Called by the *Indians, Wapacuthu,* or the Spotted
Owl. The *Europeans* fettled in the bay, reckon it a very delicate
food.

120. SOOTY.   Cinereous Owl, *Latham,* i. 134, N° 19.—BR. MUS.

O. With a whitifh bill : bright yellow irides : circlets confift of
elegant alternate lines of black and pale afh-color : head, hind
part of the neck, and coverts of wings, footy, marked with narrow
bars of dirty white : primaries deep brown, with broad bars, compofed of leffer of dufky and pale cinereous : tail moft irregularly
marked with oblique ftrokes of brown and dirty white : the breaft and
belly whitifh, greatly covered with large oblong blotches of dufky
brown : as a fingular mark, from the chin to the vent is a fpace,
about an inch in breadth, entirely naked : legs feathered to the feet.
WEIGHS three pounds : length two feet : extent four.

PLACE.

Inhabits *Hudfon's Bay* the whole year. Flies in pairs. Feeds on
Mice and Hares. Flies very low ; yet feizes its prey with fuch
force, that, in winter, it will fink into the fnow a foot deep ; and, with
great eafe, will fly away with the AMERICAN HARE, N° 38, alive in
its talons. It makes its neft in a pine-tree, in the middle of *May,*
with a few fticks lined with feathers ; and lays two eggs, fpotted
with a darkifh color. The young take wing in the end of *July.*

Great

121. Snowy.

Great White Owl, *Edw.* 61.—*Ellis's voy.* 40.—*Du Pratz,* ii. 91.—*Clayton's Virginia.*—*Ph. Tranf.* iii. 589.

Great Speckled Owl, *Egede, Greenland,* 64.

Strix Nyctea, Harfang, *Faun. Suec.* Nº 76.—*Buffon,* i. 387.—*Latham,* i. 132, Nº 17.—Bl. Mus.—Lev. Mus.

O. With a head lefs in proportion than other Owls : irides yellow : whole plumage of a fnowy whitenefs, fometimes pure, oftener marked with dufky fpots : the legs and feet covered warmly to the very claws with long fnowy feathers of the moft delicate and elegant texture : the claws are of a fine contrafting blacknefs, very large and very crooked. Its length two feet ; but it varies greatly in weight, from three pounds to one and a half.

Size.

It inhabits the coldeft parts of *America,* even as high as the remote mountains in the icy centre of *Greenland ;* from which, in intenfe cold, it migrates to the fhores. It adds horror even to that country, by its hideous cries, refembling thofe of a man in deep diftrefs.

Place.

It is rare in the temperate parts of *America,* and feldom ftrays as low as *Penfylvania* or *Louifiana.* Is very common in *Hudfon's Bay,* in *Norway,* and *Lapland.* It fears not the rigor of the feafon, but bears the cold of the northern regions the whole year. It flies by day, and is fcarcely to be diftinguifhed from the fnow : it flies pretty fwiftly, and falls perpendicularly on its prey. Feeds on the White Grous, and probably on the Hares ; for to the laft circumftance it owes its *Swedifh* name, *Harfang.* It preys alfo on Mice, and Carrion ; and in *Hudfon's Bay* is almoft domeftic, harbouring in places near the tents of the *Indians.*

Is fcarce in *Ruffia ;* grows more common on the *Uralian* mountains, and all over the north and eaft of *Sibiria,* and in its *Afiatic* empire, even in the hot latitude of *Aftrakan* * ; are very numerous in *Kamtfchatka.*

* *Extracts,* i. 91. ii. 142.

H h

Latham,

122. BARRED.   *Latham*, i. 133, N° 18.—BL. Mus.—LEV. Mus.

O. With a pale yellow bill, befet with ftrong briftles: irides yellow: circlets whitifh, barred with dufky lines: head, back, coverts of the wings, and the breaft, barred with dark brown, and white tinged with yellow; the primaries with black and white: the belly white, marked downwards with long ftripes of deep brown: tail barred with broad bands of black, and narrower of white: wings reach only half the length of the tail: feet feathered to the claws.

SIZE.   A large fpecies, two feet long; the extent four. Weight three pounds.

PLACE.   Inhabits *Hudfon's Bay*, and *New York*. Preys on Hares, Grous, Mice, &c.

123. HAWK.   Little Hawk Owl, *Edw.* 62.—*Latham*, i. 142, N° 29; 143, N° 30; 147, N° 36; 148, N° 37.—*Phil. Tranf.* lxi. 385.
Le Chat-huant de Canada, *Briffon*, i. 518.—*De Buffon*, i. 391.
Chouette a longue queue de Sibirie, *Pl. Enl.* 463.—LEV. Mus.

O. With yellow irides: head finely fpotted with dufky and pure white: back brown, with a few large white fpots: primaries of a deep brown, regularly fpotted with white on each web: upper part of the breaft white; lower part and belly barred with brown: tail very long, and cuneiform, marked with broad bars of brown, and narrow of white: feet protected with feathers to the claws.

LENGTH feventeen inches. WEIGHT twelve ounces. Never hatches above two young at a time; which, for fome months after flight, retain a rufty brown plumage.

PLACE.   This fpecies is common to *North America, Denmark*, and *Sweden*. The Savages who come down to *Hudfon's Bay*, call it *Cabetilutch*. It

1   flies

117

118

M. Griffiths del.

P. Mazell Sculp.

117 *Red Owl.* 118 *Mottled.* 120 *Barred.*

flies high, like a Hawk, and preys by day on the White Grous. Like the Short-eared Owl, will hover over the nocturnal fires. Is a bold bird; will attend the fowler, and often steal the game he has shot, before he can pick it up. Was seen by the navigators near *Sandwich* found, in *lat.* 61 north.

This bird is very frequent in all *Sibiria*, and on the west side of the *Uralian* chain, as far as *Casan* and the *Volga:* not in *Russia*.

124. WHITE.

Tuidara, *Margrave*, 205.
Barn Owl, *Clayton's Virginia.*—*Phil. Transf.* iii. 589.
White Owl, *Br. Zool.* i. N° 67.—*Latham*, i. 138.
Strix Flammea, *Faun. Suec.* N° 73.
L'Effraie, ou L'Effrasaie, *De Buffon*, i. 366. pl. xxvi.—*Pl. Enl.* 440.—LEV.
    MUS.—BL. MUS.

O. With a white bill: dusky irides: head, back, and coverts of wings, of a pale beautiful yellow, with two grey and two white spots placed alternately on each side of the shafts: breast and belly wholly white: interior sides of the feathers of the tail white; exterior marked with obscure dusky bars: legs feathered: feet covered with short hairs. LENGTH fourteen inches. WEIGHT eleven ounces.

PLACE.

This bird is common to *North* and *South America*, and to *Europe*. Was found by the navigators near *Sandwich* found, *lat.* 61 north. Is rare in *Sweden*, and, I believe, not found farther north. Inhabits *Tartary*. The *Mongol* and *Kalmuc Tartars* almost pay it divine honors; because they attribute to this species the preservation of the founder of their empire, *Cingis Khan*. That prince with his small army happened to be surprized and put to flight by his enemies, and forced to conceal himself in a little coppice: an Owl settled on the bush under which he was hid, and induced his pursuers not to search there, as they thought it impossible any man could be concealed in a place where that bird would perch. From thenceforth they held it to be sacred, and every one wore a plume of the feathers of
H h 2   this

this fpecies on his head.   To this day the *Kalmucs* continue the cuftom, on all great feftivals ; and fome tribes have an idol in form of
an Owl, to which they faften the real legs of one *.

125. BROWN.   Brown Owl, *Br. Zool.* i. N° 69.—*Latham*, i. 140.—*De Buffon*, i. 372.—*Pl. Enl.*
                 438.
                 Strix Ulula, *Faun. Suec.* N° 78.—BL. MUS.—LEV. MUS.

O. With dark hazel irides: head, wings, and back, of a deep
brown fpotted with black : coverts of the wings and fcapulars
varied with white fpots : breaft of a pale afh-color, marked with
dufky jagged ftrokes pointing downwards : feet feathered to the
claws. LENGTH about fourteen inches. WEIGHT nineteen ounces.

PLACE.   Inhabits *Newfoundland :* rare in *Ruffia :* unknown in *Sibiria :* found
in *Sweden* and *Norway* †.

126. LITTLE.   Little Owl, *Br. Zool.* i. N° 70.—*De Buffon*, i. 377.
                 Strix Pafferina, *Faun. Suec.* N° 79.—*Latham*, i. 149, N° 38, N° 39 ; 150, N° 40.
                 —BL. MUS.—LEV. MUS.

O. With pale yellow irides : bill whitifh brown : head light brown,
fpeckled with white : back, and coverts of the wings, and fcapulars, of the fame color, marked in parts with white fpots : the
breaft whitifh, varied with ruft-color : tail barred with white, and
marked regularly on each web with circular white fpots : feet feathered
to the claws.   It varies in length, from eight to feven inches.   The
fmalleft I have feen is from *Nova Scotia* ; which has white circlets
about the eyes, and fewer white fpots on its plumage.

PLACE.   Inhabits from *Hudfon's Bay* to *New York.* Called by the natives of
the firft, *Shipmofpifh.* Lives in all feafons among the pines : builds
its neft half way up the tree : lays two eggs. Are moft folitary birds.
Keep clofe in their retreat the whole day ; but are moft active
moufers during night.   Frequent in *Ruffia* ; lefs fo in *Sibiria.*

* *Extracts,* ii. 142.   † *Brunnick,* N° 19.

                                                         SCANDI-

* E A R E D.

A. Scandinavian Eared Owl, Strix Scandiaca, *Faun. Succ.* N° 70.—*Latham* i. 120.

O. With the plumage entirely white, fprinkled with black fpots.

Size of a Turky : in all refpects like the Snowy Owl, except Size. the ears.

Inhabits the *Lapland alps.* Mentioned by *Linnæus* ; who feems to Place. take his defcription from a painting of *Rudbeck*'s ; but its exiftence is confirmed by Mr. *Tonning* of *Drontheim* *.

** E A R L E S S.

B. Tawny Owl, *Br. Zool.* i. N° 68.—*Latham*, i. 139.
Strix Stridula, Skrik Uggla, *Faun. Succ.* N° 77.—*Pl. Enl.* 437.—Lev. Mus.

O. With a plain head : dufky irides : plumage of the head, and the whole upper part of the body, tawny, fpotted and powdered with dufky fpots : breaft and belly yellowifh, mixed with white, marked downward with dufky ftreaks : tail blotched, barred, and fpotted with pale ruft-color and black : toes feathered to the claws. Weight nineteen ounces.

Inhabits *Europe,* as far as *Sweden.* Frequent in the fouth of *Ruffia,* Place. and deferts of *Tartary* ; and breeds in the nefts of Rooks. None in *Sibiria :* a fufpicion that it is found in *Hudfon's Bay ?*

* *Rariora Norvegiæ,* in *Aman. Acad.* vii. 479.

ORDER.

ORDER II. PIES.

IV. S H R I K E.   *Gen. Birds* IV.

127. GREAT.

Great Shrike, *Br. Zool.* i. N° 71.
Lanius Excubitor, Warfogel, *Faun. Suec.* N° 80.—*Latham*, i. 160.
White Whisky John, *Phil. Transf.* lxii. 386.
La Pie-grieche Grise, *De Buffon*, i. 296. pl. xx.—*Pl. Enl.* 445.—LEV. MUS.

S. With a black bill and legs : cinereous crown, hind part of the neck, and back : cheeks white, crossed from the bill with a bar of black : under side, from chin to tail, white, marked with semicircular lines of a pale brown : lesser coverts black ; those on the joints of the wings ash-color : primaries black, marked with a single band of black ; secondaries tipt with white : the tail cuneiform ; the two middle feathers black, the tips of the next on each side white ; on the rest the white prevales, till the exterior, when the black almost entirely vanishes : beyond each eye of the female is a brown bar.

PLACE.

Inhabits *North America*, from *Hudson's Bay* to *Louisiana*. In *Hudson's Bay*, lives in the woods remote from shores, and is the first bird there which brings out its young in the spring. Makes its nest with dry grass or bents, and lines it thickly with feathers : lays seven eggs, of a pale blue color, blotched with brown.

Is frequent in *Russia*, but does not extend to *Sibiria* ; yet one was taken by our navigators within *Bering's* straits, in *lat.* 66, on the *Asiatic* side of the Frozen Sea. Has the same manner of transfixing and tearing its prey as the *English* kind.

128. BLACK-CROWNED.

S. With the bill, legs, crown, and sides of the head, back, and coverts of wings, black : primaries black, marked with a small spot of white, and another on the ridge of the wing : throat, cheeks, and vent, pure white : breast and belly tinged with ash-color : tail long ;

long; middle feathers black; the reft marked at their ends with white, which increafes to the exterior; in which the black almoft vanifhes. Rather inferior in fize to the laft.

Inhabits *North America.* Seems to be *La Pie Griefche de la Louifi-* PLACE. *ane, Briffon,* ii. 162; *Latham,* i. 162.

Lanius Canadenfis, *Lin. Syft.* 134.—*De Buffon,* i. 316.—*Pl. Enl.* 479. fig. 2.— 129. CRESTED. *Latham,* i. 182.
La Pie Griefche de Canada, *Briffon,* ii. 171.—LEV. MUS.

S. With black bill and legs: head adorned with a reddifh creft: cheeks dufky, fpotted with white: hind part of neck and back brown, inclining to red: throat and breaft of a yellowifh red: belly and vent of a fine afh-color: coverts of the wings black, edged with white; primaries with white on their exterior fides: tail black, bordered on each fide, and tipt with white. LENGTH fix inches and a half: EXTENT about eleven.

Inhabits *Canada.* PLACE.

S. With the bill flightly incurvated at the end, black, except the 130. NATKA. upper half of the lower mandible: crown, lower part of the upper fide of the neck, and the back, black: over each eye is a white line, extending to the very nape; beneath that one of black: from chin to vent is wholly white: a narrow white circle quite encom-paffes the neck: leffer coverts of the wings black; greater white, more or lefs dafhed down the fhafts with black: primaries dufky, fringed with yellowifh brown; fecondaries black, edged and tipped with white: tail black, a little rounded; the four outmoft feathers tipped with white: rump cinereous, the edges of the feathers grey: legs black. LENGTH feven inches one-fourth.

Brought from *Natka* found in *North America.* Communicated to PLACE. me by Mr. *Latham.*

Br.

131. RED-BACKED.

Br. Zool. i. N° 72.—*Latham,* i. 167.
Lanius Collurio, *Faun. Suec.* N° 81.
Pie-grieche de la *Louiſiane, De Buffon,* i. 307.—*Pl. Enl.* 397.—Lev. Mus.

S. With grey crown and rump : ferruginous back and coverts of wings : black line acroſs the eyes : breaſt and belly roſeate : tail black ; exterior feathers edged with white : head and upper part of the Female dirty ruſt-color ; line over the eyes the ſame color : breaſt and belly dirty white, marked with duſky ſemicircular lines. Length ſeven inches and a half.

Place. Inhabits *Ruſſia* ; not *Sibiria.* Is found in *Sweden* and *Chriſtianſoe.* The Count *De Buffon* ſays, he received one from *Louiſiana.* I imagine, that, as the *Norwegians* give the Great Shrike and this a name, that they may be found in their country. The firſt they call *Klavert,* the laſt *Hanvark.* Mr. *Ekmark* has obſerved both of them, only during ſummer, in *Eaſt Gothland* ; but is not certain whether they winter. Each ſpecies appears in *Italy* in the ſpring ; retires in autumn.

A. Grey, Lanius Nengeta, *Lin. Syſt.* 135.—*Latham,* i. 183.
Grey Pye of *Braſil, Edw.* 318.

S. With the crown, hind part of the neck, back, and coverts of the wings, deep cinereous : a black line paſſes from the bill through the eyes to the hind part of the head : greater coverts and ſe-

condaries

condaries black, tipt with dirty white; primaries black: breaſt and belly light aſh-color: tail black; ends of the outmoſt feathers white. Much larger than N° 127, the common Great Shrike; and differs ſpecifically.

Inhabits *Ruſſia*, but is more frequent in *Sibiria*; where it lives in the foreſts the whole winter. Taken and tamed by the fowlers; and kept by the *Ruſſians* for the diverſion it affords in the manner of killing its prey. They ſtick a rod with a ſharp point into the wall of a room, on which the Shrike perches. They turn looſe a ſmall bird, which the former inſtantly ſeizes by the throat, ſtrangles, and then ſpits it on the point of the ſtick, drawing it on with its claws and bill. Thus it ſerves as many as are turned to it, and afterwards eats them, thus ſuſpended, at its leiſure *. The *Germans* ſtyle it *Wurch-angel*, or the *Suffocating-angel*. The old *Engliſh*, *Wariangel*, which ſig-nifies a bird of ſome very miſchievous qualities; as is evident from *Chaucer*.

PLACE.

> This Sompnour, which that was as ful of jangles,
> As ful of venime ben thiſe *Wariangles* †.

B. Lᴇssᴇʀ Gʀᴇʏ, Pie Grieche *d'Italie*, *De Buffon*, i. 298.—*Pl. Enl.* 32.

S. With the forehead black: a black line croſſes the eyes, like as in the former: head, hind part and ſides of the neck, back, and coverts of wings, cinereous, paleſt on the rump: ridge of the wing white: primaries black, with a white ſpot near the baſe; ſecondaries black, tipt with white: throat white: breaſt and belly tinged with roſe-color: tail marked like the preceding.

Inhabits *Ruſſia*, but not *Sibiria*. Found in *Italy* and *Spain*. PLACE.

* Eᴅᴡᴀʀᴅs, *Gl.* p. 233.

† The Freres tale. *Ful of venime*, becauſe it was believed, that the thorn on which it ſtuck its prey was venomous.

V. PARROT

V. PARROT. *Gen. Birds* V.

232. CAROLINA.   Parrokeeto, *Lawfon*, 142.—*Latham*, i. 227.—LEV. MUS.
Parrot of *Carolina, Catefby*, i. 11.—*Du Pratz*, ii. 88.
Pfittacus Carolinenfis, *Lin. Syft.* 141.—*Briffon*, iv. 350.
La Perruche a tete jaune, *De Buffon*, vi. 274.
Le Papegai a tete aurore, *De Buffon*, vi. 247.

P. With the forehead, ridge of the wings, and feathers round the knees, orange: head and neck yellow: back, body, and coverts of wings and tail, green: primaries dufky, mixed with blue and green; the upper exterior fides edged with yellow: tail very long and cuneiform: legs white. LENGTH thirteen inches. WEIGHT three ounces and a half.

PLACE.   Inhabits the fouthern parts of *North America*, but never appears higher than *Virginia*. It is in general a migratory bird, even in *Carolina*; arriving at the feafon when mulberries are ripe, which they are very fond of, and which are the earlieft fruits of the country, except ftrawberries. They infeft, in autumn, the apple-orchards in vaft flocks, and make great havock by fplitting the fruit for the fake of the kernels only, being very greedy of them, and the feeds of cyprefs, and other trees. They devour too the buds of the birch.

Few of thefe tender birds continue in *Carolina* during the whole year. They breed in hollow trees, in low fwampy grounds. When taken, they eafily grow tame, but do not fpeak. Their inteftines are faid to be a fpeedy poifon to Cats.

EGGS.   The eggs of Parrots are roundifh, and generally of a pure white; thofe of the Maccaws fpotted, like the eggs of a Partridge. The number ufually two; yet the Count *De Buffon* gives an inftance of a

Perroquet,

Perroquet, in a ſtate of confinement, which laid four eggs every ſpring, during five or ſix years: one of the eggs was addle; the others productive *.

Tui-apeta-jube, *Margrave,* 206, N° 2.—*Wil. Orn.* 116.—*Raii. Syn. av.* 34.—   133. ILLINOIS.
     De *Buffon,* vi. 269.—*Latham,* i. 228.
Pſittacus Pertinax, *Lin. Syſt.* 142.
La Perruche Illinoiſe, *Briſſon,* iv. 353.
Yellow-faced Parrot, *Edw.* 234.

P. With a cinereous bill: orange-colored irides: forehead, cheeks, and ſometimes the hind part of the head, of a rich orange: crown, upper part of the body, tail, and coverts of the wings, of a fine green: primaries green, edged externally with blue: breaſt and belly of a yellowiſh green: vent yellow: tail very long and cuneiform. Of the ſame ſize with the former. Perhaps differs only in ſex?

PLACE.

Inhabits the interior parts of *North America,* in the country of the *Illinois,* ſouth of lake *Michigam :* it is alſo met with in the *Brazils.* Is a lively bird; but its voice not very articulate. Father *Charlevoix* met with ſome on the banks of the *Theatiki,* a river that riſes a little ſouth of lake *Michigam,* and runs into the *Miſſiſipi.* He ſays, that thoſe he ſaw were only ſtragglers, which migrated before winter; but that the main body paſſed the whole year on the borders of the *Miſſiſipi* †.

LATITUDES OF
PARROTS.

The Count *De Buffon* confines the whole genus of Parrots to exactly twenty-five degrees on each ſide of the Equator ‡. It always gives me pain to differ in opinion with ſo illuſtrious a character; but I muſt produce my authorities of their being common at far greater diſtances. On the continent of *America,* two ſpecies have been obſerved by the *Spaniards* about *Trinity Harbour,* in the South Seas, in

* *Oiſ.* vi. 115.   † *Journal Hiſtorique,* vi. 124.   ‡ *Oiſ.* vi. 82.

north *lat.* 41. 7 *. Dr. *Forster* faw, in the raw, rainy latitude of *Dusky Bay*, in *New Zealand*, 46 fouth, two kinds. In the neighborhood of *Botany Bay*, in *New Holland*, in fouth *lat.* 34, five fpecies were difcovered ; among which, the greater variety of the fulphur-crefted *Cockatoo* appeared in amazing multitudes. But what is moft wonderful, a fmall fpecies of this tender genus is to be met with as low as *Port Famine*, in the ftreights of *Magellan*, in north *lat.* 53. 44 †, in flocks innumerable. They inhabited the vaft forefts of the country. Their food muft be confined to buds and berries ; for no fort of fruit-trees have been obferved there. The forefts likewife were frequently bounded by mountains, probably cloathed with eternal fnow.

---

* *Barrington's Mifcellanies*, 489. 491.

† See, *Spilbergen's voy.* in *Purchas*, i. 80 ; *Wood's*, in *Dampier's voy.* iv. 112 ; and *Byron's*, in *Hawkfworth's Coll.* i. 38. Befides thefe authorities, Lieut. *Gore* (fince Captain) and Mr. *Edwards*, now furgeon at *Caernarvon*, who failed with Mr. *Byron*, confirmed to me the exiftence of thefe birds in the ftreights of *Magellan*.

VI. C R O W.

VI. C R O W. *Gen. Birds* XII.

Br. Zool. i. N° 74.
Corvus Corax, *Lin. Syſt.* 155.
Korp, *Faun. Suec.* N° 85.—*Leems*, 240.—*Faun. Greenl.* p. 62.—*Latham*, i. 367.
    —*De Buffon*, iii. 13.—LEV. MUS.

134. RAVEN.

C· With the point of the bill a little incurvated, with a ſmall tooth on each ſide, of a black color, gloſſed with blue. It varies to white, and to pied. In the *Feroe* iſles is a breed which are black and white, and are ſaid to keep in a place ſeparate from the common kind *.

The largeſt of the genus. Weighs three pounds. Length two feet two inches.

SIZE.

Very numerous as far north as *Finmark*, *Iceland*, and *Greenland*, where it frequents the huts of the natives, and feeds on the offals of the Seals †. Preys in concert with the White Bear, Arctic Fox, and Eagle. Devours the eggs of birds, eſpecially the Ptarmigan : eats ſhore-fiſh, and ſhell-fiſh : drops the laſt from on high to break them, and get at the contents. Turns round in the air, and is dexterous ; changes its prey from its bill to its feet, or from its feet to its bill, by way of eaſe. Eats alſo berries, and, when almoſt famiſhed, dried ſkins and excrements. Neſtles on high rocks, which overhang and afford a canopy. Couples in *March* ; lays in *April*. Each preſerves a diſtrict to itſelf. The male ſits in the day ; the female in the night : the former ſleeps cloſe by its mate. Have ſtrong affection to their young brood. Hearing its croaking echoed, repeats it ; as if admiring its own note. At approach of ſtorms, collects under ſhelter of rocks.

PLACE.

---

* *Brunnick*, p. 8.   † *Egede*, 64,

Caught

Caught by the natives. Its flefh is eaten. The fkins reckoned the beft for cloathing: the wings ufed for brufhes: the quils fplit, are made into fifhing-lines. They alfo inhabit *Newfoundland*, and now and then appear as low as *Virginia* and *Carolina* *.

This bird is, among the *American* favages, an emblem of return of health. Their phyficians, or rather magicians, when they vifit a fick perfon, invoke the Raven, and mimic his croaking voice †. The northern *Indians*, on the contrary, deteft this and all the Crow kind ‡. It inhabits *Kamtfchatka* and *Sibiria*; but not within the *Afiatic* Arctic regions.

CARRION.

Br. Zool. i. N° 75 ?—*Latham*, i. 370.
Blaae Raage, *Brunnick*, N° 29.
Corvus Corone, *Faun. Suec.* N° 86.
La Corbine, ou l'Corneille, *De Buffon*, iii. 45.—*Pl. Enl.* 483.—LEV. MUS.

C. With the plumage wholly black, gloffed with violet: bill ftrong, thick, and arched: noftrils covered with ftrong black briftles: ends of the feathers of the tail flightly pointed. LENGTH eighteen inches and a half. Weight from twenty to twenty-two ounces ‖.

PLACE.   Inhabits the province of *New York*, and the inland parts of *Hud-fon's Bay*. Mr. *Blackburn* obferved, that it retains there the fame manners as the *European* fpecies; and never migrates from *New York*. MR. KALM fays, that they fly in great numbers, and have a cry much refembling the Rook §. By his account, they appear of a mixed nature, feeding not only on grain, but on carrion; and are alfo very pernicious to young poultry. Like Rooks, they pull up the

---

* *Lawfon*, 139.   † *Adair's Hift. Am.* 173.   ‡ Mr. *Hutchins.*
‖ *Voyage*, i. 121.
§ See article Rook, p. 250, A. where a comparifon is made of the differences between thefe two birds.

corn

corn of the country, the new-fown maize ; and, when it ripens, do pick a hole in the leaves which furround the ears, expofing it to corruption, by letting in the rain. The inhabitants of *Penfylvania* and *New Jerfey* were wont to profcribe them, fetting three pence or four pence on the head of each Crow ; but the law was foon repealed, becaufe of the great expence it brought on the public ftock*. Mr. *Kalm* alfo remarks this agreement with the Rook fpecies, that they fettle much on trees, both in *February* and the fpring.

Thefe birds are fo rare in *Sweden*, that *Linnæus* gives only one inftance of its being killed in his country. Yet it is found in the diocefe of *Drontheim*, and in the *Feroe* iflands. They are fcarce in *Ruffia* ; and only in the north. Grow more common in *Sibiria*, and are found plentifully beyond the *Lena*, where the Hooded Crow ceafes. Was obferved about *Botany Bay*, in *New Holland* ; and is met with in the *Philippine* ifles †.

Br. Zool. i. Nº 78.—*Latham*, i. 392.—*De Buffon*, iii. 85.
Corvus Pica, Skata, Skiura, Skara, *Faun. Suec.* Nº 92.—Lev. Mus.

136. Magpie.

C. Variegated with black and white, the black moft beautifully gloffed with green and purple : the tail very long, cuneiform, black, refplendent with the fame rich colors as the body. Length eighteen inches : weight nine ounces.

Vifits *Hudfon's Bay*, where the natives call it *Oue ta-kee Afke*, or the *Heart-bird*. It migrates, and but feldom appears there ‡.

Is found in *Europe*, as high as *Wardhuys*, in *lat.* 71½. It is efteemed there an augural bird. If it perches on the church, it is fuppofed to portend the death or removal of the minifter : if on the caftle, that of the governor ‖. The Magpies fwarm in the temperate parts of *Ruffia*. Common in *Sibiria*, and even as far as *Kamtfchatka*, and the ifles.

Place.

---

* *Voyage*, ii. 65.   † *De Buffon*, iii. 66.   ‡ *Phil. Tranf.* lxii. 387.
‖ *Leems*, 241.

Corvus

137. CINEREOUS.   Corvus Canadenfis, *Lin. Syſt.* 158.—*Latham*, i. 389.
                    Le Geay Brun de Canada, *Briſſon*, ii. 54.—*De Buffon*, iii. 117.—LEV. MUS.

C. With a black bill, ſtrong, ſtrait, notched near the end of the upper mandible : noſtrils covered with a tuft of whitiſh feathers reflected downwards : the forehead, cheeks, and under part of the body, of a dirty reddiſh white : the feathers on the crown long and black, forming a ſpecies of creſt, like that of the *Engliſh* Jay : the plumage on the back brown, ſilky, looſe, and unwebbed, like that of the Jay : wings black : tail long, cuneiform, black ; the three out-moſt feathers tipt with dirty white : legs black. LENGTH near eleven inches : extent fifteen. Weight two ounces and a half.

PLACE.   Inhabits *Hudſon's Bay*, *Newfoundland*, and *Canada*, and the woods on the weſtern coaſts of *America*. Theſe birds breed early in ſpring : their neſts are made of ſticks and grafs, and built in pine-trees. They have two, rarely three, young ones at a time. Their eggs are blue. The young are quite black, and continue ſo for ſome time. They fly in pairs. The male and female are perfectly alike. They feed on black moſs, worms, and even fleſh. When near habitations or tents, they are apt to pilfer every thing they can come at, even ſalt meat. They are bold, and come into the tents to eat victuals out of the diſhes, notwithſtanding they have their hoard of berries lodged in the hollows of trees. They watch perſons baiting the traps for Martins, and devour the bait as ſoon as they turn their backs. Theſe birds lay up ſtores for the winter ; and are ſeldom ſeen in *January*, unleſs near habitations : they are a kind of mock-bird. When caught, they pine away, and die, though their appetite never fails them †. Deteſted by the natives of *Hudſon's Bay*.

† *Mr. Hutchins.*

Jay.

Jay, *Clayton's Virginia.—Phil. Tranf.* iii. 590.—*Lawfon*, 141.
Blue Jay, *Catefby*, i. 15.—*Edw.* 239.—*Latham*, i. 386.
Corvus Criftatus, *Lin. Syft.* 157.
Le Geay Bleu de Canada, *Briffon*, ii. 55.—*De Buffon*, iii. 120.—BL. MUS.—
LEV. MUS.

138. BLUE.

C. With a ftrong thick bill: head adorned with a rich blue creft: a ftripe of black from the bill extending beyond the eyes: throat and cheeks white: neck furrounded with a black collar: breaft of a pale vinaceous red: belly white: back of a pale purple: coverts of the wings and fecondaries of a rich blue, beautifully barred with black; the fecondaries, and one order of the coverts, tipt with white: tail long and cuneiform, barred with blue and black; the tips of all white, excepting thofe of the two middlemoft: legs black. LENGTH twelve inches.

Inhabits *Newfoundland, Canada,* and as far fouth as *Carolina.* Has the fame actions and jetting motion as the *Englifh* Jay, but its cry is lefs harfh. It feeds on fruits and berries, and commonly fpoils more than it eats. It is particularly fond of the berries of the bay-leaved *Smilax.* Refides in the country all the year. Lays in *May* five or fix eggs, of a dull olive, with rufty fpots.

PLACE.

C. With a crefted head: bill, neck, and back, black: leffer coverts of the wings dufky; the others of a rich refplendent blue: exterior webs of the primaries of the fame color; the inner dufky; the fecondaries of a beautiful rich blue, croffed with narrow black bars, remote from each other: the rump, belly, and breaft, of a dull blue: tail very long, cuneiform, and of a fine gloffy blue; the middle feathers flightly barred. SIZE of an *Englifh* Jay.

139. STELLER'S *

Inhabits the woods about *Natka* or *George* found, in *North America.* It had been before difcovered by *Steller,* when he landed on the fame fide of that continent. Defcribed from a fpecimen in the collection of Sir JOSEPH BANKS.

PLACE.

* *Latham,* i. 387.

K k

ROOK,

R O O K.

A.   ROOK, *Br. Zool.* i. N° 76.—*Latham*, i. 372.
Corvus Frugilegus, Roka, *Faun. Suec.* N° 87.—*De Buffon*, iii. 55.

C. Black, gloffed with purple : a tinge of dull green over part of the tail : the ends of the feathers of the tail broad, and rounded ; thofe of the Crow, acute : the bill ftraiter, flenderer, and weaker, than that of the Carrion Crow : the length two inches and a half; that of the latter only two inches and a quarter.   The bill of the CROW is of a more intenfe black.   The noftrils and bafe of the bill of the ROOK naked, and whitifh, occafioned by being often thruft under ground in fearch of food.   The WEIGHT of both nearly the fame, about twenty-one ounces : the LENGTH about eighteen inches : the EXTENT of wings in the ROOK three feet one inch and a half ; of the Crow, two inches and a half lefs *.

PLACE.   This fpecies is not found farther north than the fouth of *Sweden*. It breeds there ; but is driven away by the feverity of the winter. No mention is made of it in the *Danifh* or *Norwegian* Faunæ.   Is common in *Ruffia*, and the weft of *Sibiria* ; but there are none in the eaft.   They migrate in the beginning of *March* to the environs of *Woronefch*, and mingle with the common Crows †.

* I once had the curiofity to compare the meafurements of thefe common birds, and found them as above ; but they are often inferior in fizes to the fubjects I examined.
† *Extracts*, i. 103.

HOODED

B HOODED CR. *Br. Zool.* i. N° 77.—*Latham*, i. 374.
   Corvus Cornix, Kraka, *Faun. Suec.* N° 88.
   Krage, *Leems*, 239.—*De Buffon*, iii. 61.

C. With black head, wings, and tail; afh-colored body
   Inhabits *Europe*, as high as the *Feroe* iflands and *Lapmark*, where   PLACE.
it continues the whole year; but in the northern countries often retires to the fhores, where it lives on fhell-fifh. Is very common in
all *Ruffia* and *Sibiria*: none beyond the *Lena*. Migrates to *Woronefch*,
and paffes the winter there. Grows very large beyond the *Ob*, and
often varies to entire blacknefs. This bird, and the Raven, in *Oc-
tover* quit the *fub-alpine* woods, where they breed; and fpread all
over the plains of *Italy*. This fpecies extends to *Syria*, as do the
Raven, Crow, Jackdaw, and Magpie *.
   This fpecies, the RAVEN, CROW, JACKDAW, PIE, and JAY, pafs
their winter at *Woronefch*†, removing probably from hotter as well as
colder climates; for three of the above can endure the fevereft cold.

C. JACKDAW, *Br. Zool.* i. N° 81.—*Latham*, i. 378.
   Kaia, *Faun. Suec.* N° 89.—*De Buffon*, iii. 69.

C. With white irides: hind part of the head light grey: breaft
   and belly dufky afh: reft of the bird black. LENGTH thirteen
inches.
   Inhabits as far north as *Sondmor*: is fometimes feen in the *Feroe*   PLACE.
ifles. Migrates from *Smoland* and *Eaft Gothland* the moment that
harveft ends; and returns in the fpring, attendant on the Stares.
Winters about *Upfal*, and paffes the night in vaft flocks in ruined
towers, efpecially thofe of antient *Upfal*. Common over all *Ruffia*,
and the weft of *Sibiria*. A few are feen beyond lake *Baikal*. Are
migratory, unlefs in the fouth of *Ruffia*.

* *Ruffel's Aleppo*, 69.   † *Extracts*, i, 100.
                    K k 2   NUT-

D. NUTCRACKER, *Br. Zool.* ii. App. p. 625.—*Latham*, i. 400.—*De Buffon*, iii. 122.
Merula Saxatilis, *Aldr. av.* ii. 284.
Corvus Caryocatactes, Notwecka, Notkraka, *Faun. Suec.* Nº 91.

C. With primaries and tail black, the laſt tipt with white: vent white: reſt of the plumage of a ruſty brown: crown, and coverts of the tail, plain; every other part marked with white triangular ſpots. SIZE of à Jackdaw.

PLACE.
Is found as high as *Sondmor.* Does not migrate. Common in the pine-foreſts of *Ruſſia* and *Sibiria*, and even in *Kamtſchatka*. Lives on nuts and acorns, and on the kernels of pine-cones. Neſtles in the bodies of trees, which it perforates like the Woodpecker.

E. JAY, *Br. Zool.* i. Nº 79 —*Latham*, i. 384.—*De Buffon*, iii. 107.
Corvus Glandarius, Allonſkrika, Kornſkrika, *Faun. Suec.* Nº 90.

C. With a black ſpot on each ſide of the mouth: very long feathers on the head: body purpliſh aſh: greater coverts of wings beautifully barred with rich blue, black, and white. LENGTH thirteen inches.

PLACE.
Is met with as high north as *Sondmor.* Not migratory. Common in the woods of *Ruſſia* and *Sibiria*; but none beyond the *Lena*.

F. ROCK, Greater Redſtart, *Wil. Orn.* 197.
La Paiſſe Solitaire, *Belon, Oyſ.* 322.
Codiroſſo Maggiore, *Olina*, 47.—*Latham*, i. 176.—*De Buffon*, iii. 354.—*Aldr. av.* ii. 282.
Stein-Rotela, *Geſn. av.* 732.

C. With crown, and neck above, and coverts of wings, brown and dirty white. In the males, the middle of the back marked with a ſpot, conſiſting of a bar of blue, black, and ruſt-colored: throat, breaſt, and belly, orange, ſpotted with white, and a few duſky
ſpots:

spots: two middle feathers of the tail dufky; the reft ferruginous: has the fame loofe filky texture of feathers as the Jay. SIZE of a Stare.

Found as high as the forefts of *Lapland*. Is called by the *Swedes*, *Lappfkata* and *Olyckfugl*; by the *Norwegians*, *Gertrudsfogel*; alfo *Ulyksfuegl*, from its being fuppofed to forebode ill-luck. *Linnæus*, for the fame reafon, ftyles it *Lanius Infauftus*; and in his *Fauna*, *Corvus Infauftus* *. It is common in the woods of the north of *Ruffia* and *Sibiria*. Is a moft audacious bird. *Linnæus* relates, that in dining amidft the *Lapland* forefts, it would often fnatch away the meat before him. Breeds in crevices of rocks. Feeds on worms and infects. Sings finely, and is often preferved in cages for its fong.

ROLLER. *Gen. Birds* XIII.

G. GARRULOUS, *Br. Zool.* ii. App. p. 530 quarto, 624 octavo.—*Latham*, i. 406.—*De Buffon*, iii. 133.—*Aldr.* i. 395.
Coracias Garrula, Spanfk-kraka, Bla-kraka, *Faun. Suec.* N° 94.

R. With a naked fpot beyond each ear: head, neck, back, breaft, belly, and greater coverts of the wings, of a light bluifh green: back ferruginous: coverts of the tail, leffer coverts of the wings, and lower parts of the fecondaries, of a rich blue; primaries black above, blue beneath: middle feathers of the tail dirty green; the reft of a light blue: the exterior feathers on each fide much longer than the reft, and tipt with black: legs yellowifh. SIZE of a Jay.

This elegant bird is found not fpread, but as if it were in a ftream, from the fouthern parts of *Norway* to *Barbary* and *Senegal*: from the fouth of *Ruffia* to the neighborhood of the *Irtifh*, only, in that empire; and foutherly, to *Syria* †. In *Sweden*, it arrives with the *Cuckoo*;

* *Syft.* 138. *Faun. Suec.* N° 93. † *Ruffel's Aleppo*, 69.

retires at the conclufion of the harveft †. It makes its neft in the birch, preferably to all other trees ‡ ; and in places where trees are wanting, fuch as *Malta* and *Barbary*, it forms its neft in clayey banks. *Zinanni* fays it lays five eggs, of clear green, fprinkled with innumerable dark fpecks ‖. It feeds on fruits, acorns, and infects. Is a fhy bird ; but, at times, is feen in company with Crows and Pies on the plough lands, picking up worms, and grains of corn. *Schwenckfelt* fays, that in autumn it grows very fat, and is efteemed as a delicacy §. It is remarkably clamorous. Is migratory. *M. Adanfon* obferved them in *Senegal*, in flocks, in the month of *September*, and fuppofes they winter there ¶.

† *Amœn. Acad.* iv. 583.

‡ *De Buffon*, iii. 139 : from this circumftance, one of its *German* names is *Birckbeher*, or the *Birch Jay*.

‖ *Zinanni delle Nova, &c.* p. 68. tab. x. fig. 29.   § *Av. Silefiæ*, 244.

¶ *Voy. Senegal*, Engl. ed. 25. 107.

VII. O R I O L E.

VII. ORIOLE. *Gen. Birds* XIV.

Atolchichi, *Fernand. Nov. Hisp.* p. 14.—*Wil. Orn.* 395.—*Raii Syn. av.* 166.—   140. RED-WING.
  Latham, i. 428.
Black Bird (2d sp.) *Lawson,* 139.
Red-winged Starling, *Catesby,* i. 13.—*Du Pratz,* ii. 91.
Le Troupiale a Aisles Rouges, *Brisson,* ii. 97.
Le Commandeur, *De Buffon,* iii. 214.—*Pl. Enl.* 402.
Oriolus Phœniceus, *Lin. Syst.* 161.

O. With black bill and legs : plumage of a fine jetty blackness, except the lesser coverts of the wings, which are of a bright scarlet, with the lowest row white. LENGTH ten inches. The FE-MALES are of a dusky color.

Inhabit from the province of *New York* to the kingdom of   PLACE. *Mexico.* In *North America* they are called Red-winged Starlings, and Swamp Black-birds ; in *Mexico, Commendadores,* from their red shoulders, resembling a badge worn by the commanders of a certain *Spanish* order. That kingdom seems to be their most southern resi-dence. They appear in *New York* in *April,* and leave the coun-try in *October.* They probably continue the whole year in the   MANNERS. southern parts, at left *Catesby* and *Lawson* make no mention of their departure. They are seen in flocks innumerable, obscuring at times the very sky with their multitudes. They were esteemed the pest of the colonies, making most dreadful havock among the maize and other grain, both when new sown, and when ripe. They are very bold, and not to be terrified with a gun ; for, notwithstanding the sportsman makes slaughter in a flock, the remainder will take a short flight, and settle again in the same field. The farmers some-times attempt their destruction, by steeping the maize in a decoction

of

of white hellebore before they plant it. The birds which eat this prepared corn are feized with a vertigo, and fall down; which fometimes drives the reft away. This potion is particularly aimed againft the PURPLE GRAKLES, or PURPLE JACKDAW, which conforts in myriads with this fpecies, as if in confpiracy againft the labors of the hufbandman. The fowler feldom fhoots among the flocks, but fome of each kind fall. They appear in greateft numbers in autumn, when they receive additions from the retired parts of the country, in order to prey on the ripened maize.

Some of the colonies have eftablifhed a reward of three pence a dozen for the extirpation of the Jackdaws: and in *New England*, the intent was almoft effected, to the coft of the inhabitants; who at length difcovered that Providence had not formed even thefe feemingly deftructive birds in vain. Notwithftanding they caufed fuch havock among the grain, they made ample recompence, by clearing the ground of the noxious worms * with which it abounds. As foon as the birds were deftroyed, the reptiles had full leave to multiply: the confequence was the total lofs of the grafs, in 1749; when the *New Englanders*, late repentants, were obliged to get their hay from *Penfylvania*, and even from *Great Britain*.

The Red-winged Orioles build their nefts in bufhes, and among the reeds, in retired fwamps, in the form of a hang-neft; leaving it fufpended at fo judicious a height, and by fo wondrous an inftinct, that the higheft floods never reach to deftroy it. The neft is ftrong, made externally with broad grafs, a little plaftered; thickly lined with bent or withered grafs. The eggs are white, thinly and irregularly ftreaked with black.

Fernandez fays, that in *Mexico* they build in trees near towns; and both he and *Catefby* agree, that they fing as well in a ftate of confinement as of nature; and that they may be taught to fpeak. I agree with *M. de Buffon*, that, in cafe the manner of their nidification

USES (margin)

NEST. (margin)

* The Caterpillar of the *Bruchus Pifi*, or Peafe Beetle, in particular. See *Kalm*, i. 173. 176.

is

is as *Fernandez* afferts, the difagreement in the different countries is very wonderful.

In *Louifiana* they appear only in winter, and are taken in a clap-net, placed on each fide of a beaten path made on purpofe, and ftrewed over with rice. As foon as the birds alight, the fowler draws the net, and fometimes takes three hundred at a haul. They are alfo eaten in the *Englifh* colonies. *Fernandez* does not commend their flefh, which, he fays, is unpalatable and unwholefome.

Du Pratz fpeaks of two kinds: this, and another which is grey and black, with a red fhoulder, like the fpecies in queftion. I fufpect YOUNG, OR FEMALES? he forms out of the young birds, not yet arrived at full color, a new kind; or perhaps a female bird: for I have received from Dr. *Gar-den* one under that title, which agrees with the defcription given by M. *Du Pratz*. Thefe are ftreaked with pale rufty brown: cheeks black: over each eye a white line: breaft and belly black, fpotted with pale brown: leffer coverts of the wings rich orange.

White-backed Maize Thieves, *Kalm*, ii. 274. 141. WHITE-BACKED.

A Species mentioned barely as above by Mr. *Kalm*, with the addi-tion of their being lefs than the laft: that they fing finely, and appeared flying now and then among the bufhes near *Saratoga*; but that he faw them for the firft time near *New York*. As Mr. *Kalm* PLACE. feems not to have had a diftinct fight of thefe birds, it is poffible that they are the WHITE-WINGED ORIOLES of Mr. *Latham*, ii. 440: the coverts of whofe wings are white; the reft of the plumage entirely black. His fpecies came from *Cayenne*.

Baltimore bird, *Catefby*, i. 48.—*Latham*, i. 432. 142. BALTIMORE.
Le Baltimore, *Briffon*, ii. 109.—*De Buffon*, iii. 231.—*Pl. Enl.* 506.
Oriolus Baltimore, *Lin. Syft.* 162.—BL. MUS.—LEV. MUS.

O. With the head, throat, neck, and upper part of the back, black: MALE. leffer coverts of the wings orange; the greater black tipt

L l with

with white : breaſt, belly, lower part of the back, and coverts of the tail, of a bright orange : primaries duſky, edged with white : two middle feathers of the tail black ; the lower part of the reſt of the ſame color, the remaining part orange : legs black.

FEMALE.   Head and back of the female olive, edged with pale brown : coverts of the wings of the ſame color, marked with a ſingle bar of white : under ſide of the body, and coverts of the tail, yellow : tail duſky, edged with yellow.   LENGTH of this ſpecies ſeven inches.

PLACE.   Inhabits from *Carolina* * to *Canada* †. Suſpends its neſt to the horizontal forks of the Tulip or Poplar trees, formed of the filaments of ſome tough plants, curiouſly woven, mixed with wool, and lined with hairs.   It is of a pear ſhape, open at top, with a hole on the ſide, through which the young diſcharge their excrements, and are fed. In ſome parts of *North America*, this ſpecies, from its brilliant color, is called the *Fiery Hang-neſt*. It is called the *Baltimore* bird, from its colors reſembling thoſe in the arms of that nobleman.

It quits *North America* before winter, and probably retires to *Mexico*, the *Xochitototl* of *Fernandez* ‡ ſeeming to be the ſame ſpecies.

143. BASTARD.   Baſtard Baltimore, *Cateſby*, i. 49.—*Latham*, i. 433.
Le Baltimore Batard, *Briſſon*, ii. 111.—*De Buffon*, iii. 233.—*Pl. Enl.* 506.
Oriolus Spurius, *Lin. Syſt.* 162.—BL. MUS.—LEV. MUS.

O.   With the head, neck, and upper part of the back, of a full gloſſy black : breaſt and belly of a fine orange bay : lower part of the back, and coverts of the tail, of the ſame color : the leſſer coverts of the wings light bay ; the greater black, edged with dirty white : the quil feathers duſky, edged with white : tail cuneiform and black.

The head of the female, and hind part of the neck, deep olive : throat black : coverts of wings duſky, edged with white ; primaries

* *Lawſon*, 145.   † *De Buffon*.   ‡ *Av. Nov. Hiſp.* 39.

and

M. Griffiths del.

Baltimore Oriole № 142.

P. Mazell Sculp.

and fecondaries of the fame colors: under fide of the body of a greenifh yellow: tail dufky, edged with yellow.

Inhabits *North America*. Arrives in *New York* in *May*. Lays five eggs; and ufually hangs its neft in an apple-tree.   PLACE.

Latham, ii. 445, N° 37.
Le Troupiale Noir, *Briffon*, ii. 103. tab. x.—*De Buffon*, iii. 320.—*Pl. Enl.* 534.—
Br. Mus.   144. BLACK.

O. With a black bill, an inch long: legs of the fame color: whole plumage black and gloffy. LENGTH near ten inches. EXTENT one foot. WEIGHT two ounces and a quarter.

FEMALE. With head, breaft, and belly, dufky, tinged with cinere-ous; the reft of the plumage of as greenifh brown.

Inhabits *North America*, even a far as *Hudfon's Bay*. Arrives there in the beginning of *June*, as foon as the ground is thawed fufficiently for them to get food, which is Worms and Maggots. They fing with a fine note till the time of incubation, when they defift, and only make a chucking noife till the young take their flight; when they refume their fong. They build their nefts in trees, about eight feet from the ground; and form them with mofs and grafs. Lay five eggs, of a dark color, fpotted with black. Gather in great flocks, and retire foutherly in *September*. A bird, which I apprehend to be only a leffer variety, is defcribed by the *Comte de Buffon*, iii. 221. *Pl. Enl.* 606. *Latham*, ii. 446.   PLACE.

O. With the head of a rufty brown: the body and wings black, gloffed with green: the tail of a dufky color. SIZE of a com-mon Blackbird.—Br. Mus. Lev. Mus.   145. BROWN-HEADED.

Inhabits *New York*, and appears there in fmall flocks during fum-mer. Perhaps migrates to *St. Domingo*, where it is alfo found, and is called there, according to Mr. *Kuchan*'s account, *Siffleur*, or Whiftler; but differs from that defcribed by M. *De Buffon*, iii. 230, which is entirely yellow beneath.   PLACE.

L l 2   O. With

146. RUSTY.　　O. With dusky bill and legs: head, and hind part of the neck, of a blackish purplish hue, with the edges of the feathers rust-colored: from the bill, over and beneath the eyes, extends a black space, reaching to the hind part of the head: throat, under side of the neck, the breast, and back, black, edged with pale rust: belly dusky: wings and tail black, glossed with green. LENGTH between seven and eight inches.

PLACE,　　Appears in *New York* in the latter end of *October*, and makes a very short stay there: it probably is on its way southerly from *Hudson's Bay*, where it is also found.

147. WHITE-HEADED.　　Le Caffique de la Louisiane, *De Buffon*, iii. 242.—*Pl. Enl.* 646.

O. With the head, neck, belly, and rump, white: the rest of the plumage changeable violet, bordered with white, or in some parts intermixed. LENGTH ten inches *French*.

PLACE.　　Inhabits *Louisiana*.

148. HUDSONIAN WHITE-HEADED.　　O. With a dusky bill: head and throat pure white: ridge of the wing, some of the under coverts, first primary, and thighs, of the same color: all the rest of the bird dusky, in parts glossed with green: on the breast a few oblong strokes of white: legs dusky. LENGTH eight inches and a half. EXTENT thirteen and a half. WEIGHT an ounce and three quarters.

PLACE.　　Inhabits *Hudson's Bay*. A very rare species. Quere, if only differing in sex from the last.—LEV. MUS.

149. OLIVE.　　Le Carouge Olive de la Louisiane, *De Buffon*, iii. 251.—*Pl. Enl.* 607.

O. With the head olive, tinged with grey: hind part of the neck, the back, wings, and tail, of the same color, tinged with brown,

brown, brighteſt on the rump and the beginning of the tail: the ſides alſo olive, daſhed with yellow; the ſame color edges the greater coverts and primaries: the throat is orange-colored: the under ſide of the body yellow: legs a browniſh aſh-color. LENGTH ſix or ſeven inches *French*. EXTENT from ten to twelve.

Inhabits *Louiſiana*.

PLACE.

O. With a bright yellow ſtroke over each eye: cheeks and throat of the ſame color: all the reſt of the plumage tinged with green, only ſome of the coverts of the wings are tipt with white: bill and legs duſky. LENGTH nine inches. EXTENT fifteen and a half.

Was ſhot in *Hudſon's Bay*.

150. YELLOW-
THROATED.

PLACE.

Latham, ii. 447, N° 40.

151. UNALASCH-
KA.

O. With a brown bill; between its baſe and the eyes a white mark: plumage above, brown; the middle of each feather clouded: chin white, bounded on each ſide by a dark diverging line: fore part of the neck and breaſt of a ruſty brown: coverts of the wings, the ſecondaries, and tail, brown, edged with ruſt: primaries and belly plain: ſides duſky: legs brown. LENGTH eight inches.

Brought by the late navigators from *Unalaſchka*.

PLACE.

Latham, i. 448.

152. SHARP-
TAILED.

O. With the crown brown and cinereous: cheeks brown, ſurrounded by a border of light clay-color, commenced at the baſe of each mandible of the bill: throat white: breaſt, ſides, and vent, of a dull pale yellow, ſpotted with brown: belly white: back

varied

varied with afh-color, black, and white: greater and leffer coverts of the wings dufky, deeply bordered with ruft-color; primaries black, flightly edged with ruft: the feathers of the tail flope off on each fide to a point, not unlike thofe of a Woodpecker; are of a dufky color, and obfcurely barred: the legs of a pale brown. Size of a Lark.

PLACE.   Inhabits the province of *New York*.—From Mrs. *Blackburn*'s collection.

VIII. G R A K L E.

VIII. GRAKLE. *Gen. Birds,* XV.

Tequixquiacatzanatl *, *Fernandez Mex.* 21.
La Pie de la Jamaique, *Briſſon,* ii. 41.—*De Buffon,* iii. 97.—*Pl. Enl.* 538.
Merops Nigér iride ſub-argentea, *Brown's Jamaica,* 476.
Purple Jackdaw, *Cateſby,* i. 12.—*Latham,* i. 462.
Black Bird, *Lawſon,* ſp. 2d, 139.—*Sloane Jamaica,* ii. 299.
Gracula Quiſcula, *Lin. Syſt.* 165.—BL. MUS.—LEV. MUS.

G. With a black bill: ſilvery irides: head and neck black, gloſſed over with a moſt reſplendent blue, variable as oppoſed to the light: back and belly, with green and copper-color, growing more duſky towards the vent: tail long, and cuneiform: legs black: wings and tail rich purple. Female entirely duſky; darkeſt on the back, wings, and tail.

LENGTH of the male thirteen inches and a half: the WEIGHT about ſix ounces. LENGTH of the female eleven inches and a half.

Theſe birds inhabit the ſame countries as the Red-wing *Orioles,* and generally mingle with them. They ſometimes keep ſeparate; but uſually combine in their ravages among the plantations of maize. After that grain is carried in, they feed on the ſeeds of the Water Tare Graſs, or *Zizania aquatica.* Their good qualities, in clearing the country from noxious inſects, have been recited before, in page mixed with the hiſtory of their congenial companions.

They appear in *New York* and *Philadelphia* in *February,* or the beginning of *March;* and ſit perched on trees near the farms, and give a tolerably agreeable note. They alſo build in trees, uſually in retired places, making their neſts externally with coarſe ſtalks, inter-

* i. e. The *Salt Starling,* becauſe in *Mexico* it frequents the ſalt lakes.

nally

nally with bents and fibres, with plaifter at the bottom. They lay
five or fix eggs, of a pale plue color, thinly fpotted and ftriped.with
black. After the breeding-feafon, they return with their young from
their moft diftant quarters, in flights continuing for miles in length,
blackening the very fky, in order to make their depredations on the
ripening maize. It is unfortunate that they increafe in proportion as
the country is more cultivated; following the maize, in places they
were before unknown, wherefoever that grain is introduced.

They migrate from the northern colonies at approach of winter;
but continue in *Carolina* the whole year, feeding about the barn-
door. Their flefh is rank, and unpalatable; and is only the food of
birds of prey. The fmall Hawks dafh among the flocks, and catch
them in the air.

They are alfo found in *Mexico*, and in the ifland of *Jamaica*.
They are fometimes eaten; but their flefh is hard, rank, and of bad
nourifhment.

154. **Boat-tail.**   Gracula Barrita, *Lin. Syft.* 165.—*Latham*, i. 460.
Le Troupiale Noir, Icterus Niger, *Briffon*, ii. 105.—*De Buffon*, iii. 220.—*Pl.
Enl.* 534.
Monedula tota nigra, *Sloane*, 299.—*Raii Syn. av.* 185.—**Lev. Mus.**

G.   With the bill an inch and a half long, fharp, and black:
plumage black, gloffed with purple: tail cuneiform, expanded
when walking; in flight, or on the perch, folded, fo as to form an
oblong cavity in its upper part. Length about thirteen inches.

Place.   Inhabits not only the greater *Antilles*, but the warmer parts of
North America; conforting with the *Purple Grakles*, and *Red-winged
Orioles*. Feeds on *maize* and infects; in the iflands on *Bananas*.

IX. C U C K O O.

IX. CUCKOO. *Gen. Birds,* XIX.

Cuckoo of Carolina, *Cateſby,* i. 9.—*Lawſon,* 143.   155. CAROLINA.
Le Coucou de la *Caroline, Briſſon,* iv. 112.
Cuculus Americanus, *Lin. Syſt.* 170 —*Latham,* i. 537.—LEV. MUS.—BL. MUS.

C. With the upper mandible of the bill black, the lower yellow : head, and whole upper part of the body, and coverts of
the wings, cinereous ; under ſide entirely white : primaries brown
on their exterior, orange on their interior ſides : tail long ; two
middle feathers entirely cinereous, the others tipt with white : legs
duſky. LENGTH twelve inches.

Inhabits *North America.* Arrives in *New York* in *May.* Makes its   PLACE.
neſt in *June,* uſually in apple-trees ; and lays four eggs, of a bluiſh
white color. The neſt is made of ſmall ſticks and roots, and reſembles greatly that of the *Engliſh* Jay ; but is ſmaller. It retires from
North America in autumn.

This bird, as well as all the foreign Cuckoos, have only the generical character of the well-known *European* ſpecies. They differ in
their œconomy, nor have the opprobrious notes of that bird.

M m   EUROPEAN

A. Europᴇᴀɴ Cᴜᴄᴋᴏᴏ, *Br. Zool.* i. N° 82. tab. xxxvi. fem.—*Latham*, i. 509.
Cuculus Canorus, Gjok, *Faun. Suec.* N° 96.
Le Coucou, *De Buffon*, vi. 305.—Lᴇᴠ. Mᴜs.

C. With dove-colored head, hind part of the neck, back, rump,
and coverts: throat, and under fide of the neck, of a pale grey:
breaſt and belly white, barred with black : primaries duſky ; inner
webs marked with white oval ſpots : tail cuneiform ; middle feathers
black, tipped with white ; the reſt marked with white ſpots on each
web. Fᴇᴍᴀʟᴇ. Neck of a browniſh red : tail barred with ruſt-color
and black, and ſpotted with white.

Pʟᴀᴄᴇ.   Inhabits all parts of *Europe*, as high as *Saltens Fogderie*, in *Norway* *, within the Arctic circle ; and even at *Loppen*, in *Finmark* †.
It is found equally high in *Aſia* ; and extends as far eaſt as *Kamtſchatka*. In all places it retains its ſingular note, and its more ſingular nature of laying its eggs in the neſts of ſmall birds, and totally
deſerting them ‡. Of the above circumſtance I beg leave to add a
proof, which fell under my own notice in *June* 1778 ; when I ſaw a
young Cuckoo, almoſt full grown (when I firſt diſcovered it) in the
neſt of a white Wagtail, beneath ſome logs in a field adjacent to my
houſe. The Wagtail was as ſolicitous to feed it, as if it had been
its own offspring ; for, many days after the Cuckoo fled, it was ſeen
often perched on the adjacent walls, ſtill attended and fed by the
Wagtail.

It arrives in the northern and eaſtern parts of *Aſia*, about the tenth
of *June*.

      * *Pontop.* ii. 75.   † *Leems*, 291.   ‡ Dr. Pᴀʟʟᴀs.

WRYNECK.

W R Y N E C K.   *Gen. Birds,* **XX.**

B. WRYNECK, *Br. Zool.* i. N° 83.—*Latham*, i. 548:
   Jynx Torquilla Gjoktyta, *Faun. Suec.* N° 97.
   Le Torcol, *De Buffon*, vii. 84.—*Pl. Enl.* 698.—LEV. MUS.

W. With a black and colored lift dividing lengthways the crown
and back : upper part of the body elegantly pencilled with
grey, black, white, and ferruginous : tail confifts of ten feathers,
grey, fpeckled with black, and marked equidiftant with four broad
black bars.

Extends over all *Ruffia* and *Sibiria*, and even to *Kamtfchatka*.   PLACE.
Found in *Sweden*, and as high as *Drontheim*, in *Norway* ; and probably
migrates as far as the *Cuckoo*. The *Swedes* call this bird *Gjoktyta*, or
the bird which *explains the Cuckoo* : probably for the fame reafon as
the *Welfh* and *Englifh* ftyle it the *Cuckoo's Man*, as it feems its attendant, and to point out its arrival.

X. WOODPECKER. *Gen. Birds,* XXI.

156. WHITE-
BILLED.

Quatotomomi, *Fernand. Mex.* 50.—*Wil. Orn.* 390.
Ipecu, *Marcgrave,* 207.—*Wil. Orn.* 138.—*Raii Syn. Quad.* 43.—*Latham,* ii. 553.
Picus principalis, *Lin. Syst.* 173.
Largeft White-bill Woodpecker, *Catefby,* i. 16.—*Lawfon,* 142.—*Barrere Fr.*
     Equin. 143.—*Kalm,* ii. 85.
Grand Pic noir a bec blanc, *De Buffon,* vii. 46.—*Pl. Enl.* 690.

W. With a bill of ivory whitenefs; great ftrength; three inches
long: irides yellow: a conic creft, of a rich fcarlet color,
on the hind part of the head: head, throat, neck, breaft, and belly,
black: beneath each eye is a narrow ftripe of white, crooked at its beginning, running afterwards ftrait down the fides of the neck: upper
part of the back, primary feathers, and coverts of the wings, black;
lower part of the back, and the fecondaries, white: tail black.

This is a gigantic fpecies, weighing twenty ounces; and in bulk
equal to a Crow.

PLACE.

Inhabits the country from *New Jerfey* to the *Brafils.* Is in *North
America* a fcarce bird; in *South America* more common. It breeds
in the kingdom of *Mexico* in the rainy feafon; for which reafon
Nieremberg ftyles it *Picus Imbrifœtus* *. The *Spaniards* call them
Carpenteros, Carpenters, on account of the multitude of chips which
they hew out of the trees, either in forming their nefts, or in fearch
of food, infects, and worms, which lurk beneath the bark. They are
very deftructive to trees; for they have been known to cut out a
meafure of chips in an hour's time †. Inftinct directs them to form
their holes in a winding form, in order the better to protect their
nefts from the injury of the weather ‡.

* *Eufeb. Nieremberg.*   † *Catefby.*   ‡ *Barrere.*

10   *Canada*

Canada is deftitute of thefe birds. The *Indians* of that fevere climate purchafe the bills from the favages of the more fouthern parts, at the rate of two or three Buck fkins apiece, in order to form the coronets * of their fachems and warriors. Thefe coronets were made with feveral materials. Gay plumes formed the rays; the beaks of birds, claws of rare animals, and the little horns of their Roes, were the other ornaments. They were never worn but on high folemnities; either when a warrior fung the fong of war, or was fetting forward on his march to meet the enemy. He went forth like a *Spartan* hero, dancing, and crowned †.

Larger Red-crefted Woodpecker, *Catefby*, i. 17.
Le Pic noir hupe de Virginie, *Briffon*, iv. 29.
Picus Pileatus, *Lin. Syft.* 173.—*Latham*, i. 554.
Le Pic noir a huppe rouge, *De Buffon*, vii. 48.—*Pl. Enl.* 718.—Lev. Mus.— Bl. Mus.

157. PILEATED.

W. With a bill two inches long, of a dufky color on the upper, and whitifh on the lower mandible: irides of a gold-color: a tuft of light brown feathers reflected over the noftrils: the crown adorned with a rich fcarlet creft, bounded by a narrow buff-colored line; beneath that is a broad band of black, reaching from the eyes to the hind part of the head; under this is another line of buff-color, commencing at the bill, and dropping down on each fide of the neck to the pinions of the wings: from the lower mandible a line of fcarlet extends along the lower part of the cheeks: chin and throat white: fore and hind part of the neck, back, breaft, belly, and tail, black: the wings black, marked with a double line of white: legs dufky. Length eighteen inches. Weight nine ounces.

Inhabits the forefts of *Penfylvania* and *New York*. When the maize begins to ripen, this and the other kinds make great havock, by

PLACE.

* *Catefby.* † *Lafitau Mœurs de Sauvage*, ii. 60.

fettling

settling on the heads, and picking out the grain; or making holes in the leaves, and letting in the wet, to the deſtruction of the plant*. It breeds and reſides the whole year in the country. It extends as high as lat. 50. 31. north; being found near the banks of *Albany* river, near four hundred miles from its diſcharge into *Hudſon's Bay*. Lays ſix eggs, and brings forth its young in *June*. The *Indians* deck their Calumets with the creſt of this ſpecies.

158. GOLDEN-
WING.

Golden-winged Woodpecker, *Cateſby*, i. 18.
Lo Pic Rayè de Canada, *Briſſon*, iv. 70.
Picus Auratus, *Lin. Syſt.* 174.—*Latham*, i. 597.
Le Pic aux ailes dorees, *De Buffon*, vii. 39.—*Pl. Enl.* 693.—LEV. Mus.—BL.
Mus.

W. With a black bill, bending like that of a Cuckoo: crown cinereous; on the hind part a ſcarlet ſpot: cheeks and under ſide of the neck of a pale red: from each corner of the mouth a black line extends along the cheeks: the upper part of the breaſt is marked with a black creſcent; the remainder and the belly whitiſh, ſpotted with black: back and coverts of wings of a fine pale brown, barred with black: the primaries cinereous; their ſhafts of a moſt elegant gold-color; the under ſide of the webs of a gloſſy yellow: rump white, ſpotted with black: tail black, edged with white: the ſhafts of all the feathers gold-colored, except thoſe of the two middle feathers: legs duſky. LENGTH twelve inches. WEIGHT five ounces. The FEMALE wants the black on each ſide of the throat.

PLACE.

Inhabits from *Hudſon's Bay* to *Carolina*, and again on the weſtern ſide of *North America*. In the firſt is migratory, appearing in *April*, and leaving the country in *September*. All the *American* Woodpeckers agree with thoſe of *Europe* in building in hollow trees, and in laying ſix white eggs. The natives of *Hudſon's Bay* call this ſpecies, *Ou-thee-*

* *Kalm.*

quan-

XIII

Ferruginous Woodpecker N.º 159.   *Nuthatch N.º 170.*

P. Mazell sculp.

quan-nor-ow, from the golden color of the ſhafts and under ſide of the wing feathers *.

The *Swediſh Americans* call it *Hittock*, and *Piut* † ; words formed from its notes. It is almoſt continually on the ground ; and never picks its food out of the ſides of trees, like others of the genus : nei-ther does it climb, but ſits perched, likc the Cuckoo ; to which it has ſome reſemblance in manners, as well as form. It feeds on in-ſects. Grows very fat, and is reckoned very palatable. It inhabits the *Jerſies*, and other provinces to the ſouth, the whole year.

Latham, i. 592.
Le Pic Mordore, *De Buffon*, vii. 34.—*Pl. Enl.* 524.

159. FERRUGI-
NOUS.

W. With a duſky bill : the crown and pendent creſt of a pale yellow : a crimſon bar extends from the mouth along the lower part of the cheek : the cheeks, back, and coverts of the wings, of a deep ferruginous color : lower part of the back of a pale yel-low : primaries ferruginous, barred on their inner webs with black. SIZE of the Green Woodpecker.

This new ſpecies was ſent to me by Dr. *Garden*, of *Charleſtown*, *South Carolina*.

PLACE.

Red-headed Woodpecker, *Cateſby*, i. 20.—*Lawſon*, 3d ſp. 143.—*Du Pratz*, 92.—
Latham, i. 561.
Picus Erythrocephalus, *Lin. Syſt.* 174.
Le Pic a teſte rouge, de la Virginie, *Briſſon*, iv. 53.—*Pl. Enl.* 117.
Le Pic noir a domino rouge, *De Buffon*, vii. 55.—*Pl. Enl.* 117.—LEV. MUS.—
BL. MUS.

160 RED-HEADED,

W. With a lead-colored bill : head and neck of the moſt deep and rich ſcarlet : back, coverts of wings, primaries, and tail, of a gloſſy blackneſs : the ſecondaries white, marked with two black

* *Phil. Tr.* lxii. 387.   † *Kalm*, ii. 36.

bars :

bars: breaſt and belly white: legs black. The head of the FEMALE is brown. LENGTH nine inches and a half. WEIGHT two ounces.

PLACE.

Inhabits *Penſylvania*, and the neighboring provinces. Feeds on maize and apples; and is a moſt deſtructive ſpecies. They pick out all the pulp, and leave nothing but the mere rind. They feed alſo on acorns. They were formerly proſcribed; a reward of two pence was put on their heads: but the law was repealed. They migrate ſouthward at approach of winter. When they are obſerved to linger in numbers in the woods, in the beginning of winter, the inhabitants reckon it a ſign of a mild ſeaſon *.

This ſpecies extends acroſs the continent to the weſtern coaſt of *America.*.

161. CAROLINA.

Red-bellied Woodpecker, *Cateſby*, i. 19.
Picus Carolinus, *Lin. Syſt.* 174.—*Latham*, i. 570.
Le Pic varié de la Jamaique, *Briſſon*, iv. 59.—*De Buffon*, vii. 72.
Woodpecker of Jamaica, *Edw.* 244.—BL. MUS.

W. With the forehead, crown, and hind part of the head, of an orange red; under ſide of a light aſh-color, tinged with yellow: the vent ſpotted with black: the back and wings cloſely barred with black and white: middle feathers of the tail black, the outmoſt barred with black and white. The crown of the female is light grey: hind part of the head red. LENGTH eleven inches. WEIGHT two ounces eleven penny-weights.

PLACE.

Inhabits *North America*, and the greater *Antilles*.

162. SPOTTED.

Great Spotted Woodpecker? *Br. Zool.* i. N° 85.—*Latham*, i. 564.
Le Pic varié, *Briſſon*, iv. 34.—*De Buffon*, vii. 57.—*Pl. Enl.* 196. 595.
Picus Major, *Faun. Suec.* N° 100.—LEV. MUS.—BL. MUS.

W. With buff forehead; black crown, bounded behind with a crimſon band: vent feathers crimſon: back black: ſcapulars white: wings and tail barred with black and white: breaſt and belly

* *Kalm*, ii. 87.

white,

white, tinged with yellow. LENGTH nine inches. EXTENT fixteen. WEIGHT two ounces three quarters. FEMALE wants the crimfon marks.

Sent to Mrs. *Blackburn* from *New York*. Inhabits *Europe*, as high as *Lapmark*. Extends to the moft eaftern part of *Sibiria*.

PLACE,.

L'Epeiche de Canada, *De Buffon*, vii. 69.—*Pl. Enl.* 347.—*Briffon*, iv. 45.

163. CANADA SPOTTED.

W. With white forehead, throat, breaft, and belly: crown, black; beneath is a band of white, encircling the head; from each eye another of black, uniting behind, and running down the hind part of the neck; each fide of this bounded by white; that again bounded by black, commencing at the bafe of the bill, and uniting with the fcapulars: the back black; fcapulars of the fame color, mixed with a few white feathers: wings fpotted with black and white: middle feathers of the tail black; the outmoft black and white. SIZE of the laft.

Inhabits *Canada*.

PLACE.

Hairy Woodpecker, *Catefby*, i. 19.—*Latham*, i. 572.
Picus Villofus, *Lin. Syft.* 175.
Le Pic varié de la Virginie, *Briffon.* iv. 48.
L'Epeiche ou Pic Chevelù de Virginie, *De Buffon*, vii. 75.—LEV. MUS.—BL. MUS.

164. HAIRY,

W. With the crown black: the hind part of the head marked with a crimfon fpot; the cheeks with two lines of white and two of black: whole under fide of the body white: back black, divided in the middle lengthways with a line of white unconnected feathers, refembling hairs: the wings black, fpotted in rows with

* *Phil. Tranf.* lxii. 388.

N. n

white ::

white: two middle feathers of the tail black; the two outmoft entirely white; the reft black, marked croffways with white. The female wants the red fpot on the head. LENGTH nine inches. WEIGHT two ounces.

PLACE. Inhabits from *Hudfon's Bay* * to *Carolina.* In the laft very deftructive to apple-trees.

165. DOWNY.

Smalleft Spotted Woodpecker, *Catefby*, i. 21.
Picus Pubefcens, *Lin. Syft.* 175.—*Latham*, i. 573.
Le Petit Pic varié de la Virginie, *Briffon*, iv. 50.
Fourth Woodpecker, *Lawfon*, 143.
L'Epeiche ou Petit Pic varie de Virginie, *De Buffon*, vii. 76.—LEV. MUS.—
BL. MUS.

W. Of the fize of a Sparrow. In all refpects refembles the laft, except in fize; and in having the outmoft feather of the tail marked with a fingle white bar.

PLACE. Inhabits *Penfylvania* and *Carolina*, and is very numerous. It is alfo found, but more rarely, near *Albany* fort, in *Hudfon's Bay.* The Woodpecker tribe is the moft pernicious of all the birds of *America*, except the PURPLE GRAKLE; but this little fpecies is the moft deftructive of its whole genus, becaufe it is the moft daring. It is the peft of the orchards, alighting on the apple-trees, running round the boughs or bodies, and picking round them a circle of equidiftant holes. It is very common to fee trees encircled with numbers of thefe rings, at fcarcely an inch's diftance from each other; fo that the tree dries and perifhes.

* *Phil. Tranf.* lxii. 388.

Yellow-bellied Woodpecker, *Catesby*, i. 21.
Picus Varius, *Lin. Syst.* 176.—*Latham*, i. 574.
Le Pic Varie, *Brisson*, iv. 62.
Le Pic Varie de Carolina, *De Buffon*, vii. 77.—Lev. Mus.—Bl. Mus.

166. YELLOW-
BELLIED.

W. With a crimson crown, surrounded by a line of black:
cheeks white, with two lines of black: chin crimson: breast
and belly light yellow; the first spotted with black: coverts black,
crossed by two bars of white: primaries spotted with black and white:
tail black; interior webs of the two middle feathers barred with
white; the two outmost feathers edged with the same color. The
FEMALE wants the red on the crown. LENGTH nine inches. WEIGHT
one ounce thirteen penny-weights.

Inhabits the same country with the former. Is very numerous, and
very destructive to the fruits.

PLACE.

THIS is inserted on the suspicious authority of *Albin* *. He says,
that it is of the size of the Little *English* Spotted Woodpecker;
that the hind part of the head is black; the ridges of the wings, and
the lower part of the belly, white; the rest of the plumage, and the
tail, black; the legs yellow.

167. YELLOW-
LEGGED.

Three-toed Woodpecker, *Edw.* 114.—*Phil. Transf.* lxii. 388.—*Latham*, i. 600, 601.
Picus Tridactylus, *Lin. Syst.* 177.—*Faun. Suec.* N° 103.
Le Pic varie de la Cayenne, *Brisson*, iv. 55.—Lev. Mus.

168. THREE-
TOED.

W. With black feathers reflected over the nostrils: crown of a
bright gold color: irides blue: cheeks marked lengthways
with three black and two white lines: hind part of the neck and back

* Vol. iii. 9.—*Brisson*, iv. 24, who follows *Albin*, calls it, *Le Pic noir de la Nouvelle Angleterre.*

black;

black; the laſt ſpotted on the upper part with white: coverts of the wings black; primaries black, ſpotted with white: all the under ſide of the body white; the ſides barred with black: the middle feathers of the tail black; the outmoſt ſpotted with white: legs duſky: toes, two before, only one behind; which forms the character of this ſpecies. LENGTH eight inches. EXTENT thirteen. WEIGHT two ounces.

PLACE.

Inhabits *Hudſon's Bay*, and *Norton Sound*, lat. 64. Is frequent in *Sibiria*, and common as far as *Moſcow*, in the alps of *Dalecarlia* in *Sweden*, and in thoſe of *Switzerland* *.

A. BLACK W. Picus Martius, *Lin. Syſt.* 173.
    Spillkraka, Tillkraka, *Faun. Suec.* No 93.—*De Buffon*, vii. 41.—*Wil. Orn.* 135.—
       Latham, i. 552.—LEV. MUS.

W. With the crown of the head of a rich crimſon: the reſt of the plumage of a full black: the head of the female marked with red only behind. LENGTH eighteen inches. EXTENT twenty-nine. WEIGHT near eleven ounces.

PLACE.

Inhabits the foreſts of *Germany, Switzerland,* and the north, from *Peierſbourg* to *Ochotſk,* on the eaſtern ocean, eaſtward, and to *Lapmark* weſtward. It migrates to *Woroneſch,* about the third of *March,* and continues coming in greateſt numbers in *April.* Is called there *The*

* *M. Sprunglin's* collection at *Stettlin,* near *Bern,* who told me it was common among the *Alps.*

Fuſilier;

Fufilier; and is the moſt cunning, and difficult to be ſhot, of all the tribe.

It does vaſt damage to trees, by making holes of a great depth in the bodies to neſtle in. A buſhel of duſt and chips, a proof of its labors, are often found at the foot of the tree. Makes as much noiſe in the operation, as a woodman does with an axe. Rattles with its bill againſt the ſides of the orifice, till the woods reſound. Its note very loud. Lays two or three white ſemi-tranſparent eggs. Feeds on caterpillars and inſects, eſpecially Ants.

B. GREEN, *Br. Zool.* i. N° 84.—*Latham,* i. 577.
  Picus Viridis, Wedknar, Gronſpik, Grongjoling, *Faun. Succ.* N° 99.—*De Buffon,* vii. 7.—LEV. MUS.

W. With crimſon crown: green body; lighteſt below. LENGTH thirteen inches.

Inhabits *Europe*, as high north as *Lapmark*, where it is called *Zbi-aine* *. Is found in *Ruſſia*; but diſappears towards *Sibiria.*

C. GREY-HEADED, *Edw.* 65.—*Latham,* i. 583.

W. With a grey head, and neck of a bluiſh grey: noſtrils covered with harſh black feathers, extending in a line to the eyes: a black line, beginning at the baſe of the lower mandible, points beneath the cheeks towards the hind part of the neck: under ſide of the body of the color of the head, daſhed with green: all other parts ſo exactly like the laſt, that I ſhould ſuppoſe it to have been a variety, had not my very ſcientific friend, PALLAS, aſſured me that it was a diſtinct ſpecies, and inferior in ſize to the common GREEN.

* *Leems*, 292.

It

PLACE.

It is found in *Norway*, and among the *alps* of *Switzerland* * ; and common in the north of *Ruffia*, and ftill more in *Sibiria*. The *Tun-gufi*, of *Nijmaia Tungoufka*, roaft this fpecies, bruife the flefh, and mix it with any greafe, except that of the Bear, which diffolves too readily. They anoint their arrows with it, and pretend, that the animals, which are ftruck with them, inftantly fall †.

D. **MIDDLE SPOTTED W.** *Br. Zool.* i. N° 86.—*Latham*, i. 565.
Picus Medius, *Faun. Suec.* N° 101.—*Briffon*, iv. 38.

W. With a crimfon crown and vent: in all other refpects like the GREAT SPOTTED, N° 162, except in fize, being rather lefs.

E. **LEST SPOTTED W.** *Br. Zool.* i. N° 87.
Picus Minor, *Faun. Suec.* N° 102.
Le Petit Epeiche, *De Buffon*, vii. 62.—*Pl. Enl.* 598.—*Briffon*, iv. 41.—LEV. MUS.

W. With a crimfon crown: the reft of the head, breaft, and belly, like thofe of the former: back barred with black and white: the white on the wings diffufed in broad beds. WEIGHT under an ounce. LENGTH fix inches. EXTENT eleven.

PLACE.

The MIDDLE is only found in *Ruffia*. This, and the GREAT SPOTTED, extend to the eaftmoft parts of *Sibiria*; but all three are found as high as *Lapmark* ‡, the extremity of northern *Europe*, far within the polar circle; a country which is one vaft foreft of pines, firs, and birch ‖. Innumerable infects, or their *larvæ*, lurk in all feafons in the bark of the trees; fo that this tribe of birds is never compelled, for want of food, to fhun even the moft rigorous winters of that fevere climate. It alfo bears the heats of the torrid zone; for I difcovered it among the drawings in the collection of Governor *Loten*, made in the ifland of *Ceylon*.

* Catalogue of *Swifs* birds in M. *Spruuglin*'s cabinet, which that gentleman favored me with. This fpecies was not unnoticed by the great GESNER. See his *Hift. av.* ed. p. 710, line 20.
† *Gmelin. voy. Sibirie*, ii. 113. ‡ *Leems*, 292. ‖ *Flora Lapp. Prolog.* 21.

XI. KING-

XI. KINGFISHER. *Gen. Birds*, XXIII.

Kingfiſher, *Cateſby*, i. 69.
American Kingfiſher, *Edw.* 115.
Le Martin peſcheur hupè de la Caroline, *Briſſon*, iv. 512. & de *St. Domingue*, 515.
Alcedo Alcyon, *Lin. Syſt.* 180.—*Latham*, i. 637.
Le Jaguacati, *De Buffon*, vii. 210.—LEV. MUS.

K. With a black bill, two inches and a half long: head creſted
with long bluiſh grey feathers: above the upper mandible of
the bill, on each ſide, is a white ſpot; beneath each eye is another:
chin and throat white: the upper part of the breaſt croſſed by a
broad grey belt; the lower part, and belly, white: the ſides of a ver-
milion color; in ſome croſſing the breaſt: upper part of the neck, the
back, and coverts of the wings, of a pleaſant bluiſh grey: the ſecondaries of the ſame color; their ends, and thoſe of the lower order of
coverts, tipt with white: primaries black, barred with white: tail grey;
the two middle feathers plain; the reſt barred with white: the legs
orange. LENGTH thirteen inches. WEIGHT three ounces and a half.

Inhabits *Hudſon's Bay*, *Norton Sound*, and other parts of *North Ame-* PLACE.
rica. The *Achalalaɛti*, i. e. the Devourer of fiſh, of the *Mexicans* *,
ſeems to be the ſame bird. It has the ſame cry, manners, and ſolitary
diſpoſition, with the *European* ſpecies; and feeds not only on fiſh, but
Lizards. It makes its neſt in the face of high banks, penetrating
deep into them in an horizontal direɛtion. Lays four white eggs,
which diſcharge the young in *June*. It migrates in *Mexico*; is there
eaten, but is obſerved to have the ſame ranknefs as other piſcivorous
birds.

* *Fernandez, Nov. Hiſp.* 13.

EUROPEAN

A. European Kingfisher, *Br. Zool.* i. N° 88.—*Latham,* i. 626.
Le Martin-Pecheur, *Buffon,* vii. 164.—*Pl. Enl.* 77.
Alcedo Ifpida, *Lin. Syft.* 179.—Lev. Mus.

PLACE.

K. With the crown, and coverts of the wings, of a deep green, fpotted with cærulean: fcapulars and back bright cærulean: tail rich deep blue: breaft and belly orange red.

Said by *Du Pratz* to be found in *North America*; but, as I never faw it in any collection, doubt the fact. Inhabits the temperate parts of *Ruffia* and *Sibiria,* and is frequent about the *Jenefei,* but not farther eaft. It does not extend to *Sweden,* and it even feems a rarity in *Denmark ***.

The *Tartars* and *Oftiaks* ufe the feathers of this bird as a love-charm. They fling them on water, and preferve thofe which fwim; believing, that the woman, whom they touch with one of thefe feathers, will immediately become enamoured with them. The *Oftiaks* preferve the bill, feet, and fkin, in a purfe, and imagine them to be prefervatives againft all forts of misfortunes †.

The moft fingular northern philtre, is a fort of mufhroom, worn by the youth of *Lapland* in a purfe, *ante pubem pendulo.* LINNÆUS's apoftrophe is very diverting.

"O ridicula VENUS, tibi, quæ in exteris regionibus uteris *caffea et choco-*
"*lata,* conditis et faccharatis, vinis et bellariis, gemmis et margaritis, auro
"et argento, ferico et cofmetico, faltationibus et conventiculis, mufica et
"comœdiis, tibi fufficit hic folus exfuccus fungus." *Flora Lappon.* 368.

*· *Muller, Prod. Zool. Dan.* 13.   † *Gmelin, voy.* ii. 112.

XII. N U T-

XII. N U T H A T C H. *Gen. Birds,* XXIV.

NUTHATCH, *Br. Zool.* i. N° 89 ?—*Latham,* i. 648. 651.   170. CANADA.
Le Torchepot de Canada, *Briſſon,* iii. 592.
Sitta Europea Notwacka, *Faun. Suec.* N° 104.
La Sittelle, *De Buffon,* v. 460.—LEV. MUS.

N. With the crown, hind part of the neck, and ſhoulders, black: back and rump of a light blue grey: over each eye a white line: cheeks white: primaries duſky, edged with grey: breaſt and belly of a pure white: two middle feathers of the tail grey. the others black, with a white ſpot at the end: vent ruſt-colored. SIZE of the *European;* of which it ſeems a mere variety.

Inhabits *Canada,* and as far ſouth as *New York;* and extends to   PLACE.
the weſtern ſide of *America, Kamtſchatka*, Sibiria,* and *Ruſſia; Sweden,* and *Sondmor* † in *Norway:* and does not migrate.

Nuthatch, *Cateſby,* i. 22, lower figure.—*Latham,* i. 650. B.   171. BLACK-
Le Torchepot de la Caroline, *Briſſon,* iii. 22.   HEADED.

N. With the bill, head, and hind part of the neck, black: over each eye is a white line: back of a fine grey: wings duſky, edged with grey: breaſt and belly, and vent feathers, red: two middle feathers of the tail grey; the reſt black, marked with a white ſpot. Leſs than the *European.*

Inhabits the temperate parts of *America.*   PLACE.

---

* Among a ſmall collection of drawings made in that country by one of our voyagers.
† *Strom.* 247.

172. Lev.   Small Nuthatch, *Catefby*, i. 22.—*Briffon*, iii. 958.—*Latham*, i. 651. C.
                 La Petite Sittelle à tête Brune, *De Buffon*, v. 474.

N.   With a brown head, marked behind with a white fpot : back grey : wings of a deep brown : under fide of the body of a dirty white : two middle feathers of the tail grey ; the others black.

Place.   Inhabits *Carolina*, and other parts of *North America*.

XIII. T O D Y.

XIII. TODY. *Gen. Birds*, XXV.

Todi Sp. quarta, *Pallas Spicil.* vi. 17.—*Latham*, ii. 661, N° 9.—Br. Mus.   173. DUSKY.

T. With a bill half an inch long, broad at the bafe, flightly indented above the noftrils, and a little bent near the point; bafe befet with briftles; upper mandible brown, lower white: colors above dufky; below yellowifh white: primaries and tail of the fame color with the back, edged with dirty white: legs dark. SIZE of a Hedge Sparrow.

Inhabits *Rhode Ifland*. Has the actions of a Flycatcher. Frequents   PLACE. decayed trees, and feeds on infects. Has a brief agreeable note, which it repeats twice or thrice.—Br. Mus.

HOOPOE. *Gen. Birds*, XXVII.

A. HOOPOE, *Br. Zool.* i. N° 90.—*Latham*, i. 687.—*De Buffon*, vi. 439.
Upupa Epops, Harfogel, Popp, *Faun. Suec.* N° 105.—LEV. Mus.

H. With a high creft, of pale orange tipt with black: back and wings barred with black and white: neck reddifh brown: breaft and belly white: only ten feathers in the tail; black, with a

white

white crefcent * acrofs the middle : legs black. LENGTH twelve inches.

Inhabits *Europe*, as far as *Sweden*, where it is called *Harfugl*, or Soldier-bird, not only on account of its plumed head, but becaufe the common people believe its appearance to be an omen of war. The *Norwegians* ftyle it *Ærfugl*; it is therefore likely that it may fometimes vifit their country. It is properly a fouthern bird, and extends even to *Egypt* and *India*. Is common in the fouthern deferts of *Ruffia* and *Tartary*; grows fcarcer beyond the *Ob*; yet fome are feen beyond lake *Baikal*. Dr. *Pallas* confirms to me its filthy manners †. He affures, that it breeds, in preference, in putrid carcafes; and that he had feen the neft of one in the privy of an uninhabited houfe, in the fuburbs of *Tzaritfyn*. Lays from two to feven cinereous eggs. Ufually has no neft of its own. Breeds fometimes in hollow trees, holes in walls, or on the ground. Migratory.

* Correct the defcription of this part in the *Britifh Zoology*.
† See *Br. Zool.* i. 258.—Is rarely feen in *Britain*.

XIV. CREEPER.

XIV. CREEPER. *Gen. Birds*, XXVIII.

Br. *Zool.* i., N° 91.—*Catefby*, App. xxxvi.
Certhiu Familiaris Krypare, *Faun. Suæc.* N° 106.—*Latham*, i. 701.
Le Grimpereau, *De Buffon*, v. 481.—Lɛv. Mus.

174. Eʋropean.

C. With head and neck brown, ftreaked with black: rump tawny: coverts of wings varied with brown and black: primaries dufky, edged with white, and edged and barred with ferruginous marks: breaft and belly filvery: tail very long, confifting of twelve fharp-pointed feathers of a tawny hue.

Inhabits *North America*. Is found, but very rarely, in *Ruffia* and *Sibiria*. Found in *Sweden*, and never quits the country; and extends as far north as *Sondmor* *.

Place.

Bahama Titmoufe, *Catefby*, i. 59.
Yellow-bellied Creeper, *Edw.* 362.
Certhia Flaveola, *Lin. Syft.* 187.—*Latham*, i. 737.
Le Grimpereau de Martinique, ou le Sucrier, *Briffon*, iii. 611.
Le Sucrier, *De Buffon*, v. 542.

175. Bahama.

C. With a dufky bill head, and back: cheeks black: above each eye is a yellow line: rump yellow: wings dufky; the primaries croffed with a bar of white: neck, breaft, and belly, yellow: tail black; the exterior feathers tipt with white.

The female hath the fame marks, but the colors are more obfcure.

Inhabits the *Bahama Iflands*, and the *Antilles*; in the laft it lives among the fugar-canes, and fucks the fweet juice which exudes from them †.

Place.

* *Strom*, 244.   † *De Buffon*, v. 542.

XV. HONEY-

XV. HONEYSUCKER. *Gen. Birds,* XXIX*.

176. RED-
THROATED.

Paſter Muſcatus, *Geſner, av.* 655.
Ouriſſia five Tomineio, *Cluſ. Exot.* 96.
Guainumbi Prima, (fœm.) *Marcgrave,* 296.
Colibry, Viamelin, or Riſing Bird, *Joſſelyn's voy.* 100.—*Rarities,* 6.—Lᴇᴠ. Mᴜs.
Trochilus Colubris, *Lin. Syſt.* 191.—*Latham,* i. 769.
L'Oyſeau Mouche a rouge gorge, *Briſſon,* iii. 716.
Humming Bird, *Cateſby,* i. 65.—*Lawſon,* 146.—*Edw.*
Le Rubis, *De Buffon,* vi. 13.

H. With a black bill, three quarters of an inch long : crown, upper part of the neck, back, and coverts of the wings, of a moſt reſplendent variable green and gold : chin and throat of a ſhining rich ſcarlet, changing, as oppoſed to the light, from gold to a full black ; theſe feathers lie nearly as compaĉtly as ſcales : breaſt and belly white ; the ſides green : middle feathers of the tail green ; the exterior purple.

The chin, throat, and whole under ſide, of the female, is white : the exterior feathers of the tail tipt with white.

MANNERS. This bird, ſo admirable for its minuteneſs, vaſt ſwiftneſs of flight, food, and elegance of form and colors, gave riſe to numbers of romantic tales. They were not the *Europeans* alone, who were ſtruck with its great beauty ; the natives of *America,* to whom it was ſo familiar, were affeĉted with its gemmeous appearance, and beſtowed on it titles expreſſive of its reſplendent colors. Some nations called it *Ouriſſia,* and *Guaracyaba,* or the Sun-beam ; others, *Guaraeygaba,* or Hairs of the Sun ; others again named it *Huitzitzil,* or *Vicililin,*

* This genus may be divided into thoſe with ſtrait and thoſe with incurvated bills ; but there being none of the laſt in *North America,* the diſtinĉtion is omitted.

or the *Regenerated*; becaufe they believed it died annually, and was re-animated at the return of the flowers it fed on : that it ftuck its bill into the trunk of a tree, and remained lifelefs for fix months ; when the vital powers re-migrated, and reftored to nature one of its moft brilliant wonders.

It flies with a fwiftnefs which the eye is incapable of following. **SWIFTNESS.** The motion of the wings is fo rapid as to be imperceptible to the niceft obferver. Lightning is fcarcely more tranfient than its flight, nor the glare more bright than its colors. It never feeds but upon **FOOD.** wing, fufpended over the flower it extracts nourifhment from ; for its only food is the honied juice lodged in the nectarium, which it fucks through the tubes of its curious tongue. Like the Bee, having exhaufted the honey of one flower, it wanders to the next, in fearch of new fweets. It admires moft thofe flowers which have the deepeft tubes. Thus the female *Balfamine*, and the Scarlet *Monarda*, are particular favorites. Whofoever fets thofe plants before the window is fure to be vifited by multitudes of thefe diminutive birds. It is a moft entertaining fight to fee them fwarming around the flowers, and trying every tube of verticillated plants, by putting their bills into every one which encircles the ftalk. If they find that their brethren have been beforehand, and robbed the flower of the honey, they will, in'rage, pluck off, and throw it on the ground.

The moft violent paffions animate at times their little bodies. **RAGE.** They have often dreadful contefts, when numbers happen to difpute poffeffion of the fame flower. They will tilt againft one another with fuch fury, as if they meant to transfix their antagonifts with their long bills. During the fight, they frequently purfue the conquered into the apartments of thofe houfes whofe windows are left open, take a turn round the room, as Flies do in *England*, and then fuddenly regain the open air. They are fearlefs of mankind ; and in feeding will fuffer people to come within two yards of them ; but on a nearer approach, dart away with admirable fwiftnefs.

Fernandez Oviedo, an author of great repute, fpeaks from his own knowlege of the fpirited inftinct, even of this diminutive bird, in

defence

defence of its young: " So that when they fee a man clime yᵉ tree
" where they have their nefts, they flee at his face, and ftryke hym
" in the eyes, commyng, goying, and returnyng, with fuch fwyft-
" nefs, that no man woulde lyghtly beleeve it, that hath not feene
" it *."

Father *Charlevoix* gives a more apocryphal inftance of the courage
of this bird, in its attack on its difproportioned enemy the Raven.
As foon as the laft appears, the Honeyfucker flies up like lightning,
beds itfelf beneath the Raven's wing, and, piercing him with his
needle-like bill, till the bird is heard to croak with agony, at length
tumbles to the ground dead, either from the fall or the wound.
This relation feems of a piece with the combat of the Wren with the
Eagle, mentioned by *Ariftotle* †: but, to do juftice both to the *French*
voyager and *Grecian* philofopher, I muft add, that each of them de-
livered their reports from oral evidence.

Note.

Many fables have been related of the melody of the fong of thefe
birds. In fact, their only note is *fcreep, fcreep, fcreep*; but the noife
which they make with their wings, efpecially in the morning, when
numbers are in motion, is a fort of buzz or found refembling that
of a fpinning-wheel. Their note is chiefly emitted when they happen
to ftrike againft each other in their flight.

Nest.

Their nefts are found with great difficulty, being built in the
branch of a tree, amidft the thick foliage. It is of elegance fuitable
to the architects; formed on the outfide with mofs; in the infide lined
with the down or goffamer collected from the Great Mullein, or
Verbafcum Thapfus; but it is alfo fometimes made of flax, hemp, hair,
and other foft materials. It is of an hemifpherical fhape. Its inner
diameter an inch: its depth half an inch. The female is faid to be
the builder; the male fupplying her with materials. Each affifts in
the labor of incubation, which continues during twelve days. They
lay only two eggs, white, and as fmall as peafe. The firft is very fin-

* *Hift. of Weft Indies*, tranflated by *Richard Eden*, p. 199.
† *Hift. An.* lib. ix. c. 11. vol. i. 931.—*Charlevoix*, v. 232.

9   gular,

gular, and contrary to the general rule of nature; which makes, in all other inftances, the fmalleft and moft defencelefs- birds the moft prolific. The reafons of the exception in this cafe are double. The fmallnefs of their bodies caufes them commonly to efcape the eyes of birds of prey; or if feen, their rapid flight eludes purfuit: fo that the fpecies is preferved as fully as if they had been the moft numerous breeders.

The *Indians* of *Mexico*, *Peru*, and *Maynas*, make moft exquifite pictures of the feathers of birds; but thofe of the Honeyfuckers form the moft brilliant part. Some ufe them as ornaments, and hang them as pendants in their ears, which give a blaze emulous of the Ruby and Emerald. In order to compofe pictures, the *Indians* draw off the feathers with fmall pincers, and with fine pafte moft artfully join them together. They difpofe them with fuch fkill, as to give the true lights and fhade to the performance, and imitate nature with the greateft fidelity. Thefe were meant to decorate the idols and temples; for, before the depreffion of the *Indian* fpirit by the tyranny of the *Spaniards*, religion was highly cultivated among the *Mexicans* and *Peruvians*; and, notwithftanding it was cruel, was attended with great fplendor.

The generical name (in the *Brafilian* tongue) of thefe birds, is *Guianumbi*. There are feveral fpecies, but only one which is found in *North America*. This kind is found from *Canada*, through that great **PLACE.** continent, as low as *Louifiana*, and from thence to the *Brafils*. It breeds even in the northern climate of *Canada*; but retires not only **MIGRATES.** from thence, but even from the warm provinces of *Carolina*, at approach of winter. In *Hifpaniola*, the mountains of *Jamaica*, and the *Brafils*, countries where there are a perpetual fucceffion of flowers, they refide throughout the year.

177. Ruffed.　　　*Latham,* i. 785.

H. With long ſtrait ſlender bill : head of a rich variable green and gold : the feathers on the neck long, and diſpoſed on each ſide in form of a ruff, and of a moſt brilliant crimſon and copper color : back, and coverts of the tail, ruſt-colored : breaſt and belly white, the laſt daſhed with red : feathers of the tail pointed ; the ends brown, bottoms ferruginous : coverts of wings green : primaries deep blue.

Female.　　Crown, upper part of the neck, back, and coverts of wings and tail, green and gold : throat white, ſpotted with brown and variable copper : belly white, daſhed with ruſt : primaries deep blue : middle feathers of the tail green ; thoſe on the ſide ferruginous at their bottoms, black in the middle, and tipped with white.

Place.　　Inhabit in great numbers the neighborhood of *Natka Sound.* The *Indians* brought them to our navigators alive, with a long hair faſtened to one of their legs.

　　　　　　　　　　　　　　　　　　ORDER III.

ORDER III. GALLINACEOUS.

XVI. T U R K E Y. *Gen. Birds*, XXXI.

Turkey, *Joffelyn's voy.* 99.—*Rarities,* 8.—*Clayton's Virgin.*—*Ph. Tr. Abridg.* iii.   178. **WILD.**
    590.—*Lawfon,* 149.—*Catefby,* App. xliv.
Le Coc d'Inde, *Belon,* 248.
Gallo-pavus, *Gefner, av.* 481.—*Icon.* 56.
Gallo-pavo, *Aldrov. av.* ii. 18.
Gallo-pavo, the Turkey A. 3.
Gallo-pavo Sylveftris *Novæ Angliæ,* a *New England* Wild Turkey, *Raii. Syn.*
    av. 51.
Meleagris Gallo-pavo, M. capite caruncula frontali gularique, maris pectore bar-
    bato, *Lin. Syft.* 268.
Le Dindon, *De Buffon,* ii. 132.—*Briffon,* i. 158. tab. xvi.—*Pl. Enl.* 97.

T. With the characters defcribed in the definition of the genus.   **DESCRIPTION.**
  Color of the plumage dark, gloffed with variable copper co-
lor and green : coverts of the wings, and the quil-feathers, barred
with black and white. Tail confifts of two orders ; the upper, or   **TAIL.**
fhorter, very elegant ; the ground color a bright bay ; the middle
feathers marked with numerous bars of fhining black and green ;
the greateft part of the exterior feathers of the fame ground with the
others, marked with only three broad bands of mallard green, placed
remote from each other ; the two next are colored like thofe of the
middle ; but the end is plain, and croffed with a fingle bar, like the
exterior.

The longer, or lower order, were of a rufty white color, mottled
with black, and croffed with numerous narrow waved lines of the
fame color, and near the end with a broad band.

Wild Turkies preferve a famenefs of coloring. The tame, as ufual with domeftic animals, vary. It is needlefs to point out the diffe-rences, in fo well-known a bird. The black approach neareft to the original ftock. This variety I have feen nearly in a ftate of nature, in *Richmond* and other parks. A moft beautiful kind has of late been introduced into *England*, of a fnowy whitenefs, finely contrafting with its red head, and black pectoral tuft. Thefe, I think, came out of *Holland*, probably bred from an accidental white pair; and from them preferved pure from any dark or variegated birds.

WHITE VARIETY.

SIZE.

The fizes of the wild Turkies have been differently reprefented. Some writers affert, that there have been inftances of their weighing fixty pounds; but I find none who, fpeaking from their own know-lege, can prove their weight to be above forty. *Joffelyn* fays, that he has eaten part of a Cock, which, after it was plucked, and the en-trails taken out, weighed thirty *. *Lawfon*, whofe authority is un-queftionable, faw half a Turkey ferve eight hungry men for two meals †; and fays, that he had feen others, which, he believed, weighed forty pounds. *Catefby* tells us, that out of the many hundreds which he had handled ‡, very few exceeded thirty pounds. Each of thefe fpeak of their being double that fize, merely from the reports of others.

MANNERS.

The manners of thefe birds are as fingular as their figure. Their attitudes in the feafon of courtfhip are very ftriking. The males fling their heads and neck backwards, briftle up their feathers, drop their wings to the ground, ftrut and pace moft ridiculoufly; wheel round the females, with their wings ruftling along the earth, at the fame time emitting a ftrange found through their noftrils, not unlike the *grurr* of a great fpinning-wheel. On being interrupted, fly into great rages, and change their note into a loud and guttural gobble; and then return to dalliance.

NOTES.

The found of the females is plaintive and melancholy.

* *New England Rarities*, 8.   † *Hift. Carolina*, 149 and 27.
‡ App. xliv. The greateft certain weight is given by Mr. *Clayton*, who faw one that reached 38 lb.—*Ph. Tranf.*

The

The paffions of the males are very ftrongly expreffed by the change of colors in the flefhy fubftance of the head and neck, which alters to red, white, blue, and yellowifh, as they happen to be affected. The fight of any thing red excites their choler greatly.

IRASCIBLE.

They are polygamous, one cock ferving   or   hens. They lay in the fpring; and will lay a great number of eggs. They will perfift in laying for a great while. They retire to fome obfcure place to fit, the cock, through rage at lofs of its mate, being very apt to break the eggs. The females are very affectionate to the young, and make great moan on the lofs of them. They fit on their eggs with fuch perfeverance, that, if they are not taken away when addle, the hens will almoft perifh with hunger before they will quit the neft.

POLYGAMOUS.

Turkies greatly delight in the feeds of nettles; but thofe of the purple Fox-glove prove fatal to them *.

They are very ftupid birds; quarrelfome, and cowardly. It is diverting to fee a whole flock attack the common Cock; who will for a long time keep a great number at bay.

They are very fwift runners, in the tame as well as the wild ftate. They are but indifferent flyers. They love to perch on trees; and gain the height they wifh, by rifing from bough to bough. In a wild ftate, they get to the very fummit of the loftieft trees, even fo high as to be beyond the reach of the mufquet †.

SWIFT.
PERCH HIGH.

In the ftate of nature they go in flocks even of five hundred ‡. Feed much on the fmall red acorns; and grow fo fat in *March*, that they cannot fly more than three or four hundred yards, and are then foon run down by a horfeman. In the unfrequented parts bordering on the *Miffifipi*, they are fo tame as to be fhot with even a piftol ‖.

GREGARIOUS.

They frequent the great fwamps § of their native country; and leave them at fun-rifing to repair to the dry woods, in fearch of

HAUNTS.

---

* *De Buffon*.   † *Lawfon*, 45.   ‡ *Lawfon*, 149.   ‖ *Adair's Amer.* 360.
§ It is in the fwamps that the loftieft and moft bulky trees grow; the wet, with which they are environed, makes them a moft fecure retreat.

acorns,

acorns, and various berries; and before fun-fet retire to the fwamps to rooft.

The flefh of the wild Turkey is faid to be fuperior in goodnefs to the tame, but redder. Eggs of the former have been taken from the neft, and hatched under tame Turkies; the young will ftill prove wild, perch feparate, yet mix and breed together in the feafon. The *Indians* fometimes ufe the breed produced from the wild, as decoy-birds, to feduce thofe in a ftate of nature within their reach *.

Wild Turkies are now grown moft exceffively rare in the inhabited parts of *America*, and are only found in numbers in the diftant and moft unfrequented fpots.

The *Indians* make a moft elegant cloathing of the feathers. They twift the inner webs into a ftrong double thread of hemp, or inner bark of the mulberry-tree, and work it like matting. It appears very rich and gloffy, and as fine as a filk fhag †. They alfo make fans of the tail; and the *French* of *Louifiana* were wont to make umbrellas by the junction of four of the tails ‡.

When difturbed, they do not take to wing, but run out of fight. It is ufual to chafe them with dogs; when they will fly, and perch on the next tree. They are fo ftupid, or fo infenfible of danger, as not to fly on being fhot at; but the furvivors remain unmoved at the death of their companions ‖.

PLACE.

TURKIES are natives only of *America*, or the *New World*; and of courfe unknown to the antients. Since both thefe pofitions have been denied by fome of the moft eminent naturalifts of the fixteenth century, I beg leave to lay open, in as few words as poffible, the caufe of their error.

MISTAKEN BY BELON.

Belon §, the earlieft of thofe writers who are of opinion that thefe birds were natives of the old world, founds his notion on the defcription of the *Guinea* Fowl, the *Meleagrides* of *Strabo, Athenæus, Pliny*, and others of the antients. I reft the refutation on the excel-

* *Lawfon*, 149. † *Lawfon*, 18. *Adair*, 423. ‡ *Du Pratz*, ii. 85.
‖ *Du Pratz*, 224. § 248. *Hift. des Oif.*

lent

lent account given by *Athenæus*, taken from *Clytus Milefius*, a difciple of *Ariftotle*, which can fuit no other than that fowl. " They want, fays he, " natural affection towards their young. Their head is " naked, and on the top is a hard round body, like a peg or nail : " from their cheeks hangs a red piece of flefh, like a beard : it has " no wattles, like the common poultry : the fe thers are black, fpot-" ted with white : they have no fpurs : and both fexes are fo like, as " not to be diftinguifhed by the fight." *Varro* * and *Pliny* † take notice of the fpotted plumage, and the gibbous fubftance on the head. *Athenæus* is more minute, and contradicts every character of the Turkey : whofe females are remarkable for their natural affection ; which differ materially in form from the males ; whofe heads are def-titute of the callous fubftance ; and whofe heels (in the male) are armed with fpurs.

Aldrovandus, who died in 1605, draws his arguments from the fame fource as *Belon* ; I therefore pafs him by, and take notice of the greateft of our naturalifts, GESNER ‡ ; who falls into a miftake of another kind, and wifhes the Turkey to be thought a native of *India*. He quotes *Ælian* for that purpofe ; who tells us, " that in *India* are " very large poultry, not with combs, but with various-colored crefts, " interwoven like flowers : with broad tails, neither bending, nor " difplayed in a circular form, which they draw along the ground, " as Peacocks do when they do not erect them : and that the " feathers are partly of a gold color, partly blue, and of an emerald " color ‖."

This, in all probability, was the fame bird with the Peacock Phea-fant of Mr. *Edwards*, *Le Paon de Tibet* of M. *Briffon*, and the *Pavo Bicalcaratus* of *Linnæus*. I have feen this bird living. It has a creft, but not fo confpicuous as that defcribed by *Ælian* ; but it has thofe ftriking colors in form of eyes : neither does it erect its tail like the

ALDROVANDUS,

AND GESNER.

---

* Lib. iii. c. 9.   † Lib. x. c. 26.   ‡ *Av.* 481.   ‖ *De Anim.* lib. xvi. c. 2.

Peacock.

Peacock *. The *Catreus* of *Strabo* † feems to be the fame bird. He defcribes it as uncommonly beautiful, and fpotted; and very like a Peacock. The former author ‡ gives a more minute account of this fpecies, and under the fame name. He borrows it from *Clitarchus*, an attendant of *Alexander* the *Great* in all his conquefts. It is evident from his defcription, that it was of this kind; and it is likewife probable, that it was the fame with his large *Indian* poultry before cited. He celebrates it alfo for its fine note; but allowance muft be made for the credulity of *Ælian*. The *Catreus*, or Peacock Pheafant, is a native of *Tibet*, and in all probability of the north of *India*, where *Clitarchus* might have obferved it; for the march of *Alexander* was through that part of *India* which borders on *Tibet*, and now known by the name of *Penj-ab*, or Five Rivers.

NOT NATIVES OF EUROPE;   I fhall now collect from authors the feveral parts of the world where Turkies are unknown in the ftate of nature. *Europe* has no fhare in the queftion, it being generally agreed, that they are exotic in refpect to our continent.

NOR OF ASIA.   Neither are they found in any part of *Afia Minor*, or the *Afiatic* TURKEY, notwithftanding ignorance of their true origin firft caufed them to be named from that empire. About *Aleppo*, capital of *Syria*, they are only met with domefticated, like other poultry ‖. In *Armenia* they are unknown, as well as in *Perfia*, having been brought from *Venice* by fome *Armenian* merchants into that empire §; where they are ftill fo fcarce, as to be preferved among other rare fowls in the Royal menagery ¶.

In *India* they are kept for ufe in our fettlements, and imported from *Europe*, as I have been more than once informed by gentlemen long refident in that country.

Du Halde acquaints us, that they are not natives of *China*; but were introduced there from other countries. He errs, from mifinformation, in faying that they are common in *India*.

---

* *Edw.* ii. 67. *Briffon*, i. 291. *Lin. Syft.* 268.   † *Lib.* xv. p. 1046.
‡ *De Anim.* lib. xvii. c. 23.   ‖ *Ruffell*, 63.   § *Tavernier*, 146.
¶ *Bell's Travels*, i. 128.

I will not quote *Gemelli Careri*, to prove that they are not found in the *Philippine* iflands, becaufe that gentleman, with his pen, travelled round the world in his eafy chair, during a very long indifpofition and confinement [*].

But *Dampier* bears witnefs that none are found in *Mindanao* [†].

The hot climate of *Africa* barely fuffers thefe birds to exift in that vaft continent, except under the care of mankind. Very few are found in *Guinea*, except in the hands of the *Europeans:* the negroes declining to breed any, on account of their great tendernefs [‡].

NOR AFRICA;

Profper Alpinus fatisfies us that they are not found either in *Nubia* or in *Egypt*. He defcribes the *Meleagrides* of the antients; and only proves that the *Guinea*-hens were brought out of *Nubia*, and fold at a great price at *Cairo* ‖, but is totally filent about the Turkey of the moderns.

Let me in this place obferve, that the *Guinea*-hens have long been imported into *Britain*. They were cultivated in our farm-yards: for I difcover, in 1277, in the grainge of *Clifton*, in the parifh of *Ambrofden*, in *Buckinghamfhire*, among other articles, vi. *mutilones*, and *fex* AFRICANÆ *feminæ* ¶; for this fowl was familiarly known by the names of *Afra Avis*, and *Gallina Africana* & *Numida*. It was introduced into *Italy* from *Africa*, and from *Rome* into our country. They were neglected here by reafon of their tendernefs and difficulty of rearing. We do not find them in the bills of fare of our antient feafts §: neither do we find the Turkey: which laft argument amounts to almoft a certainty, that fuch a hardy and princely bird had not found its way to us. The other likewife was then known here by its claffical name; for that judicious writer,

---

[*] Sir *James Porter's Obf. Turkey,* i. 1.   [†] I. 321.

[‡] *Barbot,* in *Churchill's Coll.* v. 29. *Bofman,* 229.

‖ *Hift. Nat. Ægypti,* i. 201.   ¶ *Kennet's Parochial Antiq.* 287.

§ Neither in that of *George Nevil,* archbifhop of *York,* in 1466, nor among the delicacies mentioned in the *Northumberland* Houfhold Book, in the beginning of the reign of *Henry* VIII.

Dr.

Dr. *Caius**, defcribes, in the beginning of the reign of *Elizabeth*, the *Guinea* fowl, for the benefit of his friend *Gefner*, under the name of *Meleagris*, beftowed on it by *Ariftotle* †.

Having denied, on the very beft authorities, that the Turkey ever exifted as a native of the old world, I muft now bring my proofs of its being only a native of the new; and of the period in which it firft made its appearance in *Europe*.

BUT OF AMERICA.

The firft precife defcription of thefe birds is given by *Oviedo*; who in 1525 drew up a fummary of his greater work, the *Hiftory of the Indies*, for the ufe of his monarch *Charles* V. This learned man had vifited the *Weft Indies* and its iflands in perfon, and payed particular regard to the natural hiftory. It appears from him, that the Turkey was in his days an inhabitant of the greater iflands, and of the main land. He fpeaks of them as Peacocks; for, being a new bird to him, he adopts that name, from the refemblance he thought they bore to the former: "But (fays he) the neck is bare of feathers, but " covered with a fkin which they change after their phantafie into " divers colours. They have a horn as it were on their front, and " HAIRES on the breaft ‡." He defcribes other birds, which he alfo calls Peacocks. They are of the gallinaceous genus, and known by the name of *Curaffao* birds; the male of which is black, the female ferruginous.

MEXICO.

The next who fpeaks of them as natives of the main land of the warmer parts of *America*, is *Francifco Fernandez*, fent there by *Philip* II. to whom he was phyfician. This naturalift obferved them in *Mexico*. We find by him, that the *Indian* name of the male was *Huexolofl*, of the female *Cibuatotolin:* he gives them the title of *Gallus Indicus*, and *Gallo-Pavo*. As the *Indians* as well as *Spaniards* domefticated thefe ufeful birds, he fpeaks of the fize by comparifon, faying that the wild were twice the magnitude of the tame; and that they were fhot with arrows or guns ‖. I cannot learn the time

* *Caii Opufc.* 93.   † *Hift. An.* lib. vi. c. 2.   ‡ In *Purchas,* iii. 995.
‖ *Hift. Av. Nov. Hifp.* 27.

7

when

when *Fernandez* wrote. It muſt be between the years 1555 and 1598, the period of *Philip*'s reign.

Pedro de Cieſa mentions Turkies on the Iſthmus of *Darien* *. **DARIEN.** *Lery*, a *Portugueſe* author, aſſerts that they are found in *Braſil*, and gives them an *Indian* name †; but ſince I can diſcover no traces of them in that diligent and excellent naturaliſt *Marcgrave*, who reſided long in that country, I muſt deny my aſſent. But the former is confirmed by that able and honeſt navigator *Dampier*, who ſaw them frequently, as well wild as tame, in the province of *Yucatan* ‡, **YUCATAN.** now reckoned part of the kingdom of *Mexico*.

In *North America* they were obſerved by the very firſt diſcoverers. **N. AMERICA.** When *Renè de Laudonniere*, patronized by Admiral *Coligni*, attempted to form a ſettlement near the place where *Charleſtown* now ſtands, he met with them on his firſt landing, in 1564, and by his hiſtorian, has repreſented them with great fidelity in the Vth plate of the recital of his voyage ‖. From his time, the witneſſes to their being natives of this continent are innumerable. They have been ſeen in flocks of hundreds in all parts, from *Louiſiana* even to *Canada :* but at this time are extremely rare in a wild ſtate, except in the more diſtant parts, where they are ſtill found in vaſt abundance.

It was from *Mexico* or *Yucatan* that they were firſt introduced into **WHEN FIRST IN-** *Europe*; for it is certain that they were imported into *England* as **TRODUCED INTO** **EUROPE.** early as the year 1524, the 15th of *Henry* VIII §. We probably received them from *Spain*, with which we had great intercourſe till about that time. They were moſt ſucceſsfully cultivated in our kingdom from that period; inſomuch that they grew common in every farm-yard, and became even a diſh in our rural feaſts by

---

* *Seventeen Years Travels*, 20.   † In *De Laet's Deſcr. des Indes*, 491.
‡ *Voyages*, vol. ii. part 2d. p. 65, 85, 114.   ‖ *De Bry*.
§ *Baker's Cbr.*   *Anderſon's Dict. Com.* i. 354.   *Hackluyt*, ii. 165. makes their introduction about the year 1532. *Barnaby Googe*, one of our early writers on huſbandry, ſays they were not ſeen here before 1530. He highly commends a Lady *Hales*, of *Kent*, for her excellent management of theſe fowl. p. 166.

the year 1585; for we may certainly depend on the word of old *Tuſſer*, in his account of the *Chriſtmas* huſbandlie fare *.

> Beefe, mutton, and porke, ſhred pies of the beſt,
> Pig, veale, goofe and capon, and *Turkie* well dreſt :
> Cheefe, apples, and nuts, jolle carols to heare,
> As then in the countrie, is counted good cheare.

But at this very time they were ſo rare in *France*, that we are told that the very firſt which was eaten in that kingdom appeared at the nuptial feaſt of *Charles* IX. in 1570 †.

They are now very common in all parts of *Ruſſia*, but will not thrive in *Sibiria*. Are cultivated in *Sweden*, and even in *Norway*, where they degenerate in ſize ‡.

* *Five hundred pointes of good huſbandrie*, p. 57.
† *Anderſon's Dict. Comm.* i. 410.   ‡ *Pontopp.* 78.

XVII.   G R O U S.

XVII. G R O U S. *Gen. Birds.* XXXVI.

Ruffed Heathcock, or Grous, *Edw.* 248.—*Latham.*

Morehen, *La Hontan*, i. 69.

Pheafant, *Lawfon*, 139.

Tetrao umbellus, *Lin. Syft.* 275.—Tetrao togatus, *ibid.*

La gelinote hupèe de Penfylvanie, *Briffon*, i. 214.—and, La groffe gelinote
    de Canada—207.

Le Coq de Bruyere a fraife, *De Buffon, Oif.* ii. 281.—*Pl. enl.* 104. — Lev.
    Mus.—Bl. Mus.

GR. With a great ruff on the hind part of the neck, to be
raifed or depreffed at pleafure: the head crefted: that, hind
part of the neck, the ruff, back, and coverts of the wings, prettily varied with brown, ferruginous, and black: the black on the
ruff difpofed in broad black bars: the coverts of the tail marked
with heart-fhaped fpots of white: chin white: fore part of the neck
yellowifh: breaft and belly dirty white, barred with cinereous
brown: primaries barred on their outmoft fides with black and ruft-
colour.

Tail large, expanfible like a fan; in fome of a cinereous colour,
in others orange, moft elegantly barred with narrow undulated lines
of black; near the end with a broad band of afh-color, another
of black, and tipped with white.

Legs feathered to the feet: toes naked and pectinated.

Female wants both creft and ruff. Crown dufky: back mixed
with black and ruft-colour like a Woodcock: breaft, belly, and coverts of the wings, barred with dirty white and cinereous brown:
tail fhort, brown, tipt with white; two middle feathers mottled
with red.

In

SIZE.

In fize thefe birds obferve a medium between a Pheafant and a Partridge. Length 1 foot 5 inches.

PLACE.

They inhabit *North America*, from *Hudfon's Bay* * to the *Carolinas*, and probably to *Louifiana* †.

MANNERS.

The hiftory of this fpecies is very curious: all which I beg leave to tranfcribe from Mr. *Edwards*, according to the accounts given him by Mr. *Bartram* and Mr *Brooke*, who had frequent opportunity of obferving its manners; to which I fhall add another, borrowed from the Travels of the Baron *La Hontan*.

FROM MR. BARTRAM.

" He is (fays Mr. *Bartram*) a fine bird when his gaiety is dif-
" played; that is, when he fpreads his tail like that of a Turkey-
" cock, and erects a circle of feathers round his neck like a ruff,
" walking very ftately with an even pace, and making a noife fome-
" thing like a Turkey; at which time the hunter muft fire immedi-
" ately at him, or he flies away directly two or three hundred yards,
" before he fettles on the ground. There is fomething very remark-
" able in what we call their thumping; which they do with their
" wings, by clapping them againft their fides, as the hunters fay.
" They ftand upon an old fallen tree, that has lain many years on the
" ground, where they begin their ftrokes gradually, at about two fe-
" conds of time diftant from one another, and repeat them quicker
" and quicker, until they make a noife like thunder at a diftance;
" which continues, from the beginning, about a minute; then ceaf-
" eth for about fix or eight minutes before it begins again. The
" found is heard near half a mile, by which means they are dif-
" covered by the hunters, and many of them killed. I have fhot
" many of them in this pofition; but never faw them thump, they
" moftly feeing me firft, and fo left off. They commonly exercife
" in thumping fpring and fall, at about nine or ten in the morning,

---

* *Phil. Tranf.* lxii. 393.

† The accounts given by *Boffu*, *Engl. ed.* i. 95. and by *Du Pratz*, ii. are too flight for us to determine the fpecies they mean. *Charlevoix*, in his account of *Canada*, vol. v. defcribes it very well.

" and

" and four or five in the afternoon. Their food is chiefly ber-
" ries and feeds of the country: their flesh is white, and choice
" food. I believe they breed but once a year, in the spring, and
" hatch twelve or fourteen at a brood; which keep in a company
" till the following spring. Many have attempted to raise the young
" ones, and to tame them; but to no purpose. When hatched under
" a hen, they escape into the woods soon after they are hatched,
" where they either find means to subsist, or perish."

The history of this bird is thus further illustrated by Mr. *Brooke*
of *Maryland*, in *North America*: " The ruffed Grous, or Pheasant,
" breeds in all parts of *Maryland*, some countries on the Eastern
" shore excepted. They lay their eggs in nests they make in the
" leaves, either by the side of fallen trees, or the roots of standing
" ones. They lay from twelve to sixteen eggs: the time of incu-
" bation is in the spring; but how long their eggs are hatching
" I cannot say; but probably it is three weeks, the time that a
" Dunghill Hen sits. I have found their nests when a boy, and have
" endeavoured to take the old Pheasant, but never could-succeed:
" she would almost let me put my hand upon her before she would
" quit her nest; then by artifice she would draw me off from her eggs,
" by fluttering just before me for a hundred paces or more; so that
" I have been in constant hopes of taking her. They leave their
" nests as soon as they are hatched; and I believe they live at
" first on ants, small worms, &c. When they are a few days old,
" they hide themselves so artfully among the leaves, that it is dif-
" ficult to find them: as they grow up, they feed on various berries,
" fruits, and grain of the country: grapes they likewise are fond of
" in the season; but the Pheasant is more particularly fond of the ivy-
" berry. I do not know any other animal that feeds on this berry:
" I know it is poison to many. Though the Pheasant hatches
" many young at a time, and often sits twice a year, the great num-
" ber and variety of Hawks in *Maryland* feeding on them, prevents
" their increasing fast. The beating of the Pheasant, as we term it,
" is a noise chiefly made in the spring of the year by the cock-bird;

" it

MR. BROOKE.

" it may be diftinctly heard a mile in a calm day : they fwell their
" breafts like the Powting Pigeon, and beat with their wings, which
" make a noife not unlike a drum in found ; but the Pheafant fhor-
" tens each founding note, till they run one into another undiftin-
" guifhably, like ftriking two empty bottles together."

LA HONTAN.

In order to perfect, as far as I am able, the hiftory of this bird, I
fhall give a quotation from Baron *La Hontan's Voyages to North
America*, publifhed in *Englifh*, (vol. i. p. 67.) where he fpeaks of
a bird found near the lakes of *Canada*, which, I think, can be no
other than the above-defcribed, though the names given them
difagree.

La Hontan fays, " I went in company with fome *Canadefe* on
" purpofe to fee that fowl flap with its wings : believe me, this fight
" is one of the greateft curiofities in the world ; for their flapping
" makes a noife much like a drum, for about the fpace of a minute ;
" then the noife ceafes for half a quarter of an hour ; after which
" it begins again.   By this noife we were directed to the place where
" the unfortunate More-hen fat, and found them upon rotten moffy
" trees.   By flapping one wing againft the other they mean to call
" their mates ; and the humming noife that enfues thereupon may
" be heard half a quarter of a league.   This they do in the months
" of *April*, *May*, *September*, and *October* ; and, which is very re-
" markable, the More-hen never flaps in this manner but upon
" one tree.   It begins at break of day, and gives over at nine
" o'clock in the morning, till about an hour before fun-fet, then
" it flutters again, and continues fo to do till night."

Mr. GRAHAM.

To thefe accounts I beg leave to add the following, out of the
Philofophical Tranfactions; which informs us, that this fpecies of Grous
bears the *Indian* name of *Pufkee*, or *Pufpufkee*, at *Hudfon's Bay*, on
account of the leannefs and drynefs of their flefh, which is extremely white, and of a very clofe texture ; but when well prepared, is
excellent eating.   They are pretty common at *Moofe Fort* and *Henly
Houfe*; but are feldom feen at *Albany Fort*, or to the northward
of the above places. In winter they feed upon juniper-tops, in fummer
on goofeberries, rafpberries, currants, cranberries, *&c*. They are not

5   migratory ;

migratory; ſtaying all the year at *Mooſe Fort*: they build their neſt on dry ground, hatch nine young at a time, to which the mother clucks as our common hens do; and, on the leſt appearance of danger, or in order to enjoy an agreeable degree of warmth, the young ones retire under the wings of their parent.

Urogallus minor fuſcus cervice plumis alas imitantibus donata, *Cateſby, App.* tab. i.   180. PINNATED.
Tetrao Cupido, *Lin. Syſt.* 274.—*Latham.*
La Gelinote hupèe d'Amerique, *Briſſon,* i. 212.—LEV. MUS.—BL. MUS.

G R. With head, cheeks, and neck of a reddiſh brown, marked with duſky lines: chin and throat of a pale ruſty brown: on the head is a ſmall creſt: on each ſide of the neck a moſt ſingular tuft (five feathers in each) gradually lengthening to the fifth, which is about three inches long: the upper feathers ferruginous and white; the lower black: back and ſcapulars black and pale ruſt-colour; the former ſpotted with white: breaſt and belly barred with white and pale brown: tail barred with pale brown and black.

Legs covered with ſoft brown feathers: toes naked and pectinated.

SIZE of a Pheaſant. A peculiar ſpecies, not to be confounded with the preceding *. Deſcribed from the real bird by Mr. *Cateſby*; and by myſelf from the ſpecimens in Mrs. *Blackburn*'s cabinet; which were ſent from the province of *Connecticut*. Is frequent about a hundred miles up *Albany* river, in *Hudſon's Bay*.   SIZE.

       PLACE.

The tufts, which diſtinguiſh this ſpecies from all others, are rooted high on the neck, not far from the hind part of the head. The bird has the power of erecting or dropping them at pleaſure. When diſturbed, it would ſpread them horizontally, like little wings; at other times let them fall on the ſides of the neck †. It is probable, that they aſſiſt in running or flying, or perhaps both, as the real wings are very ſhort, in proportion to the weight of the body. Theſe appendages are peculiar to the cock, and almoſt the only difference between it and the hen.

* The Comte *De Buffon*, ii. 282. falls into this miſtake.   † *Cateſby.*

Long-tailed Grous, *Edw.* 118.—*Ph. Tr.* lxii.

Tetrao Phasianellus, *Lin. Syst.* 273.—*Latham.*

Le Coq de Bruyeres à longue queue, de la Baye de *Hudson, Brisson, App.* 9.—*De Buffon,* ii. 286.

G R. With the head, cheeks, and hind part of the neck, varied with reddish brown and black: the back and coverts of the tail of the same color: the scapulars and great coverts of the wings ferruginous, spotted with black, and great spots of white: primaries black, spotted with white: breast and sides white, elegantly marked with sagittal spots of black: belly white: tail short and cuneiform; the two middle feathers two inches longer than the others: the tail is of the same color with the back, only the exterior feathers are spotted with white: the legs are covered with soft and long feathers, extending over the pectinated toes, which would be otherwise naked.

The LENGTH of this species is seventeen inches: the EXTENT of wings twenty-four: WEIGHT two pounds.

Inhabits *Hudson's Bay*; and, according to Dr. *Mitchel,* the unfrequented parts of *Virginia*; but none have been brought over to *England* from any other place than the *Bay.*

Linnæus confounds this with the Wood Grous, or Cock of the Wood *. Comparison will shew with how little reason the Comte *De Buffon* † makes it to be the female of the next species, our Spotted Grous. If the female of that was not ascertained, the difference in the form of the tail would be sufficient to establish a distinction; by which it approaches nearest to the *European* Pheasant of any bird in *North America.*

* *Br. Zool.* i. N° 92. tab. xl.   † *Ois.* ii. 279.

5

The

The *Indians* about *Hudfon's Bay* call this fpecies the *Au Kufkow.* It continues there the whole year; lives among the fmall larch bufhes, and feeds, during winter, on the buds of that plant and the birch; in the fummer, on all forts of berries. The females lay from nine to thirteen eggs. The young, like others of this genus, run as foon as hatched, and make a puling noife like a chicken. They differ chiefly from the cock, in having lefs of the red naked fkin over the eyes. The cock has a fhrill crowing note, but not very loud. When difturbed, or while flying, it makes a repeated noife of *cuck, cuk*; and makes a noife with the feathers of its tail like the cracking of a fan. The flefh of thefe birds is of a light brown color, plump, and very juicy.

Black and Spotted Heathcock (male) *Edw.* 118.
Brown and Spotted Heathcock (female) *Edw.* 71.
Tetrao Canadenfis (male) *Lin. Syft.* 274.
Tetrao Canace (female) *Lin. Syft.* 275.—*Latham.*
La Gelinote *de la Baye de Hudfon, Briffon,* i. 201. and the fame, *App.* 10. (male.)
La Gelinote de Canada, *Briffon,* i. 203. tab. xx. fig. 1. 2. (m. and fem.)—*De Buffon,* ii. 279.—*Pl. Enl.* 131, 132.

182. SPOTTED

G^{R.} With a white fpot before and behind each eye: head, neck, back, and coverts of the wings and tail, dufky brown, croffed with black: throat of a gloffy black, bounded by a white line, commencing at the external corner of each eye: breaft of the former color: belly white, marked with great black fpots: tail black, external feathers tipt with orange: legs feathered: toes naked and pectinated.

The FEMALE is of a reddifh brown, barred and fpotted with black: belly of a dirty white, fpotted with black: tail of a deep brown, barred with mottled bands of black; the tips of the exterior feathers orange.

FEMALE,

The WEIGHT is twenty-three ounces: LENGTH fifteen inches: EXTENT near two feet.

SIZE.

R r 2   Inhabit

PLACE.

Inhabits *Hudfon's Bay*, *Newfoundland*, and *Canada*.   Is called by the *Englifh* of *Hudfon's Bay*, the *Wood Partridge*, from its living in pine woods.   Thefe birds are very ftupid; fo that they are often knocked down with a ftick; and are ufually caught by the natives with a noofe faftened to a ftake.   In fummer they are very palatable; for in that feafon they feed on berries.   In winter they live on the fhoots of the fpruce-fir, which infects the flefh with a very difagreeable tafte.   If it is true, that this fpecies lays but five eggs *, it is a ftrange exception to the prolific nature of the genus.

183. WHITE.

White Partridge, *Edw.* 72.—*Ellis's Voy.* 37.
La Lagopede de la Baie de *Hudfon, De Buffon,* ii. 276. tab. ix —*Latham.*
La Gelinote blanche, *Briffon,* i. 216.—*Pl. Enl.*
Tetrao Lagopus, *fuecis* Snoripa, *Lappis* Cheruna, *Faun. Suec.* N° 203

GR. With a black bill: fcarlet eye-brows, very large in the male; in the female far lefs confpicuous.   Head, neck and

SUMMER PLUMAGE.

part of the back, coverts of the tail, and fcapulars, deep orange, croffed with numerous dufky lines, and often marked with great blotches of white: belly, legs, and middle feathers of the tail, white: the reft of the tail dufky, tipt with white: the fhafts of the quill feathers black: the legs and toes warmly clad with a very thick and long coat of foft white feathers: the claws broad and flat, adapted for digging.

WINTER PLUMAGE.

DOUBLY FEATHERED.

Such is the fummer drefs: in winter they change their color to white, or, more properly fpeaking, moult, and change their colored plumes for white ones.   By a wonderful providence, every feather, except thofe of the wings and tail, becomes double; a downy one fhooting out at the bafe of each, as expreffed in the plate, which gives an additional protection againft the cold.   In the latter end of *February,* the fummer plumage begins to appear firft about the

* *Ph. Tr.* lxii. 390.

7

rump,

rump, in form of brown ftumps *, the firft rudiments of the coat they affume in the warm feafon, when each feather is fingle, fuitable to the time. I ought to have obferved before, that the SPOTTED GROUS alfo changes its fingle for double feathers at approach of winter, notwithftanding it undergoes no change of color.

The WEIGHT of this fpecies is twenty-four ounces : its LENGTH fixteen inches and a quarter : EXTENT twenty-three. SIZE.

Thefe birds are met with round the globe, within and without the *arctic* circle, and as high as *lat.* 72, in the countries round *Hudfon's Bay*, and as low as *Newfoundland*; in *Norway*; perhaps in the N. of the *Ruffian* dominions in *Europe* +, and certainly in *Afia* all over *Sibiria*, as far as *Kamtfchatka*, and in the iflands which lie between that country and *America*. Finally, they abound in *Lapland* and *Iceland*; and I repeat, with certainty, that *Norway* has fupplied me with this fpecies, which was fent to me by the late Mr. *Fleifcher*, of *Copenhagen*, along with the leffer kind, which proved to be the fame with the White Grous of the *Alps*, and the *Ptarmigan* of the Highlands of *Scotland*. Each of the varieties of the *Norwegian* birds were in their fummer drefs; and differed moft materially in fize as well as color, the one being in all refpects like the *American* kind : the leffer agreed in every point with that which I defcribe, N° 95, vol. i. of my *Britifh Zoology*. PLACE.
NORWAY.

The natives diftinguifh the kinds. The larger, which inhabits forefts, is ftyled by them *Skorv Rype*, or the Wood Grous; the leffer, which lives in the mountains, is called *Fiæld Rype*, or the Mountain Grous ‡. They all burrow under the fnow; and form extenfive walks beneath. There they feed, efpecially in *Lapland*, on NORWEGIANS
DISTINGUISH
TWO KINDS.

* *Drage's Voy.* ii. 9.

+ The feathers of the *Ruffian* kind, whichfoever it was, in early times, about *Petchora*, were an article of commerce, and were fold for two pence of their money per *Pood*, or 38 lb. *Purchas*, iii. 536.

‡ The *Ruffian* White Grous inhabits indifferently woods, mountains, plains, and marfhes. The *Britifh* fpecies or variety is in *Ruffia* about half the fize of the *Sibirian* kind.

the

the feeds of the dwarf birch *, and in the feafon on variety of berries of mountain plants. During winter they are taken and brought to *Bergen* by thoufands; are half roafted, and put into firkins, and tranfported to other countries †.

The lefler variety is not unknown in *America.* The fort here defcribed is found in amazing quantities, efpecially about *Hudfon's Bay,* where they breed in all parts along the coafts, make their nefts on dry ridges on the ground, and lay from nine to eleven eggs, powdered with black.

This is the only fpecies of Grous in *N. America* to which Providence hath given that warm protection to its feet, evidently to fecure them againft the cold of their winter lodgings: and, as they are greatly fought after by Eagles, Owls, and other birds of prey, a fine provifion is made for their fafety, by the change of color, which renders them not to be diftinguifhed from the fnow they lie on.

Every morning they take a flight into the air directly upwards, to fhake the fnow from their wings and bodies. They feed in the mornings and evenings, and in the middle of the day bafk in the fun. In the morning they call to one another with a loud note, interrupted; feeding in the intervals, and calling again.

In the beginning of *October,* they affemble in flocks of two hundred, and live much among the willows, the tops of which they eat; whence they are called *Willow Partridges.* About the beginning of *December* they appear in lefs plenty, retiring from the flats about the fettlements on *Hudfon's Bay* to the mountains, where in that month the fnow is lefs deep than in the lowlands, to feed on cranberries and other berries ‡. In *Greenland* they refort in fummer to the mountains for the fake of the crowberries ‖, which they eat even with the leaves of the plant. In winter they defcend to the fhores, where the winds fweep the fnow off the rocks, and enable them to pick up a fuftenance.

* *Fl. Lap.* 268.　　† *Pontoppidan,* ii. 92.　　‡ *Drage's Voy.* i. 174.
‖ Empetrum Nigrum. See *Crantz. Greenl.* i. 64, 75.

They

They are an excellent food, and much fearched after by the *Europeans* in *Hudfon's Bay*. They are generally as tame as chickens, efpecially in a mild day: fometimes they are rather wild; but by being driven about, or fhot at with powder, they grow fo weary, by the fhort flights they take, as foon to become very tame again. Sometimes the hunters, when they fee the birds likely to take a long flight, imitate the crying of a Hawk, which intimidates them fo much, that they inftantly fettle. When the female is killed, the male can fcarcely be forced from the body of its mate *.

The ufual method of taking them is in nets made of twine, twenty feet fquare, faftened to four poles, and fupported in front in a perpendicular direction with fticks. A long line is faftened to thefe props, the end of which is held by a perfon who lies concealed at a diftance. Several people are then employed to drive the birds within reach of the net, which is then pulled down, and often covers at one haul fifty or feventy. At this time they are fo plentiful, that ten thoufand are taken for the ufe of the fettlement from *November* to the end of *April*. In former days, they muft have been infinitely more numerous ; for Sir *Thomas Button* relates, that when he wintered there in 1612, he took eighteen hundred dozens of thefe and other fowl †: but this is a trifle to the fuccefs of M. *Jeremie*, who afferts, that there were eaten in one winter, between himfelf and feventy-nine others, ninety thoufand Grous, and twenty-five thoufand Rabbets ‡

The *Laplanders* take them by forming a hedge with the boughs of birch-trees ; leaving fmall openings at certain intervals, and hang in each a fnare. The birds are tempted to come and feed on the buds or catkins of the birch ; and whenever they endeavour to pafs through the openings they are inftantly caught.

* *Faun. Groenl.* p. 117.   † Quoted in *North-weft Fox*, 228.
‡ Recueil de Voy. au Nord. iii. 344.

ROCK

184. ROCK.   ROCK Gr. With a black line from the bill to the eye. In all other parts of the plumage of the fame colors with the WHITE, N° 183; but inferior in fize by one third.

Differs in nature. Feeds on the tops of fmall birch. Frequents only the dry rocky grounds, and the larch plains. Makes a fingular fnoring noife, with its neck ftreched out; and feemingly with difficulty. Is very numerous in the northern parts of *Hudfon's Bay*, and never vifits the fouthern end, except in very hard weather. Never takes fhelter in the woods, but fits on the rocks, or burrows in the fnow. Is inferior in goodnefs to the preceding.

A. WOOD GROUS, *Br. Zool.* i. N° 92.
    Tetraonis alterum Genus, *Plinii*, lib. x. c. 22.
    Tetrao urogallus Kjader, *Faun. Suec.* N° 200.—*Latham.*
    La Tetras ou le grand Coq de Bruyere, *De Buffon*, ii. 191. tab. v.—*Pl. Enl.* 73, 74.

MALE.   GR. With head, neck, and back croffed with flender lines of black and grey: upper part of the breaft gloffy green: tail black; the feathers on each fide fpotted with white: legs feathered:

{ SIZE.   toes naked and pectinated. LENGTH two feet eight: WEIGHT fometimes fourteen pounds.

FEMALE.   Length of the female only two feet two: color ferruginous and black, difpofed generally in bars.

Notwith-

Notwithftanding the opinions of *Linnæus* and the Count *De Buffon*, this fpecies is unknown in *North America*. Its moft foutherly habitation, as far as I can difcover, is the *Archipelago,* it being found in the iflands of *Crete* and of *Milo*. One was fhot in the laft, perched on a palm-tree, on whofe fruit it probably fed. I fufpeſt that it does not extend into *Afia Minor;* for Doċtor *Ruffell* does not enumerate it among the *Syrian* birds. As the *Tetrao,* which *Athenæus* * calls a fort of Pheafant, was found in the antient *Media,* it may ftill be met with in the northern part of *Perfia*. If *Arifiotle* intends this fpecies by the words *Tetrix* and *Ourax*†, it was likewife found in *Greece;* but he applies thofe names only to a bird which lays its eggs on the graffy ground, and fays no more.

PLACE.

Pliny gives a far clearer defcription of the *Tetraones* of *Italy.* Decet TETRAONAS *fuus nitor, abfolutaque nigritia, in fuperciliis cocci rubor*. This certainly means only the cock of the *Black Grous;* which is diftinguifhed by the intenfe blacknefs and the briiliant glofs of its plumage, as well as by its fcarlet eyebrows, which is common to it and the Wood Grous ; which laft is the fpecies defcribed by the ancient naturalift ; truly in fome refpeċts, hyperbolically in others. He fays it is of the fize of a Vulture, and not unlike it in color ‡. Both thefe affertions approach the truth ; for the upper part of the body has a dufky or footy look, not unlike that of the Vulture of the *Alps*. But when he fpeaks of its being the heavieft bird next to the Oftrich, we fee plainly he goes beyond all bounds.

It is a fpecies found in moft parts of the wooded and mountanous countries of *Europe,* and extends even to the arċtic *Lapmark* § : is common in *Ruffia* and *Sibiria;* in the laft are found greater and leffer varieties. It is found even as far as *Kamtfchatka*.

---

* *Lib.* xiv. p. 654.   † *Hift. An.* lib. vi. c. 1.   ‡ *Hift. Nat.* lib. x. c. 22.
§ *Leems,* 241.

S f

B. Spurious Gr. Tetrao Hybridus. Racklehane. *Roslægis* Roslare, *Faun. Suec.* N° 201.

GR. With a spotted breast and forked tail. In size equal to the hen of the preceding. Is much scarcer, more timid, and its note very different. *Linnæus* says it is a mixed breed between the Wood and Black Grous; but his account of it is obscure.

C. Black Grous, *Br. Zool.* i. N° 93.
Tetrao i^us, *Plinii*.
Tetrao Tetrix, Orre, *Faun. Suec.* N° 202.
Le Petit Tetras ou Coq de Bruyere a queue forchue, *De Buffon*, ii. 210.—*Pl. Enl.* 172, 173.

GR. With a white spot on the shoulders, and white vent feathers : rest of the plumage of a full black, glossed with blue : tail much forked, exterior feathers curling outwards. Weight near four pounds. Length one foot ten inches.

Female weighs but two pounds. The tail is slightly forked and short : the colors rust, black, and cinereous.

PLACE.   Inhabits *Europe*, as high as *Lapland*: extends over *Russia* and *Sibiria* as far as birch-trees grow, of the catkins and buds of which it is very fond. Feeds much on the *populus balsamifera* *, which gives its flesh a fine flavor. In northern *Europe*, this and the last species live during summer on whortle-berries, and feed their young with gnats. In summer the males perch on trees, and animate the forests with their crowing. In winter they lie on the ground, become buried in the snows, and form walks beneath, in which they

---

* The *Taccamahacca* of *North America.* *Catesby*, i. 34.

often

often continue forty days *. They are at prefent taken in fnares; but in *Lapland* were formerly fhot with arrows †.

During winter, there is at prefent a very fingular way of taking the BLACK GROUS in *Sibiria*. In the open forefts of birch, a certain number of poles are placed horizontally on forked fticks: by way of allurement, fmall bundles of corn are placed on them; and not remote, are fet certain tall bafkets of a conic fhape, with the broadeft part uppermoft: within the mouth is placed a fmall wheel, through which paffes an axis fixed fo nicely as to admit it to play very readily, and permit one fide or the other, on the leaft touch, to drop down, and again recover its fituation. The BLACK GROUS are foon attracted by the corn on the horizontal poles; firft alight on them, and after a fhort repaft fly to the bafkets, attempt to fettle on their tops, when the wheel drops fideways, and they fall headlong into the trap, which is fometimes found half full.

D. PTARMIGAN, *Br. Zool.* i. N° 95.
Tetrao Lagopus. *Suecis* Snoripa. *Lappis* Cheruna, *Faun. Suec.* N° 203.
Le Lagopede, *De Buffon*, ii. 264. tab. ix.

G R. With the head, neck, back, fcapulars, and fome of the coverts of the wings, marked with narrow lines of black, afh-color, and ruft, intermixed with fome white: wings and belly white: outmoft feathers of the tail black; thofe of the middle cinereous, mottled with black, and tipt with white. The male has a black fpot between the bill and the eye; which in the female is fcarcely vifible. One which I weighed in *Scotland* was nineteen ounces. Another weighed by Mr. *Ray*, in the *Grifons* country, only fourteen. It regularly changes its colors at approach of winter.

* *Aman. Acad.* iv. 591.   † *Olaus Gent. Septr.* lib. xix. c. 13.

Inhabits

Inhabits *Greenland, Iceland, Lapland*, all *Scandinavia*, and *Ruſſia*; but I believe does not extend to *Sibiria* or *Kamtſchatka*. This, from its haunts, is called by the *Norwegians, Fiælde Rype*, or Mountain Grous. But in *Ruſſia* it inhabits indifferently woods, mountains, plains, and marſhes. Its feathers were formerly an article of commerce. It is taken among the *Laplanders*, by the ſame ſtratagem as the WHITE GROUS, N° 183.

The *Greenlanders* catch it in nooſes hung to a long line, drawn between two men, dropping them over the neck of this ſilly bird. They ſometimes kill it with ſtones; but of late oftener by ſhooting. It is ſaid, that when the female is killed the male unwillingly deſerts the body *.

The *Greenlanders* eat it either dreſſed, or half rotten, or raw, with ſeals lard. The inteſtines, eſpecially thoſe next to the rump, and freſh drawn, are reckoned great delicacies. They alſo mix the contents with freſh train-oil and berries; a luxury frequent among theſe people. The ſkins make a warm and comfortable ſhirt, with the feathers placed next to the body. The women formerly uſed the black feathers of the tail as ornaments to their head-dreſſes.

E. REHUSAK. *Montin*, in *Aƈt. Phyſiogr. Lund.* i. 150.

GR. With neck ruſt-colored, ſpotted with black: back and coverts of tail black, varied with ruſty ſtreaks: breaſt divided from the lower part of the neck by a dark ſhade: reſt of the breaſt and vent white; the hen ſpotted with yellow: primaries white: tail black; end whitiſh: thighs white, with ſome ruſty ſpots: legs feathered to the toes: toes naked, covered with large brown ſcales. SIZE of a ſmall Hen.

* *Faun. Groenl.* p. 117.

Inhabits

Inhabits both the woods and alps of *Lapland.* Lays thirteen or fourteen reddifh eggs, marked with large brown fpots. When difturbed, flies away with a loud noife, like a coarfe laugh. The *Keron,* or common Ptarmigan, on the contrary, is filent. The *Keron* inhabits the *Alps* only.

F. HAZEL GR. *Will. Orn.* 175.
Tetrao bonafia. Hiarpe, *Faun. Suec.* N° 204.
La Gelinotte, *De Buffon,* ii. 233. tab. vii.—*Pl. Enl.* 474, 475.

GR. With the chin black, bounded with white : head and upper part of the neck croffed with dufky and cinereous lines : behind each eye a white line : coverts of wings and fcapulars fpotted with black and ruft-color : breaft and belly white, marked with bright bay fpots : feathers of the tail mottled with afh and black ; and, except the two middlemoft, croffed with a broad fingle bar of black : legs feathered half way down. FEMALE wants the black fpot on the chin, and white ftroke beyond the eyes. Its fize fuperior to an *Englifh* Partridge.

Inhabits the birch and hazel woods of many parts of *Europe,* as high as the diocefe of *Drontheim,* and even *Lapland* * ; and is not unfrequent in the temperate parts. *Paulfen* † fays that it migrates into the fouth of *Iceland* in *April,* and departs in *September ?* It lays from twelve to twenty eggs : perches ufually in the midft of a tree : is attracted by a pipe, imitative of its voice, to the nets of the fportfmen, who lie concealed in a hovel ‡. Is excellent meat, infomuch that the *Hungarians* call it *Tfchafarmadar,* or the bird of *Cæfar,* as if it was only fit for the table of the Emperor. Is found in moft parts of the *Ruffian* dominions with the *Ptarmigan,* but grows fcarcer towards the eaft of *Sibiria.*

* *Scheffer Lapl.* 138. † *Catalogue of Iceland Birds,* MS. ‡ *Gefner Av.* 230.

XVII. PARTRIDGE.

XVII. P A R T R I D G E.   *Gen. Birds*, XXXVII.

185. MARYLAND.   *American* Partridge, Clayton, *Ph. Tr. abridg.* iii. 590.—*Lawson*, 140.—*Catesby, App.*
plate xii.—*Du Pratz*, ii. 86.
Tetrao Virginianus, *Lin. Syst.* 277.
Le Perdrix d'Amerique, *Brisson*, i. 231.—Et de la Nouvelle Angleterre, 229.—*De Buffon*, ii. 447.

P. With white cheeks and throat, bounded by a line of black on all sides, and marked with another passing beneath each eye: breast whitish, prettily marked with semicircular spots of black: upper part of the breast, coverts of wings, scapulars, and coverts of tail, bright bay, edged with small black and white spots; scapulars striped with yellowish white: primaries and tail of a light ash-color.

The head of the female agrees in the white marks of the male, but the boundaries are ferruginous. There is also more red on the breast. In other respects the colors nearly correspond. In SIZE, above half as big again as the *English* Quail.

PLACE.   Frequent from *Canada* to the most southern parts of *North America*, perhaps to *Mexico*. Are great breeders, and are seen in covies of four or five and twenty. Breed the latter end of *April*, or beginning of *May*. Collect, towards the beginning of *June*, in great flocks, and take to the orchards, where they perch when disturbed. Feed much on buck-wheat; grow fat, and are excellent meat. Migrate from *Nova Scotia*, at approach of winter, to the southern provinces; but numbers reside in the latter the whole year. The males have a note twice repeated, which they emit, while the females are

sitting,

fitting, ufually perched on a rail or gate. Make a vaſt noiſe with the wings when they ariſe.

Of late they have been introduced into *Jamaica*; are naturalized to the climate, and increaſe greatly in a wild ſtate; and, as I am informed, breed in that warm climate twice in the year.

In Jamaica.

A. COMMON PARTRIDGE. Tetrao Perdix. Rapphona. *Faun. Suec.* N° 205. La Perdrix Griſe, *De Buffon*, ii. 401.—*Pl. Enl.* 27.

INHABITS as high as *Sweden*; but has not yet reached *Norway* [*]. Found in the weſt, and all the temperate parts of *Ruſſia* and *Sibiria*, and even beyond lake *Baikal*, where it winters about ſteep rocky mountains expoſed to the ſun, and where the ſnow lies leſt.

During winter, in *Sweden* it burrows beneath the ſnow; and the whole covey retires there, leaving a ſpiracle at each end of their lodge.

[*] *Brunnieb*, N° 201.

QUAIL,

E. QUAIL, *Br. Zool.* i. N° 97.—*Wachtel, Faun. Svec.* N° 206.—LEV. MUS.—BL. MUS.

I S found no further north than *Sweden.* It appears there in the beginning of the *leafing month (May)*; and is neither heard or seen there in autumn or winter, unless it should, as *Linnæus* supposes, migrate to the southern province, or *Schonen,* or retire to the *Ukraine, Wallachia,* &c *.

Quails swarm so greatly, at the time of their migration, about the *Dnieper,* and in the south of *Russia,* that they are caught by thousands, and sent to *Moscow* and *Petersburgh* in casks. They are common in all parts of *Great Tartary*; but in *Sibiria* only in the south, as their passage is hindered by the lofty snowy mountains. It is said they winter beneath the snow; and in great frosts, to be found torpid in the *Ant-hills.* Beyond lake *Baikal,* the Quails exactly resemble those of *Europe,* but are quite mute. These are used by the *Chinese* in fighting, as we do Cocks.

* *Amæn. Acad.* iv. 592.

XVIII. B U S T A R D. *Gen. Birds.* XXXIX.

I Am forry that I have it not in my power to do more than afcertain that a bird of the BUSTARD genus is found in *North America.* Captain KING was fo obliging as to inform me, that he faw on the plains near *Norton Sound,* N. lat. 64½, great flocks of a large kind. They were very fhy; ran very faft, and for a confiderable way before they took wing; fo that he never could get one fhot.

I often meet with the word *Outarde,* or *Buftard,* among the *French* voyagers in *North America;* but believe it to be always applied to a fpecies of Goofe.

The Great Buftard, *Br. Zool.* i. N° 98, is frequent over all the defert of *Tartary,* and beyond lake *Baikal.* Is a folitary bird; but collects into fmall flocks at the time of its fouthern migration, and winters about *Aftracan*[*].

A. LESSER BUSTARD, *Br. Zool.* i. N° 99.
Tetrao Tetrax, *Faun. Suec.* N° 196.
La Petite Outarde ou la Cane-petiere, *De Buffon,* ii. 40.—*Pl. Enl.* 10. 25.—LEV. MUS.

B. With crown, back, fcapulars, and coverts of the wings, ferruginous and black; primaries black at their ends, white at their bottoms; the fecondaries quite white: neck black, marked near the top and bottom with a white circle: breaft and belly white: middle feathers of the tail croffed with ruft and black, the reft white. FEMALE entirely ferruginous and black, except wings and belly. SIZE of a Pheafant.

Appears in *Sweden* rarely in the fpring: not traced further north. Very frequent in the fouthern and fouth-weft plains of *Ruffia,* and in fmall flocks when it migrates. Continues a good way into the deferts of *Tartary;* but is never feen in *Sibiria.*

PLACE.

[*] *Extrafts,* 143.

T t   ORDER

ORDER IV.   COLUMBINE.

XIX. PIGEON.   *Gen. Birds,* XL.

187. Passenger

Pigeon, *Joſſelyn's Voy.* 99.
Wild Pigeon, *Lawſon,* 140.—*Kalm.* ii. 82.
Pigeon of Paſſage, *Cateſby,* i. tab. 23.
Wood Pigeon, *Du Pratz,* ii. 88.
Columba Migratoria, *Lin. Syſt.* 285.
Le Pigeon ſauvage d'Amerique, *Briſſon,* i. 100.—*De Buffon, Oiſ.* ii. 527.—
Lev. Mus.—Bl. Mus.

P. With a black bill: red irides: head, and hind part and ſides of the neck, of a ſlaty blue; on each of the laſt a large ſhining golden ſpot: coverts of the wings of a dark blueiſh grey, marked with a few black ſpots: quil feathers brown: tail of a great length, and cuneiform: the middle feathers duſky grey, the next paler, the outmoſt white.   Weight nine ounces.

Manners.

Theſe birds viſit the provinces of *North America* in moſt amazing numbers every hard winter.   They appear in greater or leſſer numbers, according to the mildneſs or ſeverity of ſeaſon; for when the weather proves mild, few or none are ſeen in the ſouthern parts. Neceſſity alone obliges them to change their quarters, in ſearch of

Food.

acorns, maſt, and berries, which the warmer provinces yield in vaſt abundance.   When they alight, the ground is ſoon cleared of all eſculent fruits, to the great loſs of the hog, and other maſt-eating animals.   When they have devoured every thing which has fallen on the ſurface, they form themſelves into a great perpendical column, and by rotation keep flying among the boughs of the trees, from top to bottom, beating down the acorns with their wings, and ſome

or

or other, in fucceffion, alight on the earth and eat *. The fpecies of food they are fondeft of is the fmall acorn, called the *Turkey acorn,* it being alfo a favorite food of thofe birds. In *Canada* they do vaft damage in autumn, by devouring the corn, before they begin their fouthern flight †.

They build their nefts in trees, and coo like the *Englifh* Wild Pigeon; and lay two eggs. They breed in the more northern parts, from the country fouth of *Moofe Fort,* in *Hudfon's Bay* ‡, to that between *Fort Frederick* and *Fort Anne,* and the woods about the river *Onandago* ‖. During the time of incubation and nutrition, they feed firft on the feeds of the § red maple, which ripens in *May*; and after that, on thofe of the elm ¶. It appears by thofe accurate obfervers, Mr. *Bartram* and Mr. *Kalm,* that they continue in their breeding-places till the middle at left of *July:* the firft having feen them in *June*; the laft, the 19th of *July*. Mr. *Hutchins* affures me, that they continue in the inland parts of *Hudfon's Bay* till *December*; and when the ground is covered with fnow, feed on the buds of juniper.

As foon as thefe birds find a want of food, they collect in vaft flocks, and migrate to fuch places as are likely to fupply them with fubfiftence. The multitudes which appear during the rigorous feafons are fo immenfe, that the mention of them, unlefs fupported by good authority, would feem incredible. They fly by millions in a flock; and in their paffage literally intercept the light of the fun for a confiderable fpace. As foon as one flock has paffed, another fucceeds, each taking a quarter of an hour before the whole flock is gone. This continues, in fome feafons, for three days without any intermiffion **.

The inhabitants of *New York* and *Pennfylvania* are frequent witneffes of the phænomenon, and kill numbers of thefe migrants from

---

* *Du Pratz.*   † The fame.   ‡ *Ph. Tr.* lxii. 398.   ‖ *Bartram's Journey to Onandago,* 36.—*Kalm's Travels,* ii. 311.   § Acer Rubrum, *Lin.* —*Catefby,* i. 62.   ¶ Ulmus Americana, *Lin.*   ** *Catefby.*

T t 2

their

their balconies, and the roofs of their houses. When they alight on trees to roost, they often break the limbs of stout oaks, unable to support the weight of the crowds which perch on them. The ground beneath the trees on which they have lodged a night, is covered with their dung to a considerable depth *.

Joffelyn, who observed these Pigeons in 1638, in *New England*, before they were disturbed by population, says, he has seen flights of them moving at *Michaelmas* to the southward, four or five miles long, so thick that he lost sight of the sun. He adds, that they return in spring; and that they join nest to nest, and tree to tree, by their nests, for many miles together, in the woods †. *Kalm* mentions their passage through *Jersey* in *March*.

The inhabitants of *North America* profit by this kind gift of Providence, and shoot them in their passage; for they are very fat, and excellent meat. The *Indians* watch the roosting-places; go in the night, and, knocking them down with long poles, bring away thousands. Formerly, you could not go into a little *Indian* town, in the interior parts of *Carolina*, but you would find a hundred gallons of Pigeons oil or fat, which they use with their mayz, as we do butter ‡. They scorn to obtain that useful article from the quiet employ of the dairy; but are fond of the similitude, provided it could be obtained by any means suitable to their active spirit.

M. *du Pratz* hit upon an ingenious expedient of taking them on roost, by placing under the trees vessels filled with flaming sulphur; the fumes of which ascending, brought them senseless to the ground in perfect showers.

I shall conclude this account with what was communicated to me by the late Mr. *Ashton Blackburne*, from his own observations, or those of his friends, who were eye-witnesses to the wondrous facts related of these birds.

RETURN.

THEIR OIL.

---

* *Lawson*, 44.—The Rev. Mr. *Burnaby* relates the prodigious flights he saw passing in *September*, southerly, over *New England*. He adds, he scarcely met with any other food in the inns he was at. p. 132.

† *Voy*. 99.   ‡ *Lawson*, 44.

New

" *New York, June* 21, 1770.

" I think," fays Mr. *Blackburne*, " this as remarkable a bird as
" any in *America*. They are in vaft numbers in all parts, and have
" been of great fervice at particular times to our garrifons, in fup-
" plying them with frefh meat, efpecially at the out-pofts. A
" friend told me, that in the year in which *Quebec* was taken, the
" whole army was fupplied with them, if they chofe it. The way
" was this : every man took his club (for they were forbid to ufe
" their firelocks) when they *flew*, as it was termed, in fuch quan-
" tities, that each perfon could kill as many as he wanted. They
" in general begin to fly foon after day-break, and continue till
" nine or ten o'clock ; and again about three in the afternoon,
" and continue till five or fix: but what is very remarkable, they
" always fly wefterly. The times of flying here are in the fpring,
" about the latter end of *February* or the beginning of *March,*
" and continue every day for eight or ten days ; and again in the
" fall, when they begin the latter end of *July* or the beginning of
" *Auguft*. They catch vaft quantities of them in clap-nets, with
" ftale pigeons. I have feen them brought to this market by facks-
" full. People in general are very fond of them ; and I have heard
" many fay they think them as good as our common Blue Pigeon ;
" but I cannot agree with them by any means. They tafte more
" like our Queeft, or Wild Pigeon ; but are better meat. They
" have another way of killing them —They make a hut of boughs
" of trees, and fix ftale Pigeons on the ground at a fmall diftance
" from the hut. They plant poles for the Wild Pigeons to light
" on when they come a *falting* (as they term it) which they do
" every morning in the feafon, repairing to the marfhes near the
" fea-fide ; then the perfons in the hut pull the ftale Pigeon, when
" the birds will alight in vaft numbers on the poles, and great
" multitudes are fhot. Sir *William Johnfon* told me, that he killed
" at one fhot with a blunderbufs, a hundred and twenty or thirty.

Some

" Some years paft they have not been in fuch plenty as they ufed
" to be. This fpring I faw them fly one morning, as I thought
" in great abundance ; but every body was amazed how few there
" were ; and wondered at the reafon.

" I muft remark one very fingular fact : that, notwithftanding the
" whole people of a town go out a *pigeoning*, as they call it, they
" will not on fome days kill a fingle hen bird ; and on the very
" next day, not a fingle cock (and yet both fexes always fly
" wefterly) ; and when this is the cafe, the people are always affured
" that there will be great plenty of them that feafon. I have been
" at *Niagara* when the centinel has given the word that the
" Pigeons were *flying* ; and the whole garrifon were ready to run over
" one another, fo eager were they to get frefh meat."

188. CAROLINA.   Picacuroba, *Marcgrave*, 204.
                      Turtle Dove, *Lawfon*, 142.—*Du Pratz*, ii. 88.
                      La Tourterelle de la *Caroline*, *Briffon*, i. 110.
                      Turtle of Carolina, *Catefby*, i. 24.—*De Buffon*, ii. 557.—*Pl. Enl.* 175.
                      Columba Carolinenfis, *Lin. Syft.* 286.
                      Long-tailed Dove, *Edw.* 15.—LEV. MUS.—BL. MUS.

P. With the orbits naked and blue : crown, neck, back, and
fcapulars, brown ; the laft fpotted with black : fome of the
leffer coverts of a lead-color : quil feathers dufky : beneath each
eye in the male a black fpot ; on each fide of the neck another,
variable, with green, gold, and crimfon : breaft of a pale claret-
color : belly and thighs of a dull yellow : tail very long and cuneiform ; the two middle feathers brown ; the others white, marked
in the middle with a black fpot. LENGTH fourteen inches.

Inhabits *Carolina* the whole year ; and is found as far fouth as the
Weft Indies and *Brafil*. Feeds much on the berries of the poke or
Phytolacca Decandria *, and the feeds of the mug-apple or *Podophyl-*

* Phytolacca Decandria, *Lin. Sp. Pl.* 631.

lum

XIV

P. Mazell sculp.

Passenger Pigeon, N.º 187.   *Carolina Pigeon, N.º 186.*

lum Peltatum *. *Lawfon* fays, it is a great devourer of peas; on which account the *Americans* catch as many as they can in traps; and as an additional reafon, becaufe of the delicacy of their flefh.

White-crowned Pigeon, *Catefby*, i. 25.
Bald-pate, *Brown. Jam.* 468.—*Sloane*, ii. 303.
Le Pigeon de la Roche, de la *Jamaique*, *Briffon*, i. 137.—*De Buffon*, ii. 529.
Columba Leucocephala, *Lin. Syft.* 281.

189. WHITE-CROWNED.

P. With the end of the bill white; the bafe purple: crown white, beneath that purple: hind part of the neck changeable green, edged with black.

Inhabit the *Bahama* iflands, and breed among the rocks; and prove of great ufe to the inhabitants, who take vaft numbers. This fpecies is found alfo in *Jamaica*, where *Brown* fays they feed on the feeds of the mangrove and wild coffee.

PLACE.

I do not recollect that our navigators faw any Pigeons on the weftern fide of *America*; but the *Spaniards* faw abundance in lat. 41. 7. north †.

La Tourterelle de *Canada*, *Briffon*, i. 118.—*De Buffon*, ii. 552.—*Pl. Enl.* 176.
Columba Canadenfis, *Lin. Syft.* 285.

190. CANADA.

P. With the crown, hind part of the head, and upper part of the back, of a cinereous brown: the lower part of the back and rump afh-colored: lower part of the neck and the breaft cinereous, dafhed with ruft: coverts of the wings fpotted with black: primaries dufky, the exterior edges of the greater yellowifh: the tail long and cuneiform; the middle feathers afh-colored; the exterior on each fide white, marked on their inner fide with a red fpot, and beneath that with a great black one.

The head, neck, back, breaft, and coverts of the wings, of the female have the feathers terminated with dirty white and yellow.

* *Lin. Sp. Pl.* 723.   † In *Barrington's Mifcellanies*, 492.

3

Inhabits

PLACE.

Inhabits *Canada.* Greatly refembles, in fize, form, and fome of the colors, the *Carolina* Pigeon: I guefs therefore, that it is here needlefsly feparated from that fpecies.

191. GROUND.

Picuipinima, *Marcgrave,* 204.—*Raii Syn. Av.* 62. 184.—*Sloane. Jam.* ii. 305. —*Brown,* 469.
Ground Dove, *Catefby,* i. 26.
La petite Tourterelle, *Briffon,* i. 113.
Columba Pafferina, *Lin. Syft.* 285.
Le Cocotzin, *Fernandez,* 24.—*De Buffon,* ii. 559.—*Pl. Enl.* 243.

P. With a yellow bill tipt with black: red irides: upper part of the head, body, and coverts, of a cinereous brown; the coverts fpotted with black: breaft and belly a variable purple, fpotted with a deeper: the two middle feathers of the tail cinereous brown, thofe

SIZE.

of the fides dufky: legs yellowifh. In SIZE does not exceed a Lark.

PLACE.

This diminutive fpecies is not found further north than *Carolina,* where they fometimes vifit the lower parts near the fea, where fhrubs grow, in order to feed on the berries; efpecially of the pellitory, or tooth-ach tree *, which gives their flefh a fine flavor. The fpecies is continued through the warm parts of *America,* the iflands *Mexico* and *Brafil.* The *French* iflanders call them *Ortolans,* from their exceffive fatnefs and great delicacy. They take them young, when they will become very tame.

THERE is not a fingle fpecies of Pigeon to be found in *Kamtf-chatka*; a proof that the birds of this genus do not extend far to the north-weft of *America:* otherwife the narrow fea between the two continents could never confine birds of fo fwift and ftrong a flight.   * Xanthoxylum Clava Herculis, *Lin. Sp. Pl.* 1455.—*Catefby,* i. 26.

The

A. The STOCK DOVE, *Wil. Orn.*
Columba Oenas. Skogsdufva, *Faun. Suec.* N° 207.

IS very frequent in a wild state in the south of *Russia*, breeding in the turrets of village-churches, and in steep rocky banks of rivers ; but at approach of winter, migrates southward. It does the same in *Sweden* [*]. Is among the birds of the *Feroe* isles, and sometimes strays as far *Finmark* [†].

No Pigeons are seen in *Sibiria*, till you come beyond lake *Baikal*, where a very small variety (with a white rump) breeds in great plenty about the rocks. This is the same with our ROCK PIGEON, one stock or origin of our tame Pigeons. This species breeds in the cliffs as far north as the diocese of *Bergen* in *Norway* [‡]. Haunt during winter the cliffs of the *Orknies*, by myriads.

B. RING DOVE, *Br. Zool.* i. N° 102.
Columba Palumbus. Ringdufwa, *Faun. Suec.* N° 208.
Le Ramier, *De Buffon*, ii. 531.

COMMON in the *Russian* forests : very scarce in *Sibiria* ; none in the north-east. Visits *Sweden* in summer : migrates in autumn. None in *Norway*.

None of the Pigeon tribe inhabit the *arctic* zone, by reason not only of the cold, but of defect of food.

[*] *Ekmark Migr. av.* in *Amœn. Acad.* iv. 593. [†] *Leems*, 245.

[‡] *Pontop.* ii. 69.——Since the publication of the last edition of the *British Zoology*, I have been informed, by the Rev. Mr. *Ashby*, of *Barrow*, near *Newmarket*, that multitudes of *Stock Doves* breed in the rabbet-burrows on the sandy plains of *Suffolk*, about *Brandon* ; and that the shepherds annually take the young for sale.

U u ORDER

ORDER V. PASSERINE.

XX. STARE. *Gen. Birds.* XLI.

192. CRESCENT.

Lark, *Lawson,* 144.—*Catesby,* i. 33.
Le Merle a collier d'Amerique, *Brisson,* ii. 243.
L'Etourneau de la Louisiane—449.
Le Fer a Cheval, ou Merle a collier d'Amerique, *De Buffon, Ois.* iii. 371.—
    Pl. En. 256.—*Latham,* iii. 6.
Alauda Magna, *Lin. Syst.* 289.
Sturnus Ludovicianus—290.—LEV. MUS.—BL. MUS.

ST. With a dusky head, divided in the middle by a pale brown line, bounded on the side by two others: on the corner of each eye, above the bill, is a yellow spot: whole upper part of the body, neck, and wings, reddish brown and black: breast and belly of a rich yellow; the former marked with a black crescent: primaries pale brown, barred with a darker: tail very short, the feathers sharp-pointed; the three outmost white, marked with a pale brown stripe on the exterior side; rest of the feathers light brown, marked with pointed bars of black: legs long.

SIZE.

LENGTH above ten inches: WEIGHT between three and four ounces.

PLACE.

Inhabits most parts of the continent of *North America:* lives in the *savannas,* feeding chiefly on the seeds of grasses; sits on small trees and shrubs; has a jetting motion with its tail; is reckoned excellent meat; has a musical but not a various note. Arrives in *New York* in *March,* or the beginning of *April:* lays in *June,* in the grass, five white eggs, thinly spotted with pale rust-color. Leaves the country in *September* or *October.*

Sturnus

Sturnus Ludovicianus, *Lin. Syſt.* 290.—*Latham*, iii. 6.

L'Etourneau de la Louiſiane, *De Buffon*, iii. 192.—*Briſſon*, ii. 449.—*Pl. Enl.* 256.

ST. With a whitiſh bill, tipped with brown: with the crown, back, wings, and tail of a ruſty aſh-color; the firſt marked along the middle with a white line; and another of the ſame color over each eye; inner webs of the four outmoſt feathers white: in front of the neck a large black ſpot; each feather tipt with grey: reſt of the fore part of the neck, breaſt, and belly, of a rich yellow: thighs and vent dirty white.

Inhabits *Louiſiana.*

A. STARE, *Br. Zool.* i. Nº 104.

Sturnus, vulgaris Stare, *Faun. Suec.* Nº 213.

L'Etourneau, *De Buffon*, iii. 176.—*Latham*, iii. 2.—LEV. MUS.—BL. MUS.

ST. With a yellow bill: black body gloſſed with purple, and ſpotted with yellow and white: legs black.

Inhabits *Europe* as high as *Salten,* in the dioceſe of *Drontheim,* in *Norway;* and in great numbers in *Næſne Helgeland*,* in *Feroe,* and in *Iceland†.* They migrate from *Norway,* a few excepted, which lodge in the fiſſures of the rocky iſle near *Stavanger,* at the ſouthern extremity of that kingdom, and come out to baſk in the ſunny days of winter. They are found in vaſt flocks in all parts of *Ruſſia,* and the weſt of *Sibiria;* but are very ſcarce beyond

* *Leems,* 194.　　　　　† *Brunnich,* 64.

U u 2　　　　　　the

the *Jenefei*. In many places of *England* refide the whole year: in others, migrate after the breeding-feafons by thoufands to other countries.

WATER OUZEL. B.   WATER OUZEL, *Br. Zool.* i. N° 111.
Sturnus cinclus, Watnftare, *Faun. Suec.* N° 214.
Le Merle d'Eau, *De Buffon*, viii. 134.—*Latham*, iii. 48.—LEV. MUS.—BL. MUS.

ST. Dufky above: throat and breaft white: belly ruft-colored: tail black. WEIGHT two ounces and a half: LENGTH feven inches and a half.

PLACE.   Found in *Europe* as high as *Feroe* and *Finmark* *: in the *Ruffian* empire, as far as *Kamtfchatka*. The *Tartars* believe, that the feathers of this bird, tied to their nets, produce good fortune in their fifhery †.

* *Leems,* 261.   † *Voy. en Sibirie*, ii. 112.

XXI. THRUSH.

XXI. THRUSH. *Gen. Birds.* XLII.

Mock-bird, *Catesby*, i.—*Lawson*, 143.
Le Moqueur, *De Buffon*, *Oif.* ii. 323.—*Pl. Enl.* 645.—*Briffon*, ii. 262.—
   Latham, iii. 40.
Turdus Polyglottos, T. Orpheus, *Lin. Syff.* 293.—LEV. MUS.—BL. MUS.

TH. With a black bill and legs: head, neck, back, and leffer coverts on the ridge of the wing, afh-colored: the other coverts dufky, flightly edged with white: quil feathers black; white on their lower parts: under fide of the body white: tail very long; the middle feathers dufky; two outmoft feathers white; the exterior margins black.

The breaft of the female of a dirty white.

A. The Leffer, *Edw.* 78.

DIFFERS from the former in having a white line over each eye; and in being fomewhat inferior in fize. *Jamaica* *.

B. Varied. *Tetronpan, Fernanaez*, 38.

WITH a fpotted breaft; probably a young bird of one of the others.

Thefe birds fhun the cold parts of *America*; and are found from the province of *New York* as far fouth as *Mexico* and the *Antilles*. They are fo impatient of the rigorous feafon, as to retire at approach of winter from all the provinces north of *Carolina* or *Virginia*. In the firft they inhabit the whole year. They vifit *New York* in *April*, or the beginning of *May*, but are rather fcarce in that part of *America*: they breed there in *June*, and lay five or fix blue eggs, thickly fpotted with dull red.

They build often in fruit trees †; are very familiar, and love to be converfant about dwellings; and, during fummer, ufually deliver

PLACE.

---

* *Sloane*, ii. 306.   † *Lawfon*.

their

their song perched on the chimney's top. During breeding season, are very shy, and will desert their nest if any one looks on the eggs * : but are sometimes preserved, and brought alive to *England*.

During summer they feed on berries, mulberries, and other fruits, and insects. In winter, when other food is scarce, on the berries of the *Dogwood* †. When tamed, feed on every thing.

VARIOUS SONG. These birds are perhaps the first among the choristers of the woods; and are justly famed not only for the variety, fulness, and melody of their own notes, but their imitative faculty of the notes of all other birds or animals, from the Humming-bird to the Eagle. They will even imitate the sound of other things. I have heard of one, confined in a cage, that would mimic the mewing of a cat, the chattering of a magype, and the creaking of a sign in high winds. The *Mexicans* call them *Cencontlatolli* ‡, or the birds of four hundred tongues, on account of their vast variety of notes and imitative powers. In the warmer parts of *America* they sing incessantly from *March* to *August*, day and night, beginning with their own compositions, and then finishing by borrowing from the whole feathered choir, and repeat their tunes with such artful sweetness, as to excite pleasure and surprize. The super-excellence of their songs makes ample amends for the plainness of their plumage.

They may be said not only to sing, but dance: for, as if excited by a sort of extasy at their own admirable notes, they gradually raise themselves from the place where they stand, with their wings extended, drop with their head down to the same spot, and whirl round with distended wings, accompanying their melody with variety of pretty gesticulations §. They are birds of vast courage; and will attack any large bird.

---

* *Kalm*, i. 218.   † Cornus Florida, *Lin. Sp. Pl.*   ‡ *Fernandez*, p. 20.
§ *Catesby.*

Ground Mocking-bird, *Lawſon*, 143.
Fox-coloured Thruſh, *Cateſby*, i. 28.
Turdus rufus, *Lin. Syſt.* 293.—*Latham*, iii. 39.
La Grive de la Caroline, *Briſſon*, ii. 223.
Le Moqueur François, *De Buffon*, iii. 323.—*Pl. Enl.* 645.—Lev. Mus.—Bl. Mus.

195. FERRUGI-
NOUS.

TH. With yellow irides: head, and whole upper part of the body, coverts of the wings, and the tail, of a pale ruſt-color: under part of a dirty white, ſpotted with brown: acroſs the coverts of the wings are two white lines: tail very long: legs brown. Length twelve inches.

Inhabits *North America*, from *New York* to *Carolina*. In the former, arrives in *May*, and migrates to the ſouth in *Auguſt:* continues in *Virginia* and *Carolina* the whole year: builds in low buſhes, and (in *New York)* breeds in *June*, and lays five white eggs, cloſely ſpotted with ruſt-colour.

PLACE.

It feeds on berries, eſpecially thoſe of the cluſter bird-cherry, of which all the Thruſh kind are very fond. It is called in *America* the *French* Mocking-bird, from the variety of its notes; but they are far inferior to the real.

Fieldfare of Carolina, *Cateſby*, i. 29.
La Grive de Canada, *Briſſon*, ii. 225.
La Litorne de Canada, *De Buffon*, iii. 307.—*Pl. Enl.* 558.
Turdus migratorius, *Lin. Syſt.* 295.—*Latham*, iii. 26.—Lev. Mus.—Bl. Mus.

196. RED-
BREASTED.

TH. With the bill half yellow and half black: head and cheeks black: orbits covered with white feathers: chin and throat black, ſtreaked with white: under part of the neck, the breaſt, and upper part of the belly, of a deep orange: vent white: back and rump of a deep aſh-color: coverts and quil feathers duſky, edged with white: inner coverts of the wings orange: tail black; the outmoſt feather marked with white: legs brown. Size of an *Engliſh* Throſtle.

Inhabits

PLACE. Inhabits *North America*, from *Hudfon's Bay* to *Natka Sound*, on the weftern coaft; and fouth as low as *Carolina*. Quits the warmer parts in the fpring, and retires north to breed. Appear in pairs in *Hudfon's Bay*, on *Severn River*, at the beginning of *May*. At *Moofe* Fort, thefe birds build their neft, lay their eggs, and hatch their young, in fourteen days; but at *Severn* fettlement, which lies in 55, or four degrees more north, the fame is not effected in lefs than twenty-fix days *. They are alfo very common in the woods near *St. John's*, in *Newfoundland*.

They arrive in *New York* in *February*, and lay their eggs in *May*, and quit the country in *October*: in each country where they pafs the fummer, they adapt their retreat to the time in which winter fets in.

NESTS. They make their nefts with roots, mofs, &c.; and lay five eggs, of a moft lively fea-green colour. The cock is moft affiduous in affifting its mate in making the neft and feeding the young; and in the intervals chears her with its mufical voice.

Like the *English* Fieldfare, they come and go in large flocks. They have two notes; one a loud fcream, like the Miffel-bird: the other, a very fweet fong, which it delivers from the fummit of fome lofty tree.

They feed on worms, infects, and berries. Mr. *Catefby* brings a proof, that it is not the heat of the feafon alone that forces them away. He had, in *Virginia*, fome trees of the *Aluternus*, in full berry; the firft which were known in *America*: a fingle Fieldfare was fo delighted with them, as never to quit them during the whole fummer.

They are called in *America*, the *Robin*; not only from the rednefs of the breaft, but from their actions and tamenefs, as I have obferved in thofe kept in aviaries.

* *Ph. Tranfact.* lxii. 399.

T H.

Varied Thrush Nº 197.

P. Marell sculp.

TH. With a dufky crown: upper part of the neck and back   197. VARIED.
of an iron grey: cheeks black: beyond each eye is a bright
bay line: throat, under fide of the neck, and breaft, of the fame
color; the breaft croffed by a black band: fides orange-colored:
middle of the belly white: leffer coverts of the wings iron grey:
greater, dufky, tipped with bright ferruginous: primaries dufky,
croffed and edged with bay: tail long, of a deep cinereous color:
legs pale brown.   SIZE of the former.

Inhabits the woods of *Natka* Sound. *Latham*, iii. 27.   PLACE.

TH. With the head, back, and coverts, tawny; the head the   198. TAWNY.
brighteft: cheeks brown, fpotted with white: throat, breaft,
and belly, white, with large black fpots: rump, primaries, and
tail, of a pale brown: the ends of the tail fharp-pointed: legs pale
brown.   SIZE of the Redwing Thrufh.

From the province of *New York*. BL. MUS.—*Latham*, iii. 28.   PLACE.

TH. With the head, neck, back, cheeks, coverts, and tail, of   199. BROWN.
an olive brown: primaries dufky: breaft and belly of a dirty
white, marked with great brown fpots: legs dufky. SIZE of the
former; and a native of the fame country. BL. MUS.—*Latham*, iii. 28.

Merle appellé *Tilli?* *Feuillee*, i. 126.   200. RED-LEGGED.
Red-leg'd Thrufh, *Catefby*, i. 30.
Le Merle cendré de l'Amerique, *Briffon*, ii. 288.
Turdus plumbeus, *Lin. Syft.* 294.—*Latham*, iii. 33.
Le Tilly, ou la Grive cendrée de l'Amerique, *De Buffon*, iii. 314.—*Pl. Enl.* 560.

TH. With a dufky bill: irides, edges of the eyelids, and legs,
red: throat black: whole body of a dufky blue: tail long, and
cuneiform: tail dufky, with the three exterior feathers on each fide

tipt

tipt with white. WEIGHT two ounces and a half. The hen is a third part lefs than the cock.

PLACE.   Inhabits the *Bahama* iflands, *Andros*, and *Ilathera*. Has the voice and geftures of Thrufhes. Feeds on berries, efpecially of the gum *elimy* tree *.

201. LITTLE.   Little Thrufh, *Catefby*, i. 31.—*Edw.* 296.—*Latham*, iii. 20.
Le Mauvis de la Caroline, *Briffon*, ii. 212.
La Grivette d'Amerique, *De Buffon*, iii. 289.—*Pl. Enl.* 398.—LEV. MUS.—
BL. MUS.

TH. With the head, whole upper part of the body, wings, and tail, of an uniform brown color: eyelids encircled with white: chin white: breaft, and under fide of the neck, yellowifh, marked with large brown fpots: belly white: legs long, and brown. SIZE of a Lark.

PLACE.   Inhabits *Canada*, *Newfoundland*, and the whole continent of *North America*, and even *Jamaica*. In all the cold parts, even as low as *Penfylvania*, they migrate fouthward at approach of winter. They arrive in that country in *April*, and breed there. They inhabit thick woods, and the dark recefles of fwamps: are fcarce, and not often feen †. In *Jamaica* they inhabit the wooded mountains ‡. They feed on berries, &c. like other Thrufhes, but want their melody.

202. UNALASCHA.   TH. With the crown and back brown, obfcurely fpotted with dufky: breaft yellow, fpotted with black: coverts of the wings, primaries, and tail, dufky, edged with teftaceous. SIZE of a Lark.

PLACE.   Found on *Unalafcha*. *Latham*, iii. 23.

---

* Amyris Elemifera, *Lin. Sp. Pl.* i. 495.   † *Catefby.*   ‡ *Sloane*, ii. 305.

Golden

Golden-crowned Thrush, *Edw.* 252.
Motacilla aurocapilla, *Lin. Syft.* 334.—*Latham*, iii. 21.
Le Figuier a tete d'or, *Briffon*, iii. 504.
La Grivelette de St. Domingue, *De Buffon, Oif.* iii. 317.—*Pl. Enl.* 398.—
Bl. Mus.

203. GOLDEN-CROWNED.

TH. With the crown of the head of a bright gold-color, bounded on each fide by a black line : upper part of the body, wings, and tail, of an olive brown : under fide of the neck, breaft, and fides, white, fpotted with black ; or, as the *French* expreffively call, it *grivelées :* belly in fome of a pure white ; in others, fpotted : legs of a yellowifh brown. In SIZE leffer than the laft.

Inhabits *Penfylvania,* and probably all the fouthern provinces. It builds its neft on the ground, on the fide of a bank, in the form of an oven, with leaves, lining it with dry grafs, and lays five white eggs, fpotted with brown. Migrates on approach of winter to the iflands, fuch as *St. Domingo, Jamaica,* &c. Some having been taken at fea in *November* in their paffage *.

PLACE.

TH. With a black bill : general color of the plumage deep blueifh afh : crown, nape, coverts of the wings, and primaries, more or lefs edged with pale chefnut : coverts of the tail of the fame color : tail deep afh, rounded at the end : legs black. LENGTH of the whole bird feven inches and a half.

Inhabits *Hudfon's Bay.* LEV. MUS.

204. HUDSONIAN.

PLACE.

TH. With a dufky ftrong bill, half an inch long : head, neck, and breaft, mottled with light ruft-color and black : back very gloffy : and the edges of the feathers ferruginous : from the bill, above and beneath each eye, extends to the hind part of the head

205. NEW-YORK.

* *Edwards.*

X x 2

a band

a band of black: belly dufky: wings and tail black, gloffed with green: tail rounded: legs black. Size of an *Englifh* Blackbird.

PLACE.   Appears in the province of *New York* in the latter end of *October*, in its way from its more northern breeding place. Bl. Mus.

206. LABRADOR.   *Latham*, iii. 46.—Bl. Mus.—Lev. Mus.

TH. With a black bill, rather flender, near an inch long. In one fpecimen the plumage wholly black, gloffed with variable blue and green *: in another †, the feathers on the head, neck, and beginning of the back, flightly edged with deep ruft: tail, and reft of the plumage, full black; tail *even* at the end.

PLACE.   Inhabits *Labrador*, and the province of *New York*.

A. FIELDFARE, *Br. Zool.* i. N° 106.
Turdus pilaris. *Suecis* Kramsfogel. *Uplandis* Snofkata, *Faun. Suec.* N° 215.
La Litorne, *De Buffon*, iii. 301.—*Pl. Enl.* 490.—*Latham*, iii. 24.—Lev. Mus.—Bl. Mus.

TH. With head and rump cinereous: back and wing coverts chefnut: breaft and belly of a rufty white, fpotted with black. Weight about four ounces. Length ten inches.

PLACE.   Thefe birds fwarm in the woods of *Sweden* and *Norway*: breed in the higheft trees; and continue, at left in *Sweden*, the whole year ‡. In *Norway*, I do not trace them further north than the diocefe of

* From *Hudfon's Bay*.—Lev. Mus.   † *New York*.—Bl. Mus.
‡ *Amœn. Acad.* iv. 594.

Bergen.

Bergen. They migrate in great numbers into *Britain* at *Michaelmas,* and depart about the beginning of *March* ; but I have no certainty of the place they come from. Multitudes are found in all feafons in *Poland* * : multitudes alfo migrate from other places to the *Polifh* woods in autumn. Perhaps the woods in all thofe countries may be overcharged with them, fo that annually numbers may migrate into other places, without being miffed by the inhabitants. *Pontoppidan* fays, that Fieldfares are in great flocks in autumn, when berries are moft plenty †. Poffibly, after they have exhaufted the woods, they may migrate to us, compelled both by cold and want of food. They appear conftantly in the *Orknies,* near the approach of winter, in their way fouth, and feed during their refi-dence in thofe ifles on the berries of *empetrum nigrum, arbutus alpina,* and *uva urfi,* and thofe of the *juniper.* They arrive in *England* about *Michaelmas,* and leave it early in *March.* They are frequent in the forefts of *Ruffia, Sibiria,* and even *Kamtfchatka,* as is the REDWING THRUSH. Both vifit *Syria* ‡, and both migrate into *Minorca* in the end of *October,* and winter in that climate §.

B. MISSEL TH. *Br. Zool.* i. N° 105.
Turdus vifcivorus, Biork-Traft, *Faun. Suec.* N° 216.
Le Draine, *De Buffon,* iii. 295.—*Pl. Enl.* 489.—*Latham,* iii. 16.

TH. Olive-brown above : whitifh yellow below, fpotted with black : inner coverts of wings white : tail brown ; three out-moft feathers on each fide tipt with white. WEIGHT near five ounces : LENGTH eleven inches.

Inhabits *Europe* as far as *Norway* ; but not higher than the middle part. Common in *Ruffia* ; but has not reached *Sibiria.*

PLACE.

* *Klein Migr. av.* 178.   † *Hift. Norway,* 69.   ‡ *Ruffel's Aleppo,* 65, 71.
§ *Cleghorn's Minorca,* 56.

THROSTLE,

C. THROSTLE, *Br. Zool.* i. N° 107.
Turdus muficus, *Faun. Suec.* N° 217.
La Grive, *De Buffon*, iii. 280.—*Pl. Enl.* 406.—*Latham*, iii. 18.—LEV. MUS.—
BL. MUS.

TH. Above of an olive-brown: breaft white, with large dufky
fpots: inner coverts of the wings of a deep orange: cheeks
white, fpotted with brown. WEIGHT three ounces: LENGTH nine
inches.

PLACE.   Inhabits *Europe* as far north as *Sondmor* *. Is found in moft
parts of *Ruffia*, where juniper grows, efpecially about the river *Kama:*
not in *Sibiria*.

D. REDWING, *Br. Zool.* i. N° 108.
Turdus iliacus. Klera. Kladra. Talltraft. *Faun. Suec.* N° 218.
Le Mauvis, *De Buffon*, iii. 309.—*Latham*, iii. 22.

TH. With a whitifh line above each eye: and the cheeks
bounded beneath by another: head, and upper part of body,
brownifh: on each fide of the neck a fpot of deep yellow: tail
of an uniform brown: breaft white, fpotted with brown: infide of
the wings deep orange. WEIGHT two ounces and a quarter.

PLACE.   Is met with as remote as *Sondmor*, and even in *Iceland*. In *Sweden*
fings fweetly, perched on the fummit of a tree, among the forefts of
maples: builds in hedges, and lays fix blueifh green eggs, fpotted
with black †. Appears in *England* with the Fieldfare, and has with
us only a piping note. Such numbers of thefe birds, Throftles, and
Fieldfares, are killed for the market in *Polifh Pruffia*, that excife
was payed in one feafon at *Dantzick* for thirty thoufand pairs, befides what were fmuggled or payed duty in other places ‡. Found
with the Fieldfare in the *Ruffian* dominions.

* *Strom*, 260.   † *Faun. Suec.* N° 218.   ‡ *Klein, Migr. av.* 178.

Latham,

E. *Latham*, iii. 28.—Lev. Mus.

TH. With a dufky bill: crown, upper fide of the neck, back, and wings, light brown: from the bafe of the bill, on each fide, a black line paffes to the eyes, and a little beyond; over each, a line of white: chin and throat of an elegant rofe-color: breaft and belly of a whitifh brown: tail of a light brown, and rounded. Length fix inches.

Inhabits *Kamtfchatka*.

F. Oriole, *Br. Zool.* ii. App. p. 626, 8ᵛᵒ.—4ᵗᵒ, 532. tab. iv.—*Will. Orn.* 198.
Oriolus galbula, *Lin. Syft.* 160.—*Faun. Suec.* N° 95.
Loriot, *De Buffon*, iii. 260. tab. xvii.—*Pl. Enl.* 26.

TH. With head and whole body of a rich yellow: bill red: wings black; the primaries marked with a yellow fpot: tail black; tips yellow. Female dull green: ends of the exterior feathers of the tail whitifh. Length ten inches.

Inhabits many parts of *Europe*. Has been fhot in *Finland*; but is in *Sweden* a rare bird. Seen in *England* but very feldom: affects warm climates: frequent in *India* and *China*. Found in the temperate parts of the *Ruffian* empire, as far as lake *Baikal:* none beyond the *Lena*. Is almoft conftantly flitting from place to place: makes no long refidence in any. Builds a hanging neft between the forks of a bough, ufually of fome lofty tree. Lays four or five eggs, of a dirty white, fpotted with dufky. Is reckoned very good meat. Receives its name of *Loriot* from its note. Feeds on infects, berries, and fruits.

G. ROSE-COLORED OUZEL, *Br. Zool.* ii. App. p. 627. 8ᵛᵒ.—4ᵗᵒ. tab. v.

Turdus rofeus, *Faun. Suec.* Nº 219.—*Will. Orn.* 194.—*De Buffon,* iii. 348. tab. xxii. —*Pl. Enl.* 251.—*Latham,* iii. 50.

TH. With a crefted head: head, neck, wings, and tail, black, gloffed with variable purple, blue, and green: breaft, belly, back, and leffer coverts of the wings, of a fine rofe-color. SIZE of a Stare.

Linnæus, on the authority of Mr. *Adlerheim,* fays it is found in *Lapland.* Has been fhot in a garden at *Chefter;* and twice befides near *London.* Is migratory. I cannot difcover its breeding-place in *Europe.* Is found during fummer about *Aleppo,* where it is called, from its food, the *Locuft-bird* *.

Appears annually in great flocks about the river *Don:* and in *Sibiria* about the *Irtifh,* where there are abundance of Locufts, and where it breeds between the rocks.

H. RING-OUZEL, *Br. Zool.* i. Nº 110.

Turdus torquatus, *Faun. Suec.* Nº 221.

Ring-troft, Norvegio, *Brunnich,* Nº 237.—*De Buffon,* iii. 340.—*Latham,* iii. 46.— LEV. MUS.—BL. MUS.

TH. With wings and tail black: upper part of the body dufky; lower part the fame, edged with afh-color: breaft marked with a white crefcent. LENGTH eleven inches.

PLACE.   Inhabits *Europe* as high as *Lapmark* †. Is not found in *Ruffia* and *Sibiria:* is, in the fouth of *England,* and in *France,* an errant paffenger, for a week or two, to other countries: in alpine parts, refident. Is met with about Mount *Caucafus,* and in *Perfia.*

Belon ‡ fays, that in his time they fwarmed fo in their feafon about *Embrun,* that the hofts were ufed to treat their guefts with them inftead of other game. We are told by the *Count de Buffon,* that they build their nefts on the ground at the foot of fome bufh; from which they are called *Merles Terriers.*

* *Ruffel's Aleppo,* 70.   † Gjelavælgo Lapponum.—*Leems,* 260.   ‡ *Oyfeaux,* 319.

BLACKBIRD,

I.

BLACKBIRD, *Br. Zool.* i. N° 109.
Turdus merula. Tract. Kohltraft. *Faun. Suec.* N° 220.—*Latham,* ii. N° 46.—LEV.
Mus.—Bl. Mus.

TH. With a yellow bill: plumage and legs intenfely black.
FEMALE with bill and plumage of a dufky hue. LENGTH nine
inches and a half: WEIGHT about four ounces.

Inhabits *Europe* as high even as *Drontheim.* Is uncommon in PLACE.
Ruffia, except beyond the *Urallian* chain, and in the weftern provinces. But about *Woronefch,* this bird, and the STARE, do not
make their appearance till about the 17th or 18th of *April* *, fearch-
ing for food in other places during the fevere feafon.

* *Extracts,* i. 107.

XXII. CHATTERER. *Gen. Birds*, XLIII.

207. PRIB.

Chatterer, *Catesby*, i. 46.—*Edw.* 242.—*Br. Zool.* i. Nº 112.
Le Jaseur de la Caroline, *Brisson*, ii. 337.—*De Buffon, Ois.* iii. 441.—
Latham, ii. 93.—LEV. MUS.—BL. MUS.

CH. With a black bill: black band of feathers acrofs the fore-head, extended on each fide of the eyes towards the hind part of the head: head crefted; color of that and neck a pale reddifh brown: chin black: back deep brown: rump grey: coverts of wings cinereous: quil feathers dufky; ends of the feven laft tipt with wax-like, or enameled appendages, of a bright fcarlet color: tail dufky, tipt with bright yellow: breaft whitifh: belly and thighs of a light yellow: legs black. LENGTH feven inches. FEMALE wants the fcarlet appendages to the wings.

PLACE.　Inhabits *America*, from *Nova Scotia* to *Mexico* and *Cayenne* *. The *Mexican* name is *Coquantototl*. *Fernandez* † fays, it lives in the mountanous parts of the country. Feeds on feeds; but is remarkable neither for its fong, or the delicacy of its flefh. It migrates in flocks to *New York* the latter end of *March*; breeds there in *May* and *June*; and retires fouth in flocks in *November*.

The differences between this bird, and the CHATTERERS of the old continent, are thefe:—it is about an inch inferior in length: it wants the rich yellow on the wings; but, as a recompence, has the fame beautiful color on the belly.

IN EUROPE.　The *European* varieties are found as high as *Drontheim*, and appear in great numbers during winter, about *Peterfburg* and *Mofcow*, and in all parts of *Ruffia*, and are taken in flocks. They do not breed there: retiring to the very *arctic* circle for that purpofe. It is faid, that they never have been obferved beyond the river *Lena*: and that they are much fcarcer in *Sibiria* than *Ruffia*. Mr. *Bell* faw fome about *Tobolfki* in *December* ‡. The navigators found them, *September* 1778, on the weftern coaft of *America*, in lat. 64. 30.: long. 198. 30.

* *De Buffon.*　　　† *Hift. av. Nov. Hifp.* 55.　　　‡ *Travels,* i. 198.

XXIII. GROS-

XXIII. GROSBEAK. *Gen. Birds.* XLV.

Le Bec croifé, *De Buffon,* iii. 449 —*Pl. Enl.* 218.—*Br. Zool.* i. N° 115.   208. CROSSBILL.
Loxia curviroftra. Korffnaf. Kiagelrifvare, *Faun. Suec.* N° 224.—*Latham,* ii. 106.
—LEV. MUS.—BL. MUS.

GR. With each of the mandibles crooked, and croffing each other at the tips: wings, head, neck, and body, of a full red: wings and tail dufky: the coverts croffed with two white lines.

The FEMALE is of a dirty green: rump of a deep yellow: the colors of each fex vary wonderfully ; there being fcarcely two which agree in the degree of fhades of red or green: but the bills are fufficient diftinctions from all other birds.

Inhabits the northern latitudes of *North America,* from *Hudfon's*   PLACE.
Bay to *Newfoundland.* Mr. *Edwards* mentions one taken off *Green-land*; but that individual muft have been driven there by a ftorm, fince it could never have fubfifted in that woodlefs region, its food being the kernels of pine-cones, apples, and berries.

Thefe birds arrive at *Severn* river in *Hudfon's Bay,* the latter end of *May*; but fo greatly affect a cold climate, as to proceed even more northward to breed. They return in autumn at the firft fetting-in of the froft. Their habitations are the forefts of pines.

They are found in all the evergreen forefts of *Ruffia* and *Sibiria.* In *Scandinavia,* as high as *Drontheim.* In *England* they only appear in certain years. I do not find that they migrate in any part except in *America.*

The *American* fpecies varies from the *European* in being much lefs ; and in the two white lines acrofs the wings.

209. PINE.

PINE GROSBEAK, *Br. Zool.* i. N° 114.—*Edw.* 123.

Le Dur-bec, *De Buffon*, iii. 444.—*Pl. Enl.* 135.

Loxia enucleator. L. Canadenfis, *Lin. Syſt.* 299, 304.

Tallbit, Swanſk-Papgoia, *Faun. Suec.* N° 223.—*Latham*, ii. 111.—LEV. MUS.

G R. With a very ſtrong thick bill, hooked at the end: head, and upper part of the body, of a rich crimſon; each feather marked with black in the middle: leſſer coverts incline to orange; the others duſky, croſſed by two white lines: the primaries and tail duſky: lower part of the neck, breaſt, and belly, of a pale crimſon: vent cinereous: legs black. FEMALE of a dull dirty green; twice the ſize of the *Engliſh* Bullfinch.

PLACE. Inhabits *Hudſon's Bay* *, *Newfoundland*, and *Canada*, and as far as the weſtern ſide of *North America*: viſits *Hudſon's Bay*, in *April*: frequents the groves of pines and junipers: ſings on its firſt arrival, but ſoon grows ſilent: makes its neſt on trees, at a ſmall height from the ground, with ſticks, and lines it with feathers. Lays four white eggs, which are hatched in *June*. The clerk of the *California* obſerved theſe birds firſt on the 25th of *January*: they fed on the poplar tree †. It is remarked, that birds of plain colors only inhabit the frigid climates: but this gay bird is an exception.

It is likewiſe an inhabitant of the northern parts of *Europe*, as far as *Drontheim*; and in *Aſia*, in all the pine foreſts: is frequent in *Sibiria*, and the north of *Ruſſia*: is taken in *autumn* about *Peterſburg*, and brought to market in plenty. In ſpring it retires to *Lapland*.

I have ſeen them in the pine foreſts near *Invereauld*, in the county of *Aberdeen*, in *Scotland*, in the month of *Auguſt*; therefore ſuſpect they breed there.

* *Ph. Tranſ.* lxii.   † *Voy. to Hudſon's Bay,* ii. 5.

Coccothrauſte

Coccothrauftes Indica criftata, *Aldr.* ii. 289.
Virginian Nightingale, *Will. Orn* 245.—*Raii Syn. av.* 85.
Redbird, *Lawfon*, 144.—*Catefby*, i. 38.
La Cardinal hupè, *De Buffon*, iii. 458.
Grofbec de Virginie, *Briffon*, iii. 253.
Loxia Cardinalis, *Lin. Syft.* 300.—*Latham*, ii. 118.—LEV. MUS.—BL. MUS.

210. CARDINAL.

G^{R.} With a light red bill; bafe encompaffed with black feathers: head adorned with an upright pointed creft: head,
neck, and body, of a rich fcarlet color: wings and tail of a dark
and dull red. FEMALE of a much duller hue, with brown cheeks
and back: belly of a dirty yellow. LENGTH nine inches.

Inhabits the country from *Newfoundland* to *Louifiana*. Is a hardy
and familiar bird: very docile. Lives much on the grain of mayz,
which it breaks readily with its ftrong bill. Lays up a winter provifion of that grain; and conceals it very artfully in its retreat,
firft with leaves, and then with fmall branches, with an aperture
for an entrance *.

PLACE.

Their fong is remarkably fine; fo that they are called the
Virginian Nightingale. They fit warbling in the mornings, during
fpring, on the tops of the higheft trees †. They alfo fing when
confined in cages, and are much fought on account of their melody.
In a ftate of confinement the female and male are at fuch enmity,
that they will kill one another. They feldom are feen in larger
numbers than three or four together. I have heard that their note
is toned not unlike that of a Throftle; and that when tame, they
will learn to whiftle. Arrives in the *Jerfies* and *New York* in the
beginning of *April*; and during the fummer, haunts the *magnolia*
fwamps. In autumn retires to the fouth.

* *Du Pratz,* ii. 94.
† *Kalm,* ii. 71.—He fays that they are very deftructive to Bees.

Crefted

211. **Pope.**

Crested Cardinal, *Brown's Illustr.* tab. xxiii.
Le Paroure hupè, *De Buffon*, iii. 501.—*Pl. Enl.* 103.—*Latham*, ii. 124.

G R. With a most elegant upright pointed crest : that, head, and neck, of a most rich scarlet : sides of the neck, breast, and belly, white : upper part of the neck, back, wings, and tail, dark grey : legs flesh-color.

Size a little inferior to the last. Is said to have a soft feeble note *.

Place.　　Inhabits from *Louisiana* to *Brasil.*

212. **Red-breasted.**

Loxia Ludoviciana, *Lin. Syst.* 306.—*Latham*, ii. 126.—*Brisson*, iii. 247.
Le Rose Gorge, *De Buffon*, iii. 460.—*Pl. Enl.* 153.—Bl. Mus.

G R. With the head, chin, and back, deep black : coverts of the wings black, crossed with two white lines : upper part of the primaries black ; lower white : tail black ; inner webs of the two outmost feathers marked with a large white spot : breast and inner coverts of the wings of a fine rose-color : lower part of the back, belly, and vent, white : legs black. Head of the FEMALE spotted with white : breast yellowish brown, spotted with black.

Place.　　Inhabits from *New York* to *Louisiana.* Arrives in *New York* in *May* : lays five eggs : retires in *August.* Is in that province a scarce bird.

213. **Spotted.**

G R. With the middle of the head, neck, and whole under side of the body, white, marked with narrow spots of brown : above each eye is a long bar of white, reaching from near the bill to the hind part of the head : back, wings, and tail, brown : the coverts of the wings crossed with two white lines : inner coverts of the wings of a fine yellow : on the inner side of the outmost feathers of the tail is a white spot : legs dusky.

Place.　　Inhabits *New England.*—Lev. Mus.—Bl. Mus.—*Latham*, ii. 126.

* *Du Pratz,* ii. 93.

Grosbee

Grofbec appellé queue en eventail de Virginie, *Pl. Enl.* 380.—*De Buffon*, iii. 463. **214. FANTAIL.**
—*Latham*, ii. 128.

GR. With a dufky bill: fcarlet head, neck, breaft, and belly; colors moft lively on the head: back and wings dufky, tinged with fcarlet: the coverts of the tail a rich red: tail dufky, edged with red: lower belly and thighs in fome white, perhaps females.

Inhabits *Virginia*. Mr. *Blackburne* fent one from *New York*, fhot **PLACE.** in *November*. This fpecies has a cuftom of fpreading its tail like a fan, from which arofe the name.

GR. With a yellow bill: red head; hind part of an olive brown: **215. YELLOW-** hind part of the neck, and whole under fide, of a fine red: **BELLIED.** wings, lower part of the back, and the tail, olive, the two middle feathers of the laft excepted, which are red: belly yellow.

Inhabits *Virginia*. From Mr. *Kuckahn*'s collection.—*Latham*, ii. 125. **PLACE.**

GR. With the head, neck, and back, dufky, edged with pale **216. DUSKY:** brown: coverts of the wings dufky, croffed with two bars of white: quil feathers dark; their outmoft edges of a pale yellowifh green: middle of the throat white: the breaft, and fides of the belly, white fpotted with brown.

New York. Killed in *June*.—BL. MUS.—*Latham*, ii. 127. **PLACE.**

Loxia Cærulea, *Lin Syft.* 304.—*Latham*, ii. 116. **217. BLUE.**
Pyrrhula Carolineus Cærulea, *Briffon*, iii. 223. tab. xi.
Blue Grofbeak, *Catefby*, i. 39.—*De Buffon*, iii. 454.—*Pl. Enl.* 154.

GR. With a narrow black lift round the bafe of the bill: head, whole body, and coverts of the wings, of a deep blue; the laft marked with a tranfverfe bar of red: primaries and tail brown, dafhed with green: legs dufky. FEMALE of a dark brown, with a fmall mixture of blue.

7 Inhabits

PLACE.   Inhabits *Carolina* during fummer only. Is a fcarce bird, and feen only in pairs. Has but a fingle note.

218. PURPLE.
Loxia violacea, *Lin. Syft.* 306.—*Latham*, ii. 117.
Purple Grofbeak, *Catefby*, i. 40.
Pyrrhula Bahamenfis violacea, *Briffon*, iii. 326.
La Bouvreuil ou Bec rond violet de la Caroline, *De Buffon*, iv. 395.—
LEV. MUS.

GR. With head and body entirely purple: wings and tail of the fame color: over each eye a ftripe of fcarlet: throat and vent feathers of the fame color. FEMALE wholly brown, with red marks fimilar to the cock.

PLACE.   Inhabits the *Bahama* iflands. Feeds much on the mucilage of the poifon * wood-berries. From the trunk of this tree diftils a liquid, black as ink, faid to be a poifon.

219. GREY.
Le Grifalbin, *De Buffon*, iii. 467.—*Latham*, ii. 134.
Grofbec de Virginie, *Pl. Enl.* 393, N° 1.

GR. Entirely of a light grey color, except the head and neck, which are white. SIZE of a Sparrow.

PLACE.   Inhabits *Virginia*.

220. CANADA.
Loxia Canadenfis, *Lin. Syft.* 309.—*Latham*, ii. 127.
Le Flavert, *De Buffon*, iii. 462.—*Briffon*, iii. 229.—*Pl. Enl.* 152.

GR. With the upper part of the plumage of an olive green, the lower light-colored, and inclining to yellow: chin black: bafe of the bill furrounded with feathers of the fame color: legs grey.

PLACE.   Suppofed, from the *Linnæan* name, to inhabit *Canada*: but is alfo found in *Cayenne*.

* Amyris Toxifera, *Lin. Sp. Pl.* 496.

BULFINCH,

A. BULFINCH, *Br. Zool.* i. Nº 116.
Loxia Pyrrhula. Domherre, *Faun. Suec.* Nº 225.
Le Bouvreuil, *De Buffon,* iv. 372.—*Latham,* ii. 143.—LEV. MUS.—BL. MUS.

GR. With a fhort thick bill : full black crown : whole under fide rich crimfon : tail black. Under part of the FEMALE of a light brown.

Is found in *Europe* as high as *Sondmor**. Frequent in the north of *Ruffia*; and during winter, all over *Ruffia* and *Sibiria*, where it is caught for the table. The LOXIA ATRA, *Brunnich,* Nº 244, a bird fhot at *Chriftianfoe*, and defcribed as wholly black, except a white line on the wings, and the outmoft feather in the tail, feems only a variety of this fpecies. PLACE.

B. GREEN GR. *Br. Zool.* i. Nº 113.
Loxia chloris. Swenfka, *Faun. Suec.* Nº 226.
Le Verdier, *De Buffon,* iii. 171.—*Latham,* ii. 134.—LEV. MUS.—BL. MUS.

GR. With the plumage of a yellowifh green.
Inhabits *Europe* as far north as *Drontheim*. Is rare in *Ruffia*. Seen about the *Kama*. None in *Sibiria* : yet *Steller* defcribes it among the birds of *Kamtfchatka*. Inhabits *Sweden* the whole year, as does the BULFINCH. PLACE.

* *Faun. Suec.* Nº 222.

C. Haw Gr. *Br. Zool.* i. N° 113.
　　Le Gros-bec, *De Buffon,* iii. 444. tab. xxvii.—*Pl. Enl.* 99, 100.
　　Loxia coccothrauftes. Stenkneck, *Faun. Succ.* N° 222.—*Latham,* ii. 109.—
　　Lev. Mus.—Bl. Mus.

GR. With a large conic thick bill : crown and cheeks bay : hind part of the neck of a fine grey : chin black : breaft dirty fleflh-color : back, and coverts of wings, deep brown : tail fhort; inner webs white. Weight two ounces.

PLACE.　Is a fpecies that feldom is feen far north. Newly arrived in *Schonen,* where it does much damage to cherry-orchards. Lives on the kernels of fruits, and even on walnuts and almonds, which it eafily breaks with its ftrong bill. Is migratory : appears only accidentally in *England.* Known only in the weft and fouth of the *Ruffian* empire, where fruits grow, wild or cultivated. Difappears in other parts, as far as beyond lake *Baikal;* where they come from the fouth in great plenty, and feed on the *pyrus baccata,* a tree peculiar to that country. They build their neft, like that of the Turtle, with dry fticks faftened with flender roots; and lay five blueifh eggs, fpotted with brown.

XXIV. BUNTING.

Spotted Grosbeak Nº 213.   *White Crowned Bunting Nº 221.*

P. Mazell sculp.

XXIV. BUNTING. *Gen Birds*, XLVI.

Emberiza Leucophrys, *Forſter.—Phil. Tranſ.* lxii. 403, 426.—*Latham*, ii. 200.

221. WHITE-CROWNED.

B. With a red bill: white crown: ſides of the head black: beneath the eyes a black line joining the former: ſpace between both white: front, ſides of the neck, and breaſt, cinereous: back and coverts of the wings of a ruſty brown, ſpotted with black, croſſed with two lines of white: ſcapulars edged with white: primaries brown: tail long, and of the ſame color: legs fleſh-colored. LENGTH ſeven inches and a half: EXTENT nine: WEIGHT three-quarters of an ounce.

PLACE.

Inhabits the country round *Hudſon's Bay.* Viſits *Severn* river in *June.* Feeds on graſs ſeeds, grubs, inſects, &c. Makes its neſt at the bottom of willow-trees: lays four or five eggs, of a duſky color. Appears near *Albany* Fort in *May:* breeds there, and retires in *September.* Its flight ſhort and ſilent; but when it perches, ſings very melodiouſly.

Br. Zool. i. N° 122.—Tawny B. N° 121.—*Edw.* 126.
Emberiza nivalis, *Lin. Syſt.* 308.—*Latham*, ii. 16.
Sno-ſparf, *Faun. Suec.* N° 227. tab. i.—Sneekok, vinter fugl. *Cimbris.*—
    Snee fugl. Fialſter *Norvegis, Brunnich*, N° 245.
L'Ortolan de Neige, *De Buffon*, iv. 329.—*Pl. Enl.* 497.—LEV. MUS.—
    BL. MUS.

222. SNOW.

B. With a ſhort yellow bill, tipt with black: crown tawny: neck of the ſame color: breaſt and belly of a dull yellow, declining into white towards the vent: back and ſcapulars black, edged with reddiſh brown: the coverts of the tail white on their lower, yellowiſh on their upper parts: on the wings is a large bed of white: the other parts black and reddiſh brown: tail a little

Z z 2

forked;

forked; three outmoſt feathers white; the third black, tipt with white; the reſt wholly black: legs black: hind claw long, but not ſo ſtrait as that of the Lark. WEIGHT one ounce five drams: LENGTH ſix inches and a half: EXTENT ten.

PLACE.
HUDSON'S BAY.

The earlieſt of the migratory birds of *Hudſon's Bay*. Appeared in 1771, at *Severn* ſettlement, on *April* 11th; ſtayed about a month or five weeks; then proceeded farther north to breed *. Return in *September*; ſtay till *November*, when the ſevere froſts drive them ſouthward. Live in flocks: feed on graſs ſeeds, and are converſant about dunghills: are eaſily caught, under a ſmall net baited with oatmeal, and are very delicate meat. I am not certain of the winter retreat of theſe birds out of *Hudſon's Bay*; but having ſeen one of this ſpecies among thoſe ſent to Mrs. *Blackburn* from *New York*, I imagine that they ſpread over the more ſouthern parts of *North America* in the rigorous ſeaſon, as they do over *Europe* in the ſame period.

CHANGE OF
COLOR.

Theſe birds have a ſummer and a winter dreſs. The firſt we have deſcribed. Againſt the rigorous ſeaſon they become white on their head, neck, and whole under ſide: great part of their wings, and the rump, aſſumes the ſame color: the back, and middle feathers of the tail, are black. But *Linnæus*, who was very well acquainted with this ſpecies, ſays, that they vary according to age and ſeaſon. Mr. *Graham* ſent to the Royal Society two ſpecimens; one in its ſummer feathers, which exactly anſwered to our TAWNY BUNTING, N° 121; the other, to our SNOW BUNTING, N° 122, in its winter feathers. On this evidence, I beg the readers of the *Britiſh* iſles to conſider the above as one and the ſame ſpecies.

GREENLAND.

Hudſon's Bay is not the fartheſt of their northern migrations. They inhabit not only *Greenland* †, but even the dreadful climate of *Spitzbergen*, where vegetation is nearly extinct, and ſcarcely any but *cryptogamious* plants are found. It therefore excites wonder, how

* *Phil. Tranſ.* lxii. 403.　　　　† *Crantz*, i. 77.

birds,

birds, which are graminivorous in every other than thefe froft-
bound regions, fubfift: yet are there found in great flocks, both
on the land and ice of *Spitzbergen* *. They annually pafs to this
country by way of *Norway*: for in the fpring, flocks innumerable
appear, efpecially on the *Norwegian* ifles: continue only three
weeks, and then at once difappear †. As they do not breed in
Hudfon's Bay, it is certain that many retreat to this laft of lands,
and totally uninhabited, to perform in full fecurity the duties of
love, incubation, and nutrition. That they breed in *Spitzbergen* is
very probable; but we are affured that they do fo in *Greenland*.
They arrive there in *April*, and make their nefts in the fiffures of
the rocks, on the mountains, in *May*: the outfide of their neft is
grafs; the middle of feathers; and the lining the down of the
Arctic Fox. They lay five eggs, white, fpotted with brown: they
fing finely near their neft.

They are caught by the boys in autumn, when they collect near
the fhores in great flocks in order to migrate; and are eaten
dried ‡.

In *Europe* they inhabit, during fummer, the moft naked *Lapland
Alps*; and defcend, in rigorous feafons, into *Sweden*, and fill the
roads and fields; on which account the *Dalecarlians* call them
Illwarsfogel, or bad-weather birds. The *Uplanders*, *Hardvarsfogel*,
expreffive of the fame. The *Laplanders* ftyle them *Alaipg*. *Olaus
Magnus* fpeaks of them under the name of *Aviculæ nivales* §, but
mixes much fable in his narrative: he perches them alfo on trees;
whereas they always fit upon the ground.

Leems ‖ remarks, I know not with what foundation, that they
fatten on the flowing of the tides, in *Finmark*, and grow lean on the
ebb. The *Laplanders* take them in great numbers in hair fpringes,
for the tables, their flefh being very delicate.

They feem to make the countries within the whole *arctic* circle

* *Lord Mulgrave's Voy.* 188.—*Marten's Voy.* 73.   † *Leems*, 256.
‡ *Faun. Greenl.* 118.   § *De Gent. Septentr.* lib. xix. p. 156.   ‖ *Finmark*, 255.

5   *their*

their fummer refidence; from whence they overflow the more fouthern countries in amazing multitudes, at the fetting-in of winter in the frigid zone. In the winter of 1778-9, they came in fuch multitudes into *Birfa*, one of the *Orkney* iflands, as to cover the whole barony; yet, of all the numbers, hardly two agreed in colors.

NORTH OF
BRITAIN.

Lapland, and perhaps *Iceland*, furnifhes the north of *Britain* with the fwarms that frequent thofe parts during winter, as low as the *Cheviot* hills, in lat. 55. 32. Their refting-places, the *Feroe* ifles, *Schetland*, and the *Orknies.* The highlands of *Scotland*, in particular, abound with them. Their flights are immenfe; and they mingle fo clofely together, in form of a ball, that the fowlers make great havock among them. They arrive lean, foon become very fat, and are delicious food. They either arrive in the highlands very early, or a few breed there; for I had one fhot for me at *Invercauld*, the 4th of *Auguft*. But there is a certainty of their migration, for multitudes of them often fall, wearied with their paffage, on the veffels that are failing through the *Pentland Frith* *.

In their fummer drefs they are fometimes feen in the fouth of *England* †, the climate not having feverity fufficient to affect the colors; yet now and then a milk-white one appears, which is ufually miftaken for a white Lark.

RUSSIA.
SIBIRIA.

Ruffia and *Sibiria* receive them, in their fevere feafon, annually, in amazing flocks, overflowing almoft all *Ruffia*. They frequent the villages, and yield a moft luxurious repaft. They vary there infinitely in their winter colors; are pure white, fpeckled, and even quite brown ‡. This feems to be the influence of difference of age more than of feafon.

GERMANY.

Germany has alfo its fhare of them. In *Auftria* they are caught, and fed with millet, and afford the epicure a treat equal to that of the *Ortolan* §.

* Bifhop *Pocock*'s Journal, MS.   † *Morton's Northamp* 427.   ‡ *Bell's Travels*, i. 198.   § *Kramer Anim. Auftr.* 372.

Fringilla

Fringilla Hudfonias, *Forfter.*—*Ph. Tranf.* lxii.—*Latham*, ii. 666.   223. BLACK.
Snow-bird, *Catefby*, i. 36.—LEV. MUS.—BL. MUS.

B.   With a white fhort bill: blue eye: head, neck, wings, body, and tail, of a footy blacknefs, edged with ruft: breaft, belly, and vent, of a pure white: exterior fides of the primaries edged with white; of the fecondaries, with pale brown: exterior webs of the outmoft feathers of the tail white: of the fpecimen defcribed in the Tranfactions, the two outmoft are wholly white, and the third marked with a white fpot; the reft dufky. LENGTH fix inches and a half: EXTENT nine: WEIGHT half an ounce.

Appears near *Severn* fettlement not fooner than *June*: ftays a   PLACE.
fortnight: frequents the plains: feeds on grafs feeds: retires into the *arctic* parts to breed. Returns to *Hudfon's Bay* in autumn, in its paffage to the fouth. Migrates into *New York*, where it continues the whole winter. Appears in the fouthern provinces, as low as *Carolina*, but chiefly in fnow, or when the weather is harder than ordinary *. Arrive in millions, in very rigorous feafons, and fly about the houfes and barns to pick up the corn. Frequent the gardens, and the fmall hills, to feed on the fcattered feeds of grafs. Are called by the *Swedes, Snovogel,* or *Snow-bird*; by the *Americans, Chuck-bird* †. They do not change their colors in any feafon of the year. Are efteemed very delicate meat.

Towhee-bird, *Catefby*, i. 34.—*Latham*, ii. 199.   224. TOWHEE.
Fringilla Erythrophthalma, *Lin. Syft.*—*Briffon*, iii. 169.
Le Pinfon noir, aux yeux rouges, *De Buffon*, iv. 141.—LEV. MUS.—BL. MUS.

B.   With the head, coverts of the wings, whole upper fide of the body, and breaft, black: middle of the belly white: fides orange: quil feathers black, edged with white: tail long, and

---

* *Lawfon*, 146.   † *Kalm*, ii. 51, 81.

black; exterior edge of the outmoft feathers white; and a large white fpot on the end of the three firft; middle feathers entirely black.

FEMALE of a rufty brown: belly white, bounded by dirty yellow: irides in both fexes red. LENGTH eight inches and a half.

PLACE.   Inhabits *New York* and *Carolina*. Has a pretty note.

285. RICE.   Maia *Fernandez*, 56. C. ccxix.—*Wil. Orn.* 386.—*Raii Syn. Av.*
Rice-bird, Hortulanus Carolinenſis, *Cateſby*, i. 14.—*Edw.* 291.
Emberiza oryzivora, *Lin. Syſt.* 311.—*Latham*, ii. 288, 289.
L'Ortolan de la Caroline, *Briſſon*, iii. 282.
L'Agripenne, ou l'Ortolan de Riz, *De Buffon*, iv. 337.—*Pl. Enl.* 388.—LEV. MUS.
—BL. MUS.

B. With the head, and whole under fide of the body, black: hind part of the neck in fome pale yellow; in others, white: coverts of the wings, and primaries, black; the laft edged with white: part of the fcapulars, leffer coverts of the wings, and rump, white: back black, edged with dull yellow: tail of the fame colors, and each feather fharply pointed: legs red. LENGTH feven inches and a quarter.

Head, upper part of the neck, and back, of the FEMALE, yellowifh brown, fpotted with black: under part of a dull yellow: fides thinly ftreaked with black. The bird defcribed by *le Comte de Buffon*, under the title of *l'Agripenne de la Louiſiane* *, feems to be no other than a female of this fpecies, varied by having fome of the fecondary feathers wholly white.

PLACE.   Thefe birds inhabit in vaft numbers the ifland of *Cuba*, where they commit great ravages among the early crops of rice, which precede thofe of *Carolina*. As foon as the crops of that province are to their palate, they quit *Cuba*, and pafs over the fea, in numerous flights, directly north; and are very often heard in their paffage by failors frequenting that courfe. Their appearance is in

* *Hiſt. d'Oiſ.* iv. 339.—*Pl. Enl.* 388. fig. 2.

September,

September, while the rice is yet milky; and commit such devaftations, that forty acres of that grain have been totally ruined by them in a fmall time.

They arrive very lean; but foon grow fo fat, as to fly with diffi-culty; and, when fhot, often burft with the fall. They continue in *Carolina* not much above three weeks, and retire by the time the rice begins to harden. They are efteemed to be the moft delicate birds of the country. I am informed, that the male birds have a fine note.

It is very fingular, that, among the myriads which pay their autumnal vifit, there never is found a fingle cock-bird. Mr. *Catefby* verified the fact by diffecting numbers, under a fuppofition, that there might have been the young of both fexes, which had not ar-rived at the full colors; but found them all to be females, which are properly the RICE-BIRDS. Both fexes make a tranfient vifit to *Carolina* in the fpring. It is faid, that a few ftragglers continue in that country the whole year.

RICE, the periodical food of thefe birds, is a grain of *India* * : it probably arrived in *Europe* (where it has been much cultivated) by way of *Bactria*, *Sufia*, *Babylon*, and the lower *Syria* †. The time in which it reached *Italy* is uncertain: for the *Oryza* of *Pliny* is a very different plant from the common Rice; but the laft has been fown with great fuccefs about *Verona* for ages paft; and was imported from thence, and from *Egypt* ‡, into *England*; until, by a mere accident, it was introduced into *Carolina*. It was firft planted there about 1688, by Sir *Nathaniel Johnfon*, then governor of the province; but the feed being fmall and bad, the culture made little progrefs.

Chance brought here, in 1696, a veffel from *Madagafcar*; the mafter of which prefented a Mr. *Woodward* with about half a bufhel, of an excellent kind §; and from this fmall beginning fprung an

* *Raii Hift*. *Pl*. ii. 1446.   † *Strabo*, lib. xv. p. 1014.   ‡ *Anderfon's Dict*. ii. 327.   § The fame, 238—and *Catefby*, ii. Account of *Carolina*, xvii.

immenfe

immenfe fource of wealth to the fouthern provinces of *America*; and to *Europe* relief from want in times of dearth. Within little more than half a century, a hundred and twenty thoufand barrels of Rice have been in one year exported from *South Carolina*; and eighteen thoufand * from *Georgia* : and all from the remnant of a fea ftore, left in the bottom of a fack !—Ought I not to retract the word *chance*, and afcribe to PROVIDENCE fo mighty an event from fo fmall a caufe ?

226. PAINTED.

> Painted Finch, *Catefby*, i. 44.—*Lawfon*, 144.
> Emberiza ciris, *Lin. Syft.* 313.—*Edw.* 130, 173.
> Le Verdier de la Louifiane, dit vulgairement le Pape, *Briffon*, iii. 200. App. 74.
> —*Pl. Enl.* 159.—*De Buffon*, iv. 176.—*Latham*; ii. 206.—LEV. MUS.—
> BL. MUS.

B. With the head, and hind part of the head, of an exquifite deep blue : orbits fcarlet: back, greater coverts, and fecondaries, green : primaries dufky : the upper orders of leffer coverts of a fine blue ; the lower, orange : rump, and whole under fide of the body, of a rich fcarlet ; the fides declining into yellow : tail dufky, edged with green.

This beautiful fpecies is fome years in arriving at the height of its colors. At firft is of a plain brown, like a hen Sparrow ; in the next ftage, becomes blue ; in the third, attains the perfection of its gay teints.

The FEMALE is brown, and has over its plumage a tinge of green.

---

* *American Traveller*, 95, 101.——In a news-paper of laft year, I met with the following article :—A Gentleman died lately in *Carolina*, without any nearer relation than a third coufin. He determined to leave his eftate, confifting of three fine plantations, to fome perfon whofe public deferts would juftify fuch a ftep. The Gentleman, on confideration, determined in favour of Mr. *Afhby*, a gentleman in the province, whofe anceftor had introduced the culture of rice, by which *Carolina* had increafed fo amazingly in wealth, declaring at the fame time in his will, that if there had been any living perfon to whom his country was equally obliged, in the fame line of peace, he would have preferred him. Mr. *Afhby*, on his death, which happened lately, took poffeffion of the Gentleman's eftate, in confequence of this will.—How much more rational is fuch a conduct, than endowing colleges or hofpitals !

Inhabits

Inhabits *Carolina* in the fummer-time; but migrates in winter per-haps as far as *Vera Cruz*, in *Spanifh America*, where the *Spaniards* call it *Maripofa pintada*, or the Painted Butterfly. It chufes a tree for neftling equal to its own elegance of form and color; affecting the orange for that purpofe.

Hortulanus Ludovifianus, *Briffon*, iij. 278.—*De Buffon*, iv. 325.—*Pl. Enl.* 158.   **227. LOUISIANE.**
Emberiza Ludovicia, *Lin. Syft.* 310.—*Latham*, ii. 177.

B. With the crown reddifh, furrounded with a black mark, in form of a horfe-fhoe: another black line beneath each eye: the whole upper part of the body of a rufty brown, fpotted with black: lower part of the back, leffer coverts of the wings, and rump, black: breaft and belly reddifh; towards the vent growing more faint: tail and primaries black.

Inhabits *Louifiana*. Nearly allied to the *European* fpecies: per-haps a female, or young bird.   PLACE.

B. With a large and thick bill: fore part of the head of a yel-lowifh green: hind part and cheeks cinereous: above each eye a line of rich yellow: on the corner of the mouth another: on the throat a black fpot: breaft and belly of a fine yellow: back, fca-pulars, and fecondaries, black, edged with reddifh brown: leffer coverts of a bright bay: primaries and tail of a dufky brown: vent and thighs white: legs dufky.   **228. BLACK-THROATED.**

Inhabits *New York*. LEV. MUS.—BL. MUS.—*Latham*, ii. 197.   PLACE.

B. With a yellow line from the bill, reaching over each eye: crown dufky, divided lengthways with a white line: back black, edged with pale brown: tail and primaries dufky, edged with white: throat and breaft white, fpotted with black: belly white.   **229. UNALASCHA.**

Inhabits *Unalafcha*. *Latham*, ii. 202. N° 47.   PLACE.

230. BLACK-CROWNED. B. With a deep black crown, and a rich yellow fpot on the fore part : feathers on the back black, edged with ruft-color : wings of the fame color, croffed with a double line of white : rump olive brown : throat and breaft cinereous : belly whitifh : tail long, and of a deep brown : legs yellowifh.

PLACE. Inhabits *Natka Sound.   Latham,* ii. 202.

231. RUSTY. B. With head, neck, breaft, and fides, ruft-colored : belly white : wings ferruginous, with two white marks on the primaries : tail of the fame color : the two outmoft feathers of the tail tipt with white.

PLACE. *New York.*   In Mrs. *Blackburn's* collection. Perhaps the fame with Mr. *Latham's* fpecies, ii. 197 * : if fo, it is common to *Ruffia* and *America.   Latham,* ii. 202.

232. UNALASCHA. B. With head, upper part of the neck, back, wings, and tail, brown, tinged with red : breaft and fides dirty white, clouded : middle of the belly plain dirty white.

PLACE. Inhabits the weftern fide of *North America.*

Le Bruant de Canada ? *Briffon,* iii. 296.
Le Cul-rouffet, *De Buffon,* iv. 368.—*Latham,* ii. 204.—LEV. MUS.

233. CINEREOUS. B. With a fhort bill : head, neck, back, breaft, and coverts of the wings and tail, of a pale reddifh brown, edged with afh-color : on the neck and breaft the afh-color predominates : belly white : primaries dufky, edged with white : tail pale brown, with the ends fharp-pointed.

PLACE. Inhabits *Canada.*

233. A VAR. B. With a yellow bill : head, back, and wings, ruft-colored ; each feather deeply and elegantly edged with pale grey ; fome of the greater coverts edged with paler ruft ; the primaries and

* Emberiza Rutila, *Pallas Itin.* iii. 698.

7   tertials

XVII

Black throated Bunting N.º 228.   *Cinereous Bunting N.º 333.*

P. Mazell sculp.

tertials with white : throat, breaft, and fides, white, fully fpotted with ruft : middle of the belly white : middle feathers of the tail brown ; exterior feathers white ; each feather truncated obliquely.

Inhabits *New York.* BL. MUS. PLACE.

Le Bruant bleu de Canada, *Briffon*, iii. 298. 234. BLUE.
L'Azuroux, *De Buffon*, iv. 369.—*Latham*, ii. 205.

B. With the crown of a dirty red : the upper part of the neck and body, fcapulars, and leffer coverts of the wings, of the fame color, varied with blue : the lower part of the neck, breaft, and belly, of a lighter red, mixed with blue : primaries and tail brown; with the exterior edges of a cinereous blue.

Inhabits *Canada.* Breeds in *New England*; but does not winter there. PLACE.

Tanagra cyanea, *Lin. Syft.* 315.—*Latham*, ii. 205. 235. INDIGO.
Blue Linnet, *Catefby*, i. 45.
Le Tangara bleu de la Caroline, *Briffon*, iii. 13.
Le Miniftre, *De Buffon*, iv. 86.—BL. MUS.—LEV. MUS.

B. With a dufky bill : plumage of a rich fky-blue color ; lighteft about the belly and breaft : acrofs the coverts of the wings is a row of black fpots : primaries and tail dufky, edged with blue.

FEMALE brown above; of a dirty white beneath. In SIZE lefs than the *Englifh* Goldfinch.

Inhabits (according to *Catefby*) the interior parts of *Carolina*, a PLACE. hundred and fifty miles from the fea. Has the note of a Linnet. It is found as low as *Mexico*, where the *Spaniards* call it *Azul Lexos*, or the far-fetched bird : and the *Americans* call it the *Indigo* bird. Notwithftanding *Catefby*, it appears in the province of *New York*, in *May.* Makes its neft of dead yellow grafs, lined with the down of fome plant ; and places it between the fork of an upright branch.

GOLDEN

A. GOLDEN BUNTING. Emberiza Aureola, *Pallas Itin.* ii. 711.—*Latham,* ii. 201.

B. With the crown, hind part of the neck, and back, of a deep bay: fides of the head, throat, and fpace round the noftrils, black: under part of the neck, breaft, and belly, of a moft beautiful citron-color: the middle of the neck croffed by a bar of bay: vent white: wings dufky, marked with a great bed of white: tail a little forked; two outmoft feathers on each fide croffed obliquely with white: legs pale afh-colored.

PLACE.　　Found only in *Sibiria.* Moft frequent in the eaft part; where it extends even to *Kamtfchatka.* Is converfant in the iflands, in rivers overgrown with reeds and willows. Has the note of the Reed Sparrow.

B. COMMON B. *Br. Zool.* i. N° 118.
Kornlarka, *Faun. Suec.* N° 228.
Knotter *Norvigis,* Brunnich, N° 247.
Le Proyer, *De Buffon,* iv. 355.—*Pl. Enl.* 30. 1.—*Latham,* ii. 171.—LEV. MUS.

B. With the head, and upper part of the body, light brown; lower part yellowifh white: all parts, except the belly, fpotted with black: tail fubfurcated, dufky edged with white.

PLACE.　　Inhabits *Europe* as high as *Sondmor* *. Migrates into the fouth of *Ruffia.* Unknown in *Sibiria.*

* *Strom.* 240.

YELLOW

C. YELLOW B. *Br. Zool.* i. N⁰ 119.
Groning. Goldſpink, *Faun. Suec.* N° 230.
Le Bruant de France, *De Buffon,* iv. 340.—*Pl. Enl.* 30. 2.—*Latham,* ii. 170.—
LEV. MUS.

B. With the crown of a fine light yellow: chin, throat, and
belly, yellow: breaſt marked with orange red: rump ruſt-
colored: tail brown; two outmoſt feathers marked, near the end,
obliquely with white.

Found as high as *Sondmor* *, in the *Ruſſian* empire. In *Ruſſia*, PLACE.
and the weſt of *Sibiria:* but none in the wilds of the eaſt.

D. ORTOLAN, *Wil. Orn.* 270.—*Raii Syn. Av.* 94.
Emberiza Hortulana, *Lin. Syſt.* 309.—*Faun. Suec.* N° 229.
L'Ortolan, *De Buffon,* iv. 305.—*Pl. Enl.* 247. 1.—*Latham,* ii. 166.—LEV. MUS.

B. With a cinereous crown: yellow throat: back pale brown,
ſpotted with black: rump daſhed with yellow: belly ruſt-
colored: tail duſky; inner ends of the outmoſt feathers marked
with a great ſpot of white.

Theſe are a ſouthern ſpecies; but ſometimes wander into *Sweden,* PLACE.
in *March.* Breed, and quit the country in autumn †. Are common in *Ruſſia* and *Sibiria,* but not further than the *Oby.* Arrive in
France with the Swallows ‡. In *Italy,* about *Padua,* in *May,* and
retire in *September* §. I cannot trace their winter reſidence. They
come rather lean. Make an artleſs neſt: lay four or five greyiſh
eggs: and uſually lay twice in the ſummer. Theſe birds ſing prettily, and in the night ‖: but, by epicures, are valued more as a
delicious morſel, than for their ſong. They are taken and placed in
a chamber lightened by lanthorns; ſo that, not knowing the viciſſi-
tudes of day and night, they are not agitated by the change. Are

---

* *Strom.* 230.   † *Aman. Acad.* iv. 595.   ‡ *De Buffon,* iv. 309.
§ M. SCOPOLI's Liſt of *Italian* birds, MŚ. with which he favored me.   ‖ *Kram.*
Auſtr. 371.

fed

fed with oats and millet; and grow fo fat, that they would certainly die, if not killed in a critical minute. They are a mere lump of fat; of a moſt exquiſite taſte; but apt foon to ſatiate.

Theſe birds receive both their *Greek* and their *Latin* name from their food, the millet. *Ariſtotle* calls them *Cynchromi*; and the *Latins, Miliariæ* *. The latter kept and fattened them in their *ornithones,* or fowl-yards, as the *Italians* do at preſent; which the antients conſtructed with the utmoſt magnificence, as well as conveniency †.

E.   Reed B. *Br. Zool.* i. N° 120.
      Saf. Sparf. *Faun. Suec.* N° 231.
      Ror-Spurv. *Brunnich,* N° 251.
      L'Ortolan de roſeaux, *De Buffon,* iv. 315.—*Pl. Enl.* 247. 2.—*Latham,* ii. 173.—
      Lev. Mus.

B.   With black head and throat: cheeks and head encircled with white: body above ruſty, ſpotted with black; beneath white. Female has a ruſt-colored head, ſpotted with black: wants the white ring.

PLACE.   Is found as far north as *Denmark:* and is rare in *Sweden.* Common in the ſouth of *Ruſſia* and *Sibiria.* Its ſong nocturnal, and ſweet. Makes a neſt pendulous, between four reeds.

    * *Ariſt. Hiſt. An.* lib. viii. c. 12: and *Varro de re Ruſt.* lib. iii. c. 5 —*Ficedulæ et miliariæ dictæ à cibo, quod alteræ fico: alteræ milio fiant pingues.* Varro de Ling. Lat. iv.
    † See a plan in the *Leipſic* edition of *Var. de re Ruſt.* lib. iii. v.

XXV.   T A N A G E R.

XXV. TANAGER. *Gen. Birds.* XLVII.

Summer Red-bird, *Catesby*, i. 56.—*Edw.* 239.
Muscicapa rubra, *Lin. Syst.* 326.—*Brisson*, ii. 432.
Tangara du Missisipi, *De Buffon*, iv. 252.—*Pl. Enl.* 741.—*Latham*, ii. 220.

T. Wholly red, except the wings; the ends of which are brown: bill yellow: legs reddish. FEMALE brown, with a tinge of yellow.

PLACE.

Inhabits the woods on the *Missisipi*. Sings agreeably. Collects, against winter, a vast magazine of maize, which it carefully conceals with dry leaves, leaving only the hole by way of entrance; and is so jealous of it, as never to quit its neighborhood, except to drink *.

Scarlet Sparrow, *Edw.* 343.
Tanagra rubra, *Lin. Syst.* 314.—*Latham*, ii. 217. Nº 3. A.
Tangara de Canada, *De Buffon*, iv. 250.—*Pl. Enl.* 156.

T. With a whitish bill: head, neck, and whole body, of a brilliant scarlet; the bottoms of the feathers black: primaries dusky; lower part of their inner webs white: tail and legs black; tips of the first white; but that circumstance is sometimes wanted. The supposed FEMALE is of a green color, light and yellowish beneath. SIZE of a Sparrow.

PLACE.

Inhabits from *New York* to the *Brasils*. In *New York* it appears in *May*, and retires in *August*. Is a very shy bird, and lives in the deepest woods.

T. With the head, whole upper part of the body, and coverts of the wings, of an olive green, fading into cinereous towards the rump: wings and tail brown, edged with white: throat and breast of a fine yellow: belly white: legs brown. Wings and tail

* *Du Pratz.*

3 B

of

of the FEMALE dufky, edged with olive : under fide of the body of a very pale yellow.

PLACE.   Inhabits *New York :* and as far fouth as *Cayenne *.* BL. Mus.— *Latham,* ii. 218.

239. GREY.   Le Gris-olive, *De Buffon,* 277.—*Pl. Enl.* 714.—*Latham,* ii. 236.

T. With a black bill : forehead, and fpace above the eyes, grey : back of an olive grey : wings and tail dufky, edged with grey : under part of the body an uniform grey.

PLACE.   Inhabits *Guiana* and *Louifiana.*

240. BISHOP.   *Latham,* ii. 226.
        Tanagra epifcopus, *Lin. Syft.* 316.
        L'Eveque, *Briffon,* iii.
        Le Bluet, *De Buffon,* iv. 265.—*Pl. Enl.* 176.—LEV. Mus.

T. With a black bill : whole plumage of a blueifh grey ; in fome places greenifh : on the coverts of the wings the blue predominates : legs afh-colored. Head, neck, and breaft, of the FEMALE of a blueifh green : back, and coverts of the wings, brown ; the laft croffed obliquely with a greyifh ftroke : primaries and tail black.

PLACE.   Inhabits *Louifiana* † ; and as low as *Cayenne.* Haunts the fkirts of forefts, and feeds on the fmaller fruits. Is fometimes gregarious, but ufually found in pairs. Roofts on the palm-trees. Has a very fharp and difagreeable note.

        • L'Olivet, *De Buffon,* iv. 269.   † *Du Pratz,* ii.

XXVI. FINCH.

XXVI. FINCH. *Gen. Birds.* XLVIII.

Cowpen-bird, *Catesby*, i. 34.—*Latham*, ii. 269.
Le Pinçon de Virginie, *Brisson*, iii. 165.
Le Brunet, *De Buffon*, iv. 138.—Lev. Mus.—Bl. Mus.

241. COWPEN.

F. With the head and neck of a dusky brown: back, wings, and body, of a fine black, glossed with green and blue: tail the same: legs black. Crown and upper part of the FEMALE deep brown: throat white: breast and belly light cinereous brown: wings and tail dusky, edged with brown. Bigger than the *English* Bulfinch.

Arrives in *New York* in *May:* lays five eggs in *June:* and migrates southward in *August.* Appears in flights in winter, in *Virginia* and *Carolina*, and associates with the Redwing Orioles, and Purple Grakles. It delights much to feed about the pens of cattle; which gave occasion to the name.

PLACE.

American Goldfinch, *Catesby*, i. 43.—*Edw.* 274.—*Latham*, ii. 289.
Fringilla Tristis, *Lin. Syst.* 320.
Le Chardonneret jaune, *De Buffon*, iv. 212.—*Pl. Enl.* 202.—Bl. Mus.

242. GOLDEN.

F. With a flesh-colored bill: fore part of the head black: rest of the head, neck, and whole body, of a most beautiful gold color; whitening towards the vent: wings black, with two lines of white: tail black: inner webs of the exterior feathers white: legs brown. FEMALE wants the black mark on the head: whole upper part of an olive green; lower part of a pale yellow: in other marks the sexes agree: on coverts of wings two bars of yellow.

243. NEW-YORK SISKIN.

Le Tarin de la Nouvelle York, *De Buffon*, iv. 231.—*Pl. Enl.* 292.—*Latham*, ii. 291.

F. With a black crown: neck encircled with yellow: breaſt and rump of the ſame color; the laſt fading into white: back olive-brown: wings and tail black, edged with white: belly and vent whitiſh. The crown of the FEMALE wants the black: its colors alſo in general are leſs brilliant than thoſe of the male. Superior in ſize to the *European* kind; but ſeems only a variety.

PLACE.

Inhabits *New York.*

244. ORANGE.

Bahama Finch, *Cateſby*, i. 42.—*Latham*, ii. 276.
Fringilla Zena, *Lin. Syſt.* 320.—*Briſſon*, iii. 368.—*De Buffon*, iv. 140?

F. With a yellow throat: head and neck black: above and beneath each eye a long white line: breaſt orange-colored: belly white: back greeniſh: coverts of the wings black; loweſt order white: primaries and tail duſky, edged with white: legs lead-color. Head of the FEMALE aſh-color: back of a dull green: belly of a dull yellow.

FEMALE.

PLACE.

Inhabits the *Bahama* iſles.

245. RED-BREASTED.

F. With a white bill: cheeks, throat, and under ſide of the neck and breaſt, of a rich crimſon: belly white: crown, upper part of the neck, back, wings, and tail, black: coverts croſſed with two lines of white: legs black.

Eight of theſe were driven, in a ſtorm, on *Sandy Hook,* in *April* 1779. *Latham*, ii. 272.

Br.

Br. Zool. i. N° 128.
Mountain Sparrow, fem. *Edw.* 269.—*Latham*, ii. 252, 265.
Moineau de Canada, *Briſſon*, iii. 102.—*Pl. Enl.* 223.
Le Soulciet, *De Buffon*, iii. 500.—BL. MUS.—LEV. MUS.

F. With the end of the bill duſky; baſe of the lower mandible yellow: cheeks, and under ſide of the neck, pale aſh-color: from the baſe of the bill, on each ſide, is a red line paſſing above the throat: crown, hind part of the neck, and feathers on the ridge of the wings, bay: back ferruginous, ſpotted with black: coverts of the wings black, edged with ruſt-color, and croſſed with two bars of white: belly and breaſt of a dirty white: tail duſky, edged with aſh-color. LENGTH ſix inches and a half: EXTENT ten.

PLACE.

Inhabits *Hudſon's Bay* during ſummer. Comes to *Severn* ſettlement in *May*. Advances farther north to breed; and returns in autumn, in its way ſouthward. Found alſo in *Penſylvania*. Suppoſed, by Mr. *Edwards*, to be the female of the Tree, or Mountain Sparrow, *Br. Zool.* i. N° 128; but as I have had opportunity of ſeeing ſpecimens of this bird from *Hudſon's Bay*, *Newfoundland*, and *New York* *, all of which agreed in marks and colors, I have no doubt but that it is a diſtinct ſpecies.

Bahama Sparrow, *Cateſby*, i. 37.—*Latham*, ii. 300.
Fringilla bicolor, *Lin. Syſt.* 324.
Le Verdier de Bahama, *Briſſon*, iii. 202.—LEV. MUS.

F. With the head, neck, and breaſt, black: the remaining parts of a dirty green color. SIZE of a Canary-bird.

PLACE.

Inhabits the woods of the *Bahama* iſlands. Sits perched on a buſh, and ſings, repeating one ſet tune.

Edwards, 304.—*Latham*, ii. 272.—BL. MUS.

F. With a broad bar croſſing from the bill, over each eye, towards the hind part of the head; orange-colored near the bill;

* BL. MUS.

3   white

white beyond the eyes; and bounded above and below with a dufky line: crown divided lengthways by a white ftroke: throat white: hind part of the neck, back, and coverts of the wings, prettily fpotted with black, afh-color, and ferruginous: primaries and tail dufky, edged with white: ridge of the wing pale yellow: breaft and belly of a brownifh white: legs yellowifh.

PLACE.

Inhabits *Penfylvania*. Mr. *Blackburne* faw a fmall flock of them in the province of *New York*, in *January*. I have likewife defcribed them from *Newfoundland*, where they are found during fummer: one, which I fuppofe to be the female, had the yellow fpot at the bafe of the bill very obfcure, nor had it the white fpot on the chin.

249. YELLOW-THROATED.

F. With head, and upper part of body, cinereous: primaries dufky, edged with pale brown: chin white: on the throat a pale yellow fpot: belly of a dirty white: legs and bill of a blueifh grey.

PLACE.

Inhabits the province of *New York*.

250. STRIPED.

Latham, ii. 275.

F. With a lead-colored bill: forehead, and fpace between the beak and eyes, yellow: on the crown are three black ftripes on a white ground: behind each eye is a black fpot: cheeks and chin whitifh: hind part of the neck and back brown, fpotted with dufky: coverts of the wings uniform brown: tail of the fame color, and fubcuneiform: primaries dufky: breaft light grey: belly ftill paler.

PLACE.

Shot in *New York* in *May*. In the cabinet of Major *Davies*, of the Artillery: a gentleman to whom this Work is under great obligations.

Little Sparrow, *Edw.* 354.—*Latham,* ii. 272.—Bl. Mus.

F. With the head and back cinereous, edged with ruft-color: coverts of the wings and tail of a bright ferruginous: inner webs of the primaries, and the tail, dufky; the exterior ferruginous: the cheeks, breaft, and belly, white, marked with large bright fpots of ferruginous: legs yellowifh. Size of a Houfe Sparrow.

Inhabits *Newfoundland,* and as low as *Penfylvania.* Called in *New York,* the Shepherd, from its note *fhep, fhep:* ftays there only the winter. Fond of fcraping the ground.

A bird of a plain dufky ruft-color above, and white beneath, fpotted like the former, fhot at *Unalafcha,* feems a variety.

F. With the crown, hind part of the neck, and back, ruft-colored, fpotted with black; the fpots on the back large: coverts of the wings of a plain ferruginous: primaries dufky, edged with dirty white: whole under fide white, with black ftreaks pointing downwards: tail brown, croffed by numerous dufky bars.

Inhabits *New York.* Bl. Mus.—*Latham,* ii. 273.

F. With the head, upper part of the neck, and back, cinereous, ruft-colored, and black: cheeks brown: leffer coverts of the wings bright bay: the orders below black, edged with white: primaries dufky, edged with white: lower part of the neck and fides white, fpotted with fmall white ftreaks: belly pure white: tail dufky.

Inhabits *New York.* Lays five eggs in *May,* in the grafs. Called the *Grey Grafs-bird.* Continues the whole winter. Bl. Mus.—*Latham,* ii. 273.

254. WINTER.

F. With the head, neck, and back, of a light brown, spotted with black: under part of the neck, breast, and sides, white, with small brown spots: belly white, and unspotted: primaries brown, edged with white; as are the coverts.

PLACE.

Inhabits *New York*. Seen and killed there, out of a small flock, in *January*. BL. MUS.—*Latham*, ii. 274.

255. BLACK-FACED.

Le Moineau de la Caroline, *De Buffon*, iii. 496.—*Pl. Enl.* 181. fig. 2.— *Latham*, ii. 253.

F. With the fore part of the head and chin black: hind part, neck, and rump, crimson: back, tail, and wings, black, edged with rust-color: breast crossed with a black band: belly brownish.

PLACE.

Inhabits *Carolina*, according to the Count *De Buffon*, who supposes it to be the female of a crested Finch, of a very different aspect *.

256. NORTON.

F. With the head, upper part of the neck, and secondaries, black, edged with bright bay: rump bright bay, edged with ash: lesser coverts of the wings bright bay; middle order black, crossed with a white line; primaries dusky: throat buff-colored; bounded on each side by a dusky line: belly and sides white: sides and under part of the neck spotted with rust-color: tail dusky, edged with dirty white: along the middle of the outmost feather is a pure white line, ending at the tip.

PLACE.

Discovered in *Norton Sound*.—*Latham*, ii. 274.

257. CRIMSON-HEAD.

F. With a crimson head and breast; the first faintly marked with dusky spots: space behind each eye dusky: back, coverts of the wings, primaries, and tail, black, edged with crimson: belly white, tinged with red.

* *Pl. Enl.* 183. fig. 1.

Inhabits

Inhabits *New York*. Arrives there in *April*. Is very frequent among the Red Cedars, and fhifts moft nimbly around the ftems. Bl. Mus.—A bird of this fpecies, or nearly related, is defcribed by Doctor *Pallas*, under the name of *Fringilla rofea* * ; which, he fays, frequents lake *Baikal*, and the country to the north of that water.

Purple Finch, *Catefby*, i. 41.—*Latham*, ii. 275.
Le Bouvreuil violet de la Caroline, *Briffon*, iii. 324

F. With a purple head and body, with fome dufky mixture, efpecially the inner webs of the primaries, and the tail: belly white. Female brown, with the breaft fpotted like a Thrufh.

Appears in *Carolina*, in *November*. Feeds on *juniper*-berries. In *February*, deftroys the fwelling buds of fruit trees.

Fringilla Lapponica, *Lin. Syft.* 317.—*Faun. Suec.* Nº 235.
Fringilla calcarata, *Pallas Travels*, ii. App. 710. tab. E.
Le Grand Montain, *De Buffon*, iv. 134.—*Latham*, ii. 263.

F. With a yellow bill, with a dufky point: crown black: from the bafe of the bill is a white line, paffing under each eye, defcending down the fides of the neck, bending towards the breaft: throat, and fore part of the breaft, black: its fides and belly white: hind part of the neck and back brown, mixed with ruft-color: tail forked; that, and the wings, dufky, edged with ruft-color; fome of the exterior feathers of the tail marked, near their ends, with a white fpot: legs dark brown: hind claw long, like a Lark's, and almoft ftrait. Length five inches: Extent feven: Weight half an ounce.

A bird of a hardy conftitution. Inhabits *Hudfon's Bay* during winter. Appears in *November*, and lives among the *juniper* bufhes. Is called by the natives, *Tecurmafhifh*.

* *Travels,* iii. 699.

It alſo inhabits *Greenland*, but continues there only in the ſummer. Makes an artleſs neſt of moſs and graſs, lined with a few feathers; and lays in *June* five or ſix eggs, of a clay-color, clouded: departs early *.   Is found in *Lapland*, in the *Feroe* iſles, the northern parts of *Sibiria*, and near the *Urallian* chain, where it breeds. Arrives in flocks, from the ſouth, and frequent the fields at the firſt flowering of the *Draba verna*, or *Whitlow-graſs*. Has nearly the note of a Linnet; but its flight is higher, and more laſting. It runs on the ground like a Lark: and feeds on ſeeds.

260. CINEREOUS.   F. With the head, upper part of the body, wings, and tail, deep cinereous brown, edged with obſcure ruſt-color: at the corner of the upper mandible is a light grey line; another bounds the cheeks beneath; and a duſky line bounds that: the throat is of a light grey: under ſide of the neck pale cinereous, marked with great duſky black ſpots: middle of the belly whitiſh: bill long: that and the legs duſky.

PLACE.   Inhabits *Unalaſcha*. *Latham*, ii. 274.

261. GREATER
REDPOLL.

Greater Red-headed Linnet, *Br. Zool.* i. Nº 131.—*Latham*, ii. 304.
Hampling, *Faun. Suec.* Nº 240.
La Linotte, *De Buffon*, iv. 58.—*Pl. Enl.* 485.—LEV. MUS.—BL. MUS.

F. With a blood-red ſpot on the forehead: breaſt tinged with roſe-color. In the *European* ſpecies, a ruſt-color prevales in all the upper part of the body; in this the greateſt portion is white. LENGTH five inches and a half: EXTENT nine.

PLACE.   Is found in the northern parts of *North America*. Is ſeen only in the ſouth and weſt of *Ruſſia*: yet is met with in *Scandinavia*, as high as *Drontheim*. None in *Sibiria*.

* *Fauna Greenl.* 119.

Br. Zool. i. N° 132.—*Ph. Tranf.* lxii. 405.
Grafifka, *Faun. Suec.* N° 241.
Le Sizerin, *De Buffon,* iv. 216.—*Pl. Enl.* 151. 2.—*Latham,* ii. 305.—Lev.
    Mus.—Bl. Mus.

262. **Lesser Redpoll.**

F. With a red fpot on the forehead: breaft of the fame color: back dufky, edged with rufty brown: coverts brown, with two tranfverfe bars of white.

Inhabits *Hudfon's Bay,* and probably other parts of *America :* alfo *Greenland,* where it arrives in *April,* and quits the country in autumn. Is found in *Sweden.* Is feen in prodigious flocks all over *Ruffia* and *Sibiria,* particularly in the fpring, flying about the villages. Mr. *Steller* alfo faw it in *Kamtfchatka,* and the iflands.

Place,

. Arctic F. Fringilla flaviroftris, *Lin. Syft.* 322.
Rifka, *Faun. Suec.* N° 239.—*Latham,* ii. 260.

F. With a yellow bill: body black and afh-color, lighteft in front: wings and tail black: tips of the feathers on the breaft gloffed with crimfon. Female of a dufky afh.

Appears about the *Jenefei,* and in the eaftern parts of *Sibiria,* even in the fevereft of winters: and returns to the north even before the Snow Bunting *. Is not feen in *Ruffia,* but inhabits *Sweden.*

Place.

* *Pallas's Travels,* ii, 710.

B. LULEAN F. Fringilla Lulenſis, *Lin. Syſt.* 318.—*Faun. Suec.* N° 234.—*Latham*, ii. 287.
Le Chardonneret à quatre raies, *De Buffon*, iv. 210.

F. With body and tail duſky cinereous : chin white : breaſt and ſhoulders ferruginous : belly whitiſh : primaries duſky : on part of the wings two black lines, one ruſty, and a fourth white.

PLACE.   Inhabits about *Lulea*, in *Weſt Bothnia*.

C. TWITE, *Br. Zool.* i. N° 133.
La Linotte de Montagne, *De Buffon*, iv. 74.—*Latham*, ii. 307.

F. With a ſhort yellow bill : head cinereous, and black : above each eye a ſpot of pale brown : back ruſty, ſpotted with black : coverts of the tail rich ſcarlet : tips of the greater coverts of the wings white : primaries duſky ; inner ſides white : tail duſky ; all but the two middle feathers edged with white. About the SIZE of the greater Red-headed Linnet.

PLACE.   Is ſeen in northern *Europe* as high as *Finmark* *. I diſcover it only in the *Fauna* of that country, of *Sileſia* †, and of *Great Britain*. It flits in great numbers, in ſpring and fall, in the neighborhood of *London*, to and from its breeding place.

D. FLAMING. Fringilla flammea, *Lin. Syſt.* 322.—*Faun. Suec.* N°   —*Latham*, ii. 259.
LEV. MUS.

F. With a pale brown bill : crown of a deep crimſon flame-color, ſlightly creſted : upper part of the body and wings brown : lower parts of a light roſe-color : legs pale brown. LENGTH four inches.

PLACE.   Inhabits *Norland*, in *Sweden*.

* Gran-Iriſk, *Leems :* well deſcribed, p. 256.
† Linaria Saxatilis. Stein-henffling, *Schwenckfelt, Av. Sileſia*, 294.

BRAMBLING.

E. BRAMBLING, *Br. Zool.* i. N° 126.
Norquint, *Faun. Suec.* N° 233.—*Latham*, ii. 261.
Le Pinſon d'Ardenne, *De Buffon*, iv. 123.—*Pl. Enl.* 54. 2.

F. With head and back of a gloſſy black, edged with dull yellow: breaſt, and leſſer coverts of the wings, orange: inner coverts rich yellow: primaries duſky; exterior ſides edged with yellow: tail a little forked; black, with the outmoſt webs of the outmoſt feather white.

Breeds in the woods of *Nordland* and *Drontheim*.   In hard winters deſcends into *Eaſt Gothland* *.

PLACE.

F. CHAFFINCH. *Br. Zool.* i. N° 125.
Finke. Bofinke, *Faun. Suec.* N° 232.—*De Buffon*, iv. 109.—*Pl. Enl.* 54.—*Latham*, ii. 257.—LEV. MUS.—BL. MUS.

F. With the front black: crown blueiſh-grey: cheeks, throat, and breaſt, reddiſh: upper part of the back tawny; lower, green: wings and tail black, marked with white. FEMALE of duller colors: breaſt of a dirty white.

Is found as high as *Drontheim*.   Both ſexes continue in *England* the whole year.   By admirable and unuſual inſtinct, in *Sweden* the females, to a bird, collect in vaſt flocks at the latter end of *September*, deſert their mates, and, paſſing through *Schonen, Denmark, Holſtein,* and *Halland*, viſit ſeveral parts of *Europe*.   They reach *Holland* about a fortnight after *Michaelmas*, and at that time afford great amuſement to the gentry at their country houſes, in taking them while they ſit at tea in their pavilions.   They ſpread nets among their plantations, and ſtrew the ground with hemp-ſeed, by way of bait.   The birds arrive, and perch by thouſands in the trees: then alight on the ground, hungry, and inattentive to the danger.   The nets are cloſed by the pulling of a cord by the perſons in the pavilions; and

PLACE.

* *Amœn. Acad.* iv. 596.

multitudes

multitudes are thus taken. Thofe which efcape, continue their route to *Flanders*, *France*, and *Italy*. The males continue in *Sweden*, and enliven its rigorous winter with their chearful twitter. Towards fpring, they receive additional fpirits; perch on every tree, and animate with their notes every fpray, expecting the arrival of fpring, and of their mates. The laft return invariably the beginning of *April*, in fuch numbers as almoft to darken the fkies; join their conforts, perform their nuptials, retire to the woods, increafe and multiply *

France has its refident Chaffinches, as well as *England*: many alfo winter in *Italy*: many come there in *April*, and migrate in *October* † ; perhaps into *Minorca*, where it arrives in *October*, and continues in that ifland the whole winter ‡.

C. SPARROW, *Br. Zool.* i. Nº 127.—*Latham*, ii. 248.
   Fatting. Grafparf, *Faun. Suec.* Nº 242.|
   Le Moineau, *De Buffon*, iii. 474.—*Pl. Enl.* 6. 1. 55. 1.

PLACE. INHABITS *Europe* in plenty as high as *Drontheim* §: infefts the corn, in the *Orknies*, by thoufands: is native among the rocks beyond lake *Baikal*; but it is faid, that they were unknown in the greateft part of *Sibiria* before the *Ruffians* attracted them by the cultivation of corn. By a wonderful inftinct, thefe and many other birds difcover the effects of rural œconomy, which draws various fpecies, unknown before, from diftant parts, to fhare with mankind the feveral forts of grain or feeds which are grateful to them. Partridges keep pace with the fpreading of corn over many parts of the earth, and appear where they were never feen before: and RICE-BIRDS quickly difcovered the cultivation of *rice* in *South Carolina*, and come annually fome hundreds of miles to feed on it.

* *Amœn. Acad.* iv.–595.   † *M. Scopoli*, MS. Lift, & *Av.* 148.
‡ *Clegborn*, 56.   § *Aves Nidr. Enum.* MS.

GOLDFINCH

H. GOLDFINCH, *Br. Zool.* i. Nº 124.
Stiglitza, *Faun. Suec.* Nº 236.
Le Chardoneret, *De Buffon,* iv. 187.—*Pl. Enl.* 4.—*Latham,* ii. 281.—LEV. MUS.
BL. MUS.

F. With the bafe of the bill encircled with rich fcarlet: cheeks
white: crown black: primaries dufky, marked with a rich
yellow fpot: tail black; tips white: feathers round the bill of the
FEMALE brown: other colors lefs brilliant.

This elegant bird is found as high as *Sondmor* * : whether it goes   PLACE
farther north, is rather doubtful †. In *Italy,* appears in *April:*
breeds; and retires in *October* and *November.* Is common in *Ruffia,*
and the greateft part of *Sibiria.* None beyond the *Lena,* and lake
Baikal.

L. SISKIN, *Br. Zool.* i. Nº 129.
Le Tarin, *De Buffon,* iv. 221.—*Pl. Enl.* 485.
Sifka, Groufifka, *Faun. Suec.* Nº 237.—*Latham,* ii. 289.

F. With a black crown: body yellowifh; green above: breaft the
fame: wings green, with a yellow fpot in the middle: tail
black; yellow at the bafe: head and back of the FEMALE greenifh
afh, fpotted with brown.

Found as high as *Sweden,* and perhaps *Norway* ‡. In *Sweden,*   PLACE.
during fummer, lives in woods, and among junipers: in winter,
conforts with Red-headed Linnets, and feeds on the buds of alders.
Plenty in the fouth and weft of *Ruffia,* but none towards the *Urallian*
chain, nor in *Sibiria.*

* *Strom.* 255.   † *Gunner,* in *Leems,* 256.   ‡ Siifgen? *Pontoppidan,* ii. 94.

XXVII. FLY-

XXVII. FLY-CATCHER. *Gen. Birds*, XLIX.

263. TYRANT.   Tyrant, *Catefby*, i. 55.—*Briffon*, ii. 391.
Lanius Tyrannus, *Lin. Syft.* 136 —*Latham*, i. 186.
Le Tyran de la Caroline, *De Buffon*, iv. 577.—*Pl. Enl.* 676.—LEV. MUS.—
BL. MUS.

FL. With a black bill and head; the crown divided lengthways by a ftripe of fcarlet; in fome, yellow: back afh-color: wings dufky, edged with white: tail black, tipt with white: under fide of the body white: legs black. SIZE of a Redwing Thrufh.

PLACE.   This fpecies appears in *New York* in *April*: lays five white eggs, fpotted with ruft-color: builds in low bufhes: makes its neft with wool, and fome mofs, and lines it with fmall fibres of roots: leaves the country in *Auguft*: obferves the fame time of migration in the fouthern provinces. Mr. *Catefby* gives fo very good an account of its manners, and fingular fpirit, that I beg leave to exprefs it in his own words:—" The courage of this little bird is fingular. He purfues
" and puts to flight all kinds of birds that come near his ftation,
" from the fmalleft to the largeft, none efcaping his fury; nor did I
" ever fee any that dared to oppofe him while flying, for he does not
" offer to attack them when fitting. I have feen one of them fix
" on the back of an Eagle, and perfecute him fo, that he has
" turned on his back into various poftures in the air, in order to get
" rid of him; and at laft was forced to alight on the top of the
" next tree, from whence he dared not to move, till the little
" Tyrant was tired, or thought fit to leave him. This is the
" conftant practice of the cock, while the hen is brooding: he fits
" on the top of a bufh, or fmall tree, not far from her neft; near
" which, if any fmall birds approach, he drives them away; but
" the great ones, as Crows, Hawks, and Eagles, he won't fuffer to
" come within a quarter of a mile of him without attacking them.

7

" They

" They have only a chattering note, which they utter with great
" vehemence all the time they are fighting.
" When their young are flown, they are as peaceable as other
" birds. It has a tender bill; and feeds on infects only. They are
" tame and harmless birds. They build their nest in an open
" manner, on low trees and shrubs, and usually on the *saffafras-*
" *tree.*"

Le Tyran de la Louisiane, *De Buffon*, iv. 583.—*Latham*, ii. 358.

264. LOUISIANA TYRANT.

FL. With a long flat beak, hooked at the end: head and back
cinereous brown: throat clear flate-colour: belly yellowish:
primaries bright bay: on the greater coverts fome lines of white:
tail long, of a cinereous brown. Rather inferior in fize to the laft.
Inhabits *Louisiana*.

Mufcicapa Tyrannus, *Lin. Syf.* 325 —*Latham* ii. 355.
Le Tyran a queue fourchue, *Briffon*, ii. 395.
Le Savana, *De Buffon*, iv. 557.—*Pl. Enl.* 571.—LEV. Mus.

265. FORK-TAIL.

FL. With head and cheeks black: feathers on the crown yellow at
their bottoms: upper part of the body afh-colored; lower
white: tail greatly forked; the two outmoft feathers on each fide
five inches longer than the others; color black: the lower half of
the exterior feather white.
Inhabits *Canada*, and as low as *Surinam*.

PLACE.

Yellow-breafted Chat, *Catefby*, i. 50.—*Latham*, ii. 350.
Le Merle verde de la Caroline, *Briffon*, ii. 315.—*De Buffon*, iii. 396.—*Pl.
Enl.* 627.—BL. Mus.

266. CHATTER-ING.

FL. With the crown, upper part of neck and back, and tail, of a cinereous green: each eye encircled with yellow: from the throat
to the thighs of a fine yellow: belly white: tail dufky, edged with
white: legs black. SIZE of a Sky-Lark.

3 D   Inhabits

PLACE. Inhabits the interior parts of *Carolina*, two or three hundred miles from the fea. Is fo very fhy, as to be fhot with the utmoft difficulty. Lives by the banks of great rivers; and makes fo loud a chattering, as to reverberate from rock to rock. Flies with its legs hanging down. Its mufical note is good. Often flies up perpendicular, and lights by jerks.

267. CRESTED. Crefted Fly-catcher, *Catefby*, i. 52.—*Latham*, ii. 357.
Mufcicapa crinita, *Lin. Syft.* 325.
Le Gobe-Mouche hupe de Virginie, *Briffon*, ii. 412.
Le Moucherolle de Virginie a huppe verte, *De Buffon*, iv. 565.—*Pl. Enl.* 569.
—BL. MUS.

FL. With an upright creft: head and back olive: the coverts of the fame color, croffed with two white lines: primaries dufky; the four firft edged, on their outmoft fides, with ferruginous: tail dufky; two middle feathers plain; the inner webs of the others orange: neck and breaft of a lead-color: belly and thighs yellow: legs black. I have feen one of a cinereous color on the upper parts, and white belly: perhaps a young bird, or a hen. WEIGHT one ounce.

SIZE. LENGTH eight inches. Sent from *New York*, with the name of the Large Wild *Phœby Bird*, or Bee-eater.

PLACE. Breeds in *New York* and *Carolina*. Its note extremely brawling, as if at enmity with all other birds. Makes its neft of fnake-fkins and hair, in holes of trees. Retires in *Auguft*.

268. LESSER-CRESTED. FL. With a fmall backward creft: head, neck, and back, of a dirty light cinereous green: breaft and belly whitifh, tinged with yellow: wings and tail dufky; coverts croffed with two bars of white; fecondaries edged with white: legs black.

PLACE. Inhabits *Nova Scotia*. Captain *Davies*.

5

Black-

Black-cap Fly-catcher, *Catesby*, i. 53.—*Latham*, ii. 355.
Le Gobe-Mouche brun de la Caroline, *Brisson*, ii. 367.
Le Gobe-Mouche noirâtre de la Caroline, *De Buffon*, iv. 541.

269. BLACK-
HEADED.

FL. With a black crown : back brown, wings and tail dusky,
edged with white : whole under side white, tinged with yellowish
green : legs black. Head of the hen of not so full a black as that
of the cock.
Breeds in *Carolina*. Is supposed to migrate in the winter.

PLACE.

Little brown Fly-catcher, *Catesby*, i. 54. fig. 1.
Le Gobe-Mouche cendré de la Caroline, *Brisson*, ii. 368.
Muscicapa virens, *Lin Syst.* 327.
Le Gobe-Mouche brun de la Caroline, *De Buffon*, iv. 543.—*Latham*, ii. 350.
—BL. MUS.

270. CINEREOUS.

FL. With the upper mandible black ; the lower yellow: eyes red:
head and back of a deep ash-color : over each eye a faint white
line : wings and tail brown : secondaries edged with white : whole
under side of the body dirty white, tinged with yellow : legs black.
WEIGHT nine pennyweights.
Inhabits *Carolina*, in the summer only.

PLACE.

Red-eyed Fly-catcher, *Catesby*, i. 54. fig. 2.—*Edw.* 253.
Muscicapa Olivacea, *Lin. Syst.* 327.—*Brown Jam.* 476.
Le Gobe-Mouche de la Jamaique, *Brisson*, ii. 410.
Le Gobe-Mouche olive de la Caroline, *De Buffon*, iv. 539.—*Latham*, ii. 351,
352.—LEV. MUS.

271. RED-EYED.

FL. With red irides: crown, and whole upper part of the body,
wings, and tail, of a cinereous brown : over each eye a white
line : edges of the primaries and tail whitish : under side of the
body white, dashed with olive : legs black. WEIGHT ten penny-
weights and a half.

3 D 2

Inhabits

PLACE.   Inhabits *Carolina*, and as high as *New York*; and migrates at approach of winter : probably into *Jamaica* ; the fame kind being found there, where, from its note, it is called *Whip Tom Kelly*. Has great affinity with the preceding : perhaps they

NEST.   differ only in fex. Makes a pendulous neft, ufually in apple-trees, and hangs it between the horizontal fork of fome bough, beneath the leaves. It is moft curioufly formed with cotton and wool, lined with hair and dead grafs ; and wonderfully bound to the branches by a certain thread, like mofs, twifted round them, and likewife all about the outfide of the neft. Lays five eggs, white, thinly fpotted with deep ruft-color.

272. CAT.   Cat-bird, *Catefby*, i. 66.—*Lawfon*, 143.—*Latham*, ii. 353.
Le Gobe-Mouche brun de Virginie, *Briffon*, ii. 365.
Mufcicapa Carolinenfis, *Lin. Syft.* 328.
Le Moucherolle de Virginie, *De Buffon*, iv. 562.—LEV. MUS.—BL. MUS.

FL. With a black crown : upper part of the body, wings, and tail, blueifh grey : the tail cuneiform, marked with numerous dufky bars : under fide of the body of a pale grey : vent ferruginous : legs brown. Larger than a LARK.

PLACE.   Inhabits *New York* and *Carolina*. Mews like a kitten; from which arofe its name. Lives among bufhes and thickets. Feeds on infects. Makes the outfide of its neft with leaves and matting rufhes ; the infide with fibres of roots. Lays a blue egg. Has a great fpirit, and will attack a Crow, or any large bird. Mr. *Latham* faw one which was brought from *Kamtfchatka*, which differed from this only in having no ruft-color on the vent.

273. CANADA.   Mufcicapa Canadenfis, *Lin. Syft.* 324.—*Latham*, ii. 354.
Gobe-Mouche cendrè de Canada, *Briffon*, ii. 405. tab. xxxix.—*De Buffon*, iv. 538.
—*Catefby*, i. 60.

FL. With a cinereous head, fpotted with black ; a yellow fpot between the bill and the eyes; and beneath each eye a black one :
                                                                                                the

the upper part of the body cinereous ; the lower, yellow, marked on the under fide of the neck with fmall black fpots: the tail of a cinereous brown, with the exterior webs afh-colored.

Inhabits *Canada.* PLACE.

F<small>L</small>. With a yellow fpot on each fide of the bill : head a cinereous green : back and coverts of the wings of a pale green ; crofs the laft are two bars of white : primaries and tail dufky, edged with green : throat of a pale afh-color : middle of the belly white : fides of a fine yellow. 274. G<small>REEN</small>.

Sent from *New York* by Mr. *Blackburne*, under the name of the fmall Green Hanging Bird. It comes there in *May*, breeds, and etires in *Auguft :* and is a fcarce fpecies. B<small>L</small>. M<small>US</small>. PLACE.

F<small>L</small>. With a dufky head : back of a dull cinereous olive : quil feathers and fecondaries dufky ; the laft edged with white : breaft of a pale afh-color : belly of a whitifh yellow : tail dufky ; exterior web of the exterior feather white : legs black. 275. D<small>USKY</small>.

Sent from the fame place, under the title of The Small or Common *Phæby* Bird, or Bee-eater. Appears the latter end of *March*, or beginning of *April*; lays five white fmall eggs : difappears in *Auguft*. Eats Bees. B<small>L</small>. M<small>US</small>. PLACE.

F<small>L</small>. With the crown, upper part of the neck, and body, of a dirty olive : throat and ridge of the wing of a very rich yellow : breaft and belly white, tinged with yellow : primaries and tail of a bright olive green. 276. G<small>OLDEN</small>-T<small>HROAT</small>.

Inhabits *New York.* B<small>L</small>. M<small>US</small>. PLACE.

Striped

277 STRIPED.

Striped Fly-catcher, *Forster*, Ph. Tr. lxii. 406.
Mufcicapa ftriata, *the fame*, 429.—*Latham*, ii. 349.—*Miller's Plates*, Nº 15.

FL. With a black crown; white cheeks: hind part of the head varied with black and white: throat of a yellowifh white, ftriped with brown: breaft white, ftriped on the fides with black: belly white: back of a cinereous green, marked with black: wings dufky, mixed with white: tail dufky, with the three outmoft feathers marked with a white fpot: legs yellow.

Head of the FEMALE of a yellowifh green, with fhort ftreaks of black: a fhort yellow line paffes from the bill over each eye: throat, cheeks, and breaft, of a yellowifh white, ftriped on the fides with black: in other refpects like the MALE, but greener. LENGTH five inches; EXTENT feven.

PLACE.

Arrives at *Severn* fettlement, *Hudfon's Bay*, in the fummer. Feeds on grafs-feeds.

A. DUN FL. *Faun. Ruff.*—*Latham*, ii. 351.

FL. Dufky above; afh-colored beneath: throat and vent fpotted with white.

PLACE.

Found about lake *Baikal*, and in the eaftern part of *Sibiria*: and obferved by *Steller* in *Kamtfchatka*.

B. Pied Fl. *Br. Zool.* i. Nº 135.
Mufcicapa Atricapilla, *Faun. Suec.* Nº 256, tab. 1 *.
Le Gobe-Mouche noir a Collier, *De Buffon*, iv. 520.—*Pl. Enl.* 565.
Motacilla Leucomela, *Muller*, Nº 268.—*Latham*, ii. 324.—Lev. Mus.—Bl. Mus.

FL. With white front : bill, head, back, and legs, black : coverts of tail fpotted with white : coverts of wings dufky, croffed
with a white bar : primaries dufky : exterior fides of fecondaries
white ; interior black : breaft and belly white : middle feathers of
tail black; exterior black, marked with white : head of the FEMALE
wholly brown, as is the upper part of the body : white in the wings
obfcure : breaft and belly dirty white.

Found as far north as *Sondmor.* Inhabits that diocefe the whole PLACE.
year ; and, during winter, frequently takes refuge in the very
houfes +. Feeds on the buds of birch. Is met with in *Ruffia* only
between the *Kama* and the *Samara.*

* The defcription refers to the *Black-cap Warbler.* The figure to this bird.
† *Act. Nidros,* v. 543.

XXVIII.

XXVIII. L A R K. *Gen. Birds,* L.

278. SHORE.

Alauda gutture flavo. The Lark, *Catesby,* i. 32.
Alauda alpeſtris, *Lin. Syſt.* 289.
Gelbburtige Lerch, *Klein, Av.* 72.—*Latham,* ii. 385.
Le Hauſſe-col noir, ou l'Alouette de Virginie, *De Buffon,* v. 55.—*Briſſon,* iii. 367.
LIV. MUS.—BL. MUS.

L. With yellow cheeks and forehead: breaſt and belly white: head divided by a line of black; another paſſes beneath each eye, bounding the throat, which is yellow: acroſs the upper part of the breaſt is a broad black mark; beneath that is a tinge of red: upper part of the neck, and coverts of the wings and tail, are ferruginous: back brown: primaries duſky: two middle feathers of the tail brown; the reſt black; thoſe on the outſide edged with white: legs duſky: head of the FEMALE duſky. LARGER than the common Lark.

PLACE.

Inhabit the large plains of ſeveral provinces, and breed there. They appear on our ſettlements in *Hudſon's Bay* in *May,* and proceed farther north to breed. Feed on graſs-ſeeds, and the buds of the ſprig birch. Run into ſmall holes, and keep cloſe to the ground; whence the natives call them *Chi-chup-pi-ſue.*

In winter they retire to the ſouthern provinces in great flights; but it is only in very ſevere weather that they reach *Virginia* and *Carolina.* They frequent ſand-hills on the ſea-ſhore, and feed on the *ſea-ſide oats,* or *uniola panicula.* They have a ſingle note, like the Sky-lark in winter.

They are alſo found in *Poland*; in *Ruſſia* and in *Sibiria* more frequent: in both are very common during winter; but retire to the north on approach of ſpring, except in the north-eaſt parts, and near the high mountains.

Red

Red Lark, *Edw.* 297.—*Br. Zool,* i. N° 140.—*Briſſon,* App. 94.—*Latham,* ii. 376.
L'Alouette aux joues brunes de Penſylvanie, *De Buffon,* v. 58.—Lev. Mus.

L. With a white line above and beneath each eye: thickiſh bill: chin and throat whitiſh: head, and whole upper part of the body, and coverts, pale ferruginous, ſpotted with black: breaſt whitiſh, with duſky ſpots: belly of a dirty white: ſide tinged with ruſt: tail duſky; outmoſt feathers white; the two next edged with white: legs duſky. When the wing is cloſed, ſays Mr. *Edwards,* the third quill from the body reaches to its tip; a conſtant characteriſtic of the Wagtail genus.

Inhabits *Penſylvania*; appears there in *March,* in its paſſage northward. Found alſo near *London.*

Edw. 268.—*Latham,* ii. 382.
Alauda Calandra, *Lin. Syſt.* 288.
La Calandra ou groſſe Alouette, *De Buffon,* v. 49.—*Pl. Enl.* 363.—*Briſſon,* iii. 352.

L. With a bill thicker and ſtronger than uſual to the genus: from the bill a black line paſſes to and beyond the eye; above and beneath are two others of white, faintly appearing: head, neck, back, and coverts of the wings, reddiſh brown, ſpotted with black: primaries and tail duſky, edged with ruſt-color: throat white: upper part of the breaſt croſſed by a narrow black creſcent; beneath that the breaſt is of a pale brown, ſpotted with a darker: belly and vent white: tail a little forked: legs of a pale fleſh-color. In Size rather ſuperior to the Sky-Lark; but the body thicker. It is a ſpecies allied to the common Bunting.

Brought from *North Carolina*; and firſt deſcribed as an *American* bird by Mr. *Edwards.* Is common in many parts of *Europe,* eſpecially in the ſouthern. In *Aſia* it is found about *Aleppo,* and is pretty frequent about the *Tartarian* deſerts bordering on the *Don* and *Volga.*

A. Sky-Lark, *Br. Zool.* i. N° 136.
L'Alouette, *De Buffon*, v. 1.
Alauda arvenfis. Larka, *Faun. Succ.* N° 209.—*Latham*, ii. 368.—Liv. Mus.—
Bl. Mus.

L. With the crown of a reddifh brown, fpotted with black : hind part of the head cinereous : chin white : breaft and belly pale dull yellow ; the firft fpotted with black : back and coverts of wings dufky, edged with pale reddifh brown : exterior web, and half the interior web of the outmoft feather of the tail, white : legs dufky.

Size. Length feven inches one-fourth : Extent twelve and a half : Weight an ounce and a half.

Place. Inhabits all parts of *Europe*, even as high as *Nordland* in *Norway*, beneath the *Arctic* circle. They migrate in *Scandinavia*. They are the firft birds, in *Eaft Gothland* in *Sweden*, which give notice of the return of fpring, finging with a tremulous note, and flying in flocks near to the ground. Enlivened by the warmth of fummer, they foar and fing with full voices. In *September* they collect in flocks, and retire fouth ; probably into the province of *Schonen*, where they are found in vaft multitudes during winter *. They are frequent in all parts of *Ruffia* and *Sibiria*, and reach even *Kamtfchatka.*

* *Aman. Acad.* iv. 593.

WOOD.

B. WOOD-LARK, *Br. Zool.* i. N° 137.
Alauda arborea, *Faun. Suec.* N° 211.
Le Cujelier, *De Buffon,* v. 25.—*Pl. Enl.* 660.—*Latham,* ii. 371.

L. With crown and upper part of back reddish brown : head surrounded with a whitish coronet from eye to eye : first feather
of the wing shorter than the second. In form shorter and thicker
than the Sky-Lark.

Inhabits not farther north than *Sweden.* Found in the woods of
Ruſſia and *Sibiria,* as far east as *Kamtſchatka* *.

PLACE.

C. TIT-LARK, *Br. Zool.* i. N° 138.
Alauda pratenſis, *Faun. Suec.* N° 210.
La Farlouſe, *De Buffon,* v. 31.—*Pl. Enl.* 574.—*Latham,* ii. 374.

L. With a black bill : olivaceous brown head and back, spotted
with black : breast yellow, with oblong streaks of black. Of
a slender form.

Found not higher than *Sweden.*

PLACE.

D. FIELD-LARK, *Br. Zool.* i. N° 139.
Alauda campeſtris, *Faun. Suec.* N° 212.—*Raii Syn. Av.* 70.
La Spipolette, *De Buffon,* v. 43.—*Latham,* ii. 375.

L. With head and neck pale brown, marked with dusky lines, faint-
est on the neck : rump and back of a cinereous olive ; the first
spotted with black, the last plain : legs pale brown : hind claw
shorter than usual with Larks. Leſſer than the Sky-Lark.

Extends only to *Sweden.* These three species disappear in that
kingdom in the height of winter. If the weather softens, they return in *February.* The Comte *De Buffon* † describes a variety of this,
under the name of *La Farlouzzane* ; which, he says, came from
Louiſiana.

* Mr. Latham, ii. 372.   † v. 38.

3 E 2   WAGTAIL.

WAGTAIL. *Gen. Birds.* LI.

E. WHITE, *Br. Zool.* ii. N° 142.
   M. Alba. Arla, *Faun. Suec.* N° 252.—*Latham*, ii. 395.
   La Lavandiere, *De Buffon*, v. 251.—*Pl. Enl.* 652.—LEV. MUS.—BL. MUS.

W. T. With head, back, and neck, black: cheek, front, and chin, white: belly white: primaries dusky: tail long, dusky, with part of the webs white.

PLACE.

Inhabits as high as *Iceland*, the *Feroe Isles*, and *Drontheim* *. It is a bird of augury with the *Swedish* farmers; who have a proverb relative to this and the WHEAT-EAR, which is another bird of direction: " When you see the WAGTAIL return, you may turn your sheep into " the fields; and when you see the WHEAT-EAR, you may sow your " grain †."

It is common in *Russia, Sibiria,* and *Kamtschatka*, but does not extend to the arctic regions.

F. Yellow Wagtail, *Br. Zool.* i. N° 143.
   M. Flava. Sadesarla, *Faun. Suec.* N° 253.—*Latham*, ii. 400.
   La Bergeronette grise, *De Buffon*, v. 261.—*Pl. Enl.* 674.— LEV. MUS.

W. T. With crown and upper part of the body of an olive-green: breast and lower part of the body of a rich yellow: throat spotted with black. In the FEMALE those black spots are wanting: the other colors are also much more obscure.

PLACE.

Inhabits *Sweden*; but not higher. Migrates like the former. Common in all parts of *Russia, Sibiria,* and even *Kamtschatka*.

* *Av. Nidr. Enum. MS.*
† STILLINGFLEET's *Tracts*, 2d ed. 265.

G. **Yellow-headed Wagtail.** Motacilla cifreola, *Pallas Itin.* iii. 696.—*Latham*, ii. 401.

W. T. with citron-colored head, neck, breaſt, and belly: the hind part of the neck marked with a black creſcent: the back blueiſh grey.

Common in *Sibiria*, as far as the *Arctic* circle: leſs ſo in *Ruſſia*. Migrates with the laſt.

PLACE.

H. TCHUTSCHI, *Latham*, ii. 403.

W. T. With crown and back deep olive-brown : a ſpot of white between the upper mandible and eye : coverts and primaries deep brown ; the firſt croſſed with two bars of white : breaſt and belly white, daſhed with ruſt : vent pale yellow : tail very long ; outward web, and half the inward web, of outmoſt feather, white ; all the reſt duſky : legs black.

Taken off the *Tchutſchi* coaſt, within the Streights of *Bering*, lat. 66, north.

PLACE.

XXIX.

XXIX. WARBLER. *Gen. Birds,* LII.

281. BLUE-BACKED
RED-BREAST.

Blue-bird, *Catesby,* i. 47.

Blue Red-breast, *Edw.* 24.—*Lawson.*

Motacilla Sialis, *Lin. Syst.* 336.—*Latham,* ii. 446.

Le Rouge gorge bleu, *De Buffon,* v. 212.—*Pl. Enl.* 390.—*Brisson,* iii. 423.—
Lev. Mus.—Bl. Mus.

W. With bill and legs of a jetty blackness: head, hind part of the neck, back, tail, and coverts of the wings, of a rich deep and glossy blue: primaries dusky, tipt with brown: from the bill to the tail red.   Head, and lower part of the neck, in the FEMALE, cinereous blue: breast duller than that of the MALE.

PLACE.

Frequent in most parts of *North America,* from *New York* to the *Bermuda* islands.   Is the same in the new world as the *Robin-red-breast* is in the old.   Are harmless, familiar birds.   Breed in holes of trees.   Have long wings.   Are swift of flight, therefore elude the pursuit of the Hawk.   Have a cry and a whistle.   Feed usually on insects; but, through deficiency of that food, come to the farm-houses, to pick up grass-seeds, or any thing they can meet with.

282. BLACK-
HEADED.

Redstart, *Catesby,* i. 67 —*Edw.* 80.

Muscicapa ruticilla, *Lin Syst.* 326.—*Raii Syn. Av.* 180. N° 51.

Le Gobe-Mouche d'Amerique, *Brisson,* iii. 383.—*De Buffon,* v. 178, 566.—Lev.
Mus.—Bl. Mus.

W. With the head, neck, breast, back, and wings, black: the primaries crossed with a broad bar of orange: the sides and inner coverts of the wings, belly, and vent, white, spotted with black on the upper sides: two middle feathers of the tail dusky; the rest of the same color at their ends; the lower parts orange: legs black. The FEMALE cinereous olive above; white beneath, bounded on each side by yellow: the parts of the tail which are red in the male, are in this sex yellow.

PLACE.

Inhabits the shady woods of *New York, Virginia, Hudson's Bay,*

I   and

and *Carolina*, during the summer. Retreat to *Jamaica*, and perhaps others of the *Antilles*, during winter *.

> Maryland Yellow-throat, *Edw.* 237.
> Le figuier de Maryland, *Briffon*, iii. 506.
> Le figuier a joues noires, *De Buffon*, v. 292.
> Turdus Trichas, *Lin. Syft.* 293.—*Latham*, ii. 438.—Lev. Mus.—Bl. Mus.

283. YELLOW-
BREAST.

W. With black forehead and cheeks: crown cinereous: hind
part, whole upper part of the neck, back, wings, and tail,
of a deep olive green: primaries and tail edged with yellow: under fide of the neck, breaft, and belly, of a rich yellow.

Inhabits *Penfylvania* and *Maryland*. Frequents bufhes and low
grounds, near rills of water. Quits the country in autumn.

PLACE.

> La Fauvette a poitrine jaune de la Louifiane, *De Buffon*, v. 162.—*Pl. Enl.* 709.
> —*Latham*, ii. 439.

284. ORANGE-
THIGHED.

W. With forehead and cheeks black: head croffed in the middle
with a white band, which divides the cheeks from its hind
part: nape, back, wings, and tail, deep olive: lower part of the neck,
breaft, and belly, fine yellow: thighs and vent reddifh orange: tail
rounded.

Inhabits *Louifiana*; and is a moft elegant fpecies: differs from the
laft in its rounded tail.

PLACE.

> Blue Fly-catcher, *Edw.* 252.
> Motacilla Canadenfis, *Lin. Syft.* 336.
> Le petit figuier cendrè de Canade, *Briffon*, iii. 527.—*Latham*, ii. 487.
> Le figuier bleu, *De Buffon*, v. 304.—*Pl. Enl.* 685.—Bl. Mus.

285. BLACK-
THROAT.

W With the head, upper part of the neck, back, and coverts
of the wings, of a flaty blue: throat, under part, and fides
of the neck, black: primaries dufky; white at bottom: breaft and
belly white: tail dufky.

Inhabits, during fummer, *Canada* and other parts of *America*, to
the fouth. Arrives in *Penfylvania* in *April*. Migrates in winter to
the *Antilles*, and returns in fpring.

PLACE.

* *Sloane's Jamaica*, ii. 312.

Yellow-

286. YELLOW-THROAT.

Yellow-throated Creeper, *Catesby*, i. 62.—*Latham*, ii. 437.
La Mesange grise a gorge jaune, *De Buffon*, v. 454.—*Brisson*, iii. 563.

W. With a yellow spot on each side of the upper mandible: throat of a bright yellow: from the bill, a black line extends across each eye, pointing down, and bounding the sides of the neck: forehead black: crown, hind part of the neck, and back, grey: wings dark cinereous; the coverts edged with white: middle of the breast and belly of a pure white: side spotted with black: tail black and white. The FEMALE wants both the yellow and black marks.

PLACE.

Inhabits *Carolina*; and is continually creeping about the trees in search of insects.

287. HOODED.

Catesby, i. 60.—*Latham*, ii. 462.
Le Gobe-Mouche citrin, *De Buffon*, iv. 538.—*Pl. Enl.* 666.
La Mesange a Collier, *De Buffon*, v. 452.—Bl. Mus.

W. With the forehead, cheeks, and chin, yellow, regularly encircled with black like a hood. This black is the color of the head, breast, and each side of the neck: back, wings, and tail, of a dusky green: inner webs of the exterior feathers of the tail white: breast and belly bright yellow. SIZE of a Gold-Finch.

PLACE.

Frequents the thickets and shady parts of the uninhabited places of *Carolina*.

288. YELLOW-RUMP.

Yellow-rumped Fly catcher, *Edw.* 255.
Le figuier tacheté de la Pensylvanie, *Brisson*, iii. 503.
Le figuier a tête cendrè, *De Buffon*, v. 291.—*Latham*, ii. 481.

W. With cheeks and crown of the head cinereous: hind part of the neck and back of an olive green; the last spotted with black: rump of a bright yellow: throat and breast of the same color; the breast spotted with black drops. rest of the under side white: wings dark ash-color; the coverts crossed with two bars of white:

white : inner fides of the primaries edged with white : coverts of the tail black ; two middle feathers of the tail dufky ; the middle part of the inner webs of the reft white ; the tops and bottoms black.

Inhabits *Penfylvania*.   PLACE.

Yellow Red-poll, *Edw.* 256.
Motacilla petechia, *Lin. Syft.* 334.—*Latham*, ii. 479.
·Le figuier à tete rouge de Penfylvanie, *Briffon*, iii. 488.—*De Buffon*, v. 286,
—BL. MUS.

289. RED-HEAD.

W. With the crown fcarlet : cheeks yellow : hind part of the neck, back, and rump, of an olive-green : wings and tail ·dufky, edged with yellow : all the under fide of the body of a rich yellow, fpeckled with red, except the vent, which is plain. A bird, which I fufpect to be the FEMALE, fhot in *Newfoundland*, had the ·fcarlet crown ; but the upper part of the body was dufky, edged with pale brown : coverts of the tail white : primaries and tail dufky : breaft and belly of a dirty white, and unfpotted.

Vifits *Penfylvania* in *March.* Is a lonely bird, keeping in thickets ·and low bufhes. Does not breed there ; but goes farther north to ·breed ; probably to *Canada* and *Newfoundland.* Feeds on infects.

PLACE.

W. With the crown black : cheeks white : upper part of the body afh-colored, with long black ftrokes pointing to the tail : coverts of the wings and primaries dufky ; the firft marked with two white bars : the fecondaries edged with white : tail dufky ; ends of the two outmoft feathers marked with a white fpot : throat white, ftreaked on each fide with black : breaft and belly of a dirty white, ftreaked downwards with black : legs whitifh.

290. BLACK-POLL.

Inhabits, during fummer, *Newfoundland* and *New York* ; called in the laft, *Sailor.* Arrives there in *May* ; breeds ; and retires in *Auguft*. BL. MUS.—*Latham*, ii. 460.

PLACE.

3 F   W. With

291. GREY POLL.

W. With head, fides of the neck, and coverts of the wings and tail, of a fine grey ; the coverts of wings croffed with two white bars : primaries and tail dufky, edged with grey : throat orange : chin and breaft of a fine yellow : belly whitifh afh-color.

PLACE. Sent from *New York* to Mrs. *Blackburn.—Latham,* ii. 461.

292. YELLOW-POLL.

Le figuier tachete 1 Efpece, *De Buffon*, v. 285.—*Pl. Enl.* 58.—*Latham,* ii. 514.—LEV. MUS.—BL. MUS.

W. With the forehead and whole under fide of the body of a fine yellow ; the laft ftreaked with red : the upper part, and coverts of wings, of an olive-green : the primaries brown, bordered with green : tail brown, bordered with rich yellow. FEMALE of a duller color.

PLACE. Inhabits *Canada* ; where it makes only a fhort ftay, and does not breed there. Found in *New York* ; and even *Hudfon's Bay* during fummer. Retires into *South America*, according to M. *De Buffon.* He fufpects that N° 1, plate 58, *Pl. Enl.* is the female. Till that is afcertained, I beg leave to make a new fpecies of it, in the OLIVE, N°

The neft is very elegant, compofed of down, mixed with dead, grafs ; the infide lined with fine fibres. The eggs fpotted near the larger end. Sent from *New York,* under the name of the *Swamp Bird.*

293. WHITE-POLL.

Black and white Creeper, *Edw.* 300.
Le figuier variè, *De Buffon,* v. 305.—*Latham*, ii. 488.
——————— de St. Domingue, *Briffon*, iii. 529.—BL. MUS.

W. With the crown white, bounded by a black line paffing from the corners of the bill ; beneath that is a ftripe of white : below the eyes a broad bed of black, bounded with white : chin and throat black : hind part of the neck, back, and rump, white,

white, marked with great black fpots: coverts and primaries
black; the firft croffed with two white bars; the laft edged on
their inner fides with white: belly white: fides fpotted with black:
tail black, edged with grey; inner webs of the outmoft feathers fpotted with white.

Arrives in *Penfylvania* in *April*; ftays there the whole fummer.   PLACE.
Feeds on infects, caterpillars, &c.   Probably winters in the *Antilles*, where it is likewife found *.

Golden-crowned Fly-catcher, *Edw.* 298.
Le figuier couronne d'or, *De Buffon*, v. 312.—*Latham*, ii. 486.

294. GOLDEN-
CROWNED.

W. With a golden crown, bounded on all fides with a blueifh
flate-color: above each eye, a narrow white line: from the
bill, acrofs the eyes, a broad band of black: throat and chin white:
hind part of neck and back blueifh, with dufky oblong fpots: rump
yéllow: breaft black, edged with grey; fides of the breaft yellow:
belly and vent white, fpotted with black: wings dufky; coverts
and fecondaries edged with white: tail black; three outmoft feathers on each fide marked on their inner webs with white. FEMALE
is brown on the back; wants the black ftroke through the eye, and
mark on the breaft: in other refpects agrees with the cock.

Arrives in *Penfylvania* in fpring: ftays there but three or four   PLACE.
days, proceeding northward to breed.   Appears likewife in the fame
manner in *Nova Scotia*.

Golden-wing Fly-catcher, *Edw.* 299.
Le figuier, aux ailes dorées, *De Buffon*, v. 311.—*Briffon*, App. 109.
Motacilla chryfoptera, *Lin. Syft.* 333.—*Latham*, ii. 492.

295. GOLD-WING.

W. With a golden crown: eyes inclofed in a bed of black,
reaching from the bill to the hind part of the head, and
bounded above and below with a white line: throat, and under fide

* *Sloane*, i. 309.

3 F 2   of

of neck, black : upper part, back, and leſſer coverts of wings, pale blueiſh grey : greater coverts rich yellow : primaries and tail dark cinereous : belly white.

PLACE.   Like the preceding, tranſient in the ſpring through *Penſyl-vania.*

296. YELLOW-FRONTED.   W. With the forehead and crown of a bright yellow : from the bill extends through the eyes a band of black, bounded on each ſide with white: chin, throat,. and lower ſide of the neck, black : breaſt and belly white : upper part of the neck, back, rump, and leſſer coverts of the wings, of a light blueiſh grey ; the greater co-verts, and lower order of leſſer, of a bright yellow, forming a great ſpot in each wing : primaries and tail of a deep aſh-color ; inner webs of the outmoſt feathers of the tail ſpotted with white.

PLACE.   A paſſenger, like the former, through *Penſylvania.—Latham,* ii. 461.

297. GREEN.   Green black-throated Fly-catcher, *Edw.* 300.—*Latham,* ii. 484.
Le figuier à cravate noire, *De Buffon,* v. 298.—*Briſſon,* App. 104.

W. With yellow cheeks and ſides of the neck : black throat, under ſide of the neck, and ſides under the wings : upper part of the breaſt yellowiſh ; lower, and belly, white : head, and upper ſide of the body, of an olive-green : coverts of the wings of the ſame color, marked with two bars of white : primaries and tail duſky ; the inner webs of the firſt edged with white ; of the three outmoſt feathers of the tail, ſpotted with white.

PLACE.   Appears and migrates in the ſame manner as the other.

Red-

Red-throated Fly-catcher, *Edw.* 301.
La figuier a poitrine rouge, *De Buffon*, v. 308.—*Briſſon*, Add. 105.
Motacilla Penſylvanica, *Lin. Syſt.* 333.—*Latham*, ii. 489.

W. With a yellow crown : white cheeks : a ſmall black mark paſſing under each eye : throat, and whole under ſide of the body, white, except part of the breaſt, which is of a blood-red, which color extends along the ſides under the wings : hind part of the head black : back and rump duſky, edged with yellowiſh green : coverts of the wings, and primaries, duſky ; the firſt marked with two bars of white : tail duſky, with a white mark on the exterior feathers. FEMALE wants the black ſpot on the hind part of the head, and thoſe on the back ; in other reſpects agrees with the cock.

Attends the preceding ſpecies in their ſhort paſſage through *Penſylvania.*

PLACE.

Little blue-grey Fly-catcher, *Edw.* 302.
La figuier gris de fer, *De Buffon*, v. 309.—*Briſſon*, App. 107.
M. Cærulea, *Lin. Syſt.* 337.—*Latham*, ii. 490.

W. With the head and whole upper part of the body of a blueiſh ſlate-color : wings brown ; a few of the ſecondaries edged with white : over each eye a narrow line of black : tail duſky ; two outmoſt feathers white ; the third on each ſide tipt with white. FEMALE wants the black ſtripe over the eyes : and the colors of the tail, and upper part of it, browniſh.

Appears in *Penſylvania* in *March.* Builds its neſt in *April,* with huſks from the buds of trees, down of plants, &c. coating it with lichens, and lining it with horſe-hair. It continues in the country all

PLACE.

all fummer, and retires fouth at approach of winter; perhaps to *Cayenne*, where the fame fpecies is found *.

300. **WORM-EATER.**

Worm-eater, *Edw.* 305.—*Latham*, ii. 499.
Le Demi-fin Mangeur de vers, *De Buffon*, v. 325.

W. With the crown of a reddifh yellow, bounded by a line of a lighter; beneath that, another of black; and through the eye, from the bill, a third of yellow, bounded beneath by a dufky ftroke: cheeks, throat, and breaft, of a yellowifh red, deepeft on the breaft, fading towards the belly, which is white: upper part of the neck, back, wings, and tail, of a deep olive-green: legs flefh-colored. BILL of this fpecies is much thicker than others of the genus.

PLACE. Does not appear in *Penfylvania* till *July*, in its paffage north-ward. Does not return the fame way; but is fuppofed to go be-yond the mountains which lie to the weft. This feems to be the cafe with all the tranfient vernal vifitants of *Penfylvania*.

301. **YELLOW-TAIL.**

Yellow-tail Fly-catcher, *Edw.* 257.

W. With an afh-colored crown: hind part of the neck, co-verts of the wings, and the back, of an olive-green: rump cinereous; fometimes that and the head of the fame color with the back: throat, under fide of neck, breaft, and belly, white; the fides of the breaft dafhed with ruft-color: fides, under the wings, yellow: on the lower part of the primaries a large bed of yellow: two mid-dle feathers of the tail brown; the reft yellow, tipt with brown.

PLACE. Taken on its paffage, with other birds (before defcribed) of this genus, off *Hifpaniola*, at fea, fuppofed to be on their way to their winter quarters in *Jamaica*, and other iflands.

* *Pl. Enl.* 704.

Spotted

Spotted yellow Fly-catcher, *Edw.* 257.—*Latham*, ii. 482.   302. SPOTTED.
La figuier brun de Canada, (the male) *Briſſon*, iii. 515.·
——————— de St. Domingue (the female) 513.—*De Buffon,* v. 293.

W. With the head, upper part of the body, and wings, of a
dark olive green : primaries and tail of a more duſky hue :
the interior web of the outmoſt feathers of the tail marked with a
large white ſpot : leſſer coverts of the wings, near the ridge, croſſed
with white : rump yellowiſh : all the under ſide of the body yellow :
under ſide of the neck, breaſt, and ſides, ſpotted with black : middle of the belly and vent plain.

Taken with the preceding. Inhabits alſo *Canada*, which may be   PLACE.
its place of ſummer reſidence and breeding. The FEMALE, which
has a white breaſt, and the colors of the upper part of the body
more dull than that of the cock, has been found in the iſle of *Hiſpani-
ola*; which may be one of the winter quarters of this and congenerous birds.

Le figuier à gorge jaune, *De Buffon*, v. 288.   303. LOUISIANE.
Le figuier de le Louiſiane, *Briſſon*, iii. 500.—*Latham*, ii. 480.

W. With the head and whole upper part of the body of a clear
olive-green : cheeks inclining to cinereous : coverts of the
wings of a blueiſh aſh-color, croſſed with two white bars : primaries duſky, edged externally with olive, internally with white : tail
of a duſky brown, edged like the wings; and the three outmoſt
feathers marked near their ends with a white ſpot : lower ſide of
the neck and breaſt of a fine yellow; the laſt ſpotted with red :
belly and vent white, tinged with yellow. FEMALE wants the red
ſpots on the breaſt.

Inhabits *Louiſiana* and *St. Domingo.*   PLACE

304. ORANGE-
THROAT.

Le figuier à gorge orangée, *De Buffon,* v. 290.
Le grand figuier de Canada, *Briffon,* iii. 508.

W. With the head, upper part of the neck and back, and leffer coverts of the wings, of an olive-green : the lower part of the back, rump, and greater coverts, afh-colored : primaries brown, edged on the outmoft webs with dark cinereous ; on the inner with dirty white : throat and under fide of the body orange, except the vent, which is white. FEMALE differs from the male in having its under fide of a duller and paler color.

PLACE.   Inhabits *Canada.*

305. QUEBEC.

Le figuier à téte jaune, *De Buffon,* v. 298.—*Briffon,* iii. 517.—*Pl. Enl.* 731.
Motacilla icterocephala, *Lin. Syft.* 334.—*Latham,* ii. 484.

W. With a yellow crown : fpace between the bill and the eyes black : below the eyes, and on the fides of the neck, white : hind part of the head, neck, back, and rump, black, edged with yellowifh olive : ridge coverts of the wings, and tail, of the fame color; other leffer coverts, and the greater coverts, black, marked with two tranfverfe bars of yellow : tail dufky, edged with olive ; the outmoft feathers marked half the length of their inner webs with yellowifh white : all the lower part of the body of a dirty white.

PLACE.   Inhabits *Canada.*

306. BELTED.

Le figuier a ceinture, *De Buffon,* v. 503.
Le figuier cendrè, *Briffon,* iii. 524.
Motacilla Canadenfis, *Lin. Syft.* 334.—*Latham,* ii. 486.

W. With an oblong yellow fpot on the crown : reft of the head, upper fide of the body, and coverts of wings, of a deep blueifh afh-color, almoft black ; the laft croffed with two white

9

bars :

bars: from the bill, above each eye, paſſes a white line: the under ſide of the neck, breaſt, and belly, are white; the two firſt marked longitudinally with brown ſtreaks: between the breaſt and belly is a tranſverſe belt of yellow: tail duſky, a little forked; the two outmoſt feathers on each ſide white at their ends and inner ſides: coverts of the tail yellow. FEMALE is brown on the upper ſide: the coverts of the tail are not yellow.

Inhabits *Canada.*   PLACE.

Le figuier de la Caroline, *Pl. Enl.* 58, N° 1.—*De Buffon*, v. 286.   307. OLIVE.

W. With the head, upper part of the body, and coverts of the wings, of an olive-green: primaries and tail brown; the firſt bordered with green, the laſt with yellow: under ſide of the body of a pale yellow.

Inhabits *Carolina.*   PLACE.

Le Fauvette tachetée de la Louiſiane, *De Buffon*, v. 161.—*Pl. Enl.* 752.—   308. NEW-YORK.
        Latham, ii. 436.

W. With a black bill, ſlightly bent at the end: over each eye a white line: crown, and all the upper plumage, cinereous and deep brown: lower part of the neck and body yellowiſh, ſtreaked with black: legs reddiſh brown. LENGTH near ſix inches.

Inhabits *Louiſiana,* and the hedges about *New York.* Not gre-   PLACE.
garious.

309. Dusky.   Fauvette ombrée de la Louifiane, *De Buffon*, v. 162.—*Pl. Enl.* 709.—
Latham, ii. 437.

W. With a black flender bill : upper part of the plumage greyifh brown : back marked faintly with black : wings, coverts
of the tail, and the tail itfelf, dufky ; the laft edged with white,
thinly fpeckled with black : legs dufky.

Place.   Inhabits *Louifiana*.

310. Prothono-   Le figuier protonotaire, *De Buffon*, v. 316.—*Pl. Enl.* 704.—*Latham*, ii. 494.
tary.

W. With the head, neck, throat, breaft, and belly, of a fine
jonquil yellow : vent white : back olive : rump afh-color :
wings and tail black and cinereous.

Place.   Inhabits *Louifiana*.   Called there *le Protonotaire* ; but the reafon
has not reached us.

311. Half-col-   Le figuier a demi collier, *De Buffon*, v. 316.—*Latham*, ii. 494.
lared.

W. With a yellowifh olive crown : an afh-colored band behind
the eyes : coverts of the wings brown, edged with yellow :
primaries brown, edged with white : throat and all the under fide of the
body of a clear afh color : acrofs the breaft is a half-collar of black :
belly tinged with yellow : tail afh-color : four feathers on each fide
edged with black on their inner fides.

312. Orange-   Le figuier a gorge jaune, *De Buffon*, v. 317.—*Latham*, ii. 495.
bellied.

W. With the head and upper part of the body of an olive-
brown : coverts of the wings yellow, varied with brown ?
primaries brown : fecondaries and tail brown, bordered with
olive : throat, under fide of the neck, and breaft, yellow ; part
of

of the latter tinged with brown: the reft of the lower part of the body reddifh, approaching to yellow.

Le figuier brun olive, *De Buffon*, v. 318.—*Latham*, ii. 495.   313. OLIVE-BROWN.

W. With the upper part of the head and body of a brownifh olive : the coverts of the wings, and primaries, brown ; the firft edged and tipt with white; the laft edged with grey: throat and breaft white, varied with teints of grey : belly of a yellowifh white: vent quite yellow: tail brown, bordered with clear grey; thofe of the middle tinged with yellow ; the two outmoft on each fide bordered with white.

Le figuier graffet, *De Buffon*, v. 319.—*Latham*, ii. 495.   314. GRASSET.

W. With the head and upper part of the body of a deep greyifh green and deep olive ; the middle of the head marked with a yellow fpot : back tinged with black : wings brown or dufky : throat and under fide of the neck reddifh ; the reft of the lower part white : tail black, edged with grey ; and the four outmoft feathers on each fide marked near their ends with white.

Le figuier cendre, a gorge cendré, *De Buffon*, v. 319.—*Latham*, ii. 496.   315. GREY-THROAT.

W. With the head, and upper part of the body and wings, afh-co-lor ; the laft edged with white : throat and under fide of the body of a more clear afh-color : tail black : firft feather on each fide almoft white ; the fecond half white ; the third tipt with the fame.

These five fpecies inhabit *Louifiana*, and are called there *Graffets*,   PLACE. from their exceeding fatnefs. They frequent the tulip-trees ; in particular the *magnolia grandiflora*, or the *laurel-tree* *, whofe ever-green leaves give ample fhelter to the feathered tribe.

* *Catefby*, ii. 61.

3 G 2   Motacilla

316. GUIRA.   Motacilla Guira, *Lin. Syst.* 336.—*Edw.* 351.—*Latham*, ii. 505.—*Marcgrave*, 212.—*De Buffon*, v. 343.—BL. MUS.

W. With head, hind part of neck, and back, of an olive green; lower part dashed with yellow: lesser coverts dusky, slightly edged with white; greater, and primaries, dusky, with their edges deeply marked with white: throat, and lower part of the neck, full black: breast and belly of a fine light yellow: tail brown, edged with dull yellow. The crown of the FEMALE olive green, spotted with black: hind part of the neck plain green: chin and fore part of neck black: breast and belly yellow, spotted with red: wings and tail like those of the male.

PLACE.   Inhabits *New York.* Makes its nest between the small branches of some tree. It is open at top, shallow, and formed of broad dead grafs, and some fibres. Its eggs white, thinly spotted with black.

317. BLACK-BURNIAN.   W. With the crown intensely black, divided by a line of rich yellow: from each corner of the upper mandible is another of the same color: through the eye passes one of black, reaching beyond it, bounded beneath by a narrow yellow line: sides of the neck, the throat, and middle of the breast, are of a beautiful yellow: sides spotted with black: vent and thighs white: lesser coverts black; greater white: back striped black and white: primaries dusky: middle feathers of the tail dusky; three outmost on each side marked with white.—*Latham*, ii. 461.

PLACE.   Inhabits *New York.*—BL. MUS.

318. PINE.   Pine-Creeper, *Catesby*, i. 61.—*Edw.* 277. Le figuier de fapins, *De Buffon*, v. 296.—*Latham*, ii. 483.

W. With the crown, cheeks, breast, belly, and thighs, of a bright yellow: from the bill to the eyes is a dusky line: hind part of the neck, the back, and rump, of a yellowish green, inclining

to

to olive, brighteſt on the rump : wings and tail of a blueiſh grey : coverts marked with two white lines : outmoſt feathers of the tail with their inner webs white.   FEMALES of a browniſh color.

Appears in *Penſylvania*, from the ſouth, in *April*.   Feeds on inſects and buds of trees.   Continues there the whole ſummer.   Inhabits the ſofter climate of *Carolina* the whole winter ; and is ſeen creeping about the trees, eſpecially the firs and pine, with other congenerous birds, which aſſociate during that ſeaſon in ſmall flights.

PLACE.

Yellow Titmouſe, *Cateſby*, i. 63.
Yellow Wren, *Br. Zool.* i. N° 151.—*Edw.* 278..
Le figuier brun & jaune, *De Buffon*, v. 295.
Le Pouillot, ou le Chantre, *Ib.* 344.—*Briſſon*, iii. 479.
Le figuier de Caroline, *Ib.* 486.—*Latham*, ii. 512.
M. Trochilus, *Faun. Suec.* N° 264.—LEV. MUS.—BL. MUS.

319.   YELLOW.

W. With the head and upper part of the body, wings, and tail, of a deep olive : cheeks yellow : through the eyes paſſes a duſky line, and beneath them another : whole under ſide and inner coverts of the wings, of a fine yellow ; but in ſome much paler than others.

Inhabits *North Carolina* ; breeds there, and diſappears in winter, retiring to *Jamaica* and other iſlands.   Is almoſt an univerſal bird. Found in moſt parts of *Europe*.   Bears all climates, from the *Eaſt Indies* to the rugged *Kamtſchatka*.   Is one of the ſmalleſt birds of *Europe*.   Feeds on inſects.

PLACE..

Ruby-crowned Wren, *Edw.* 254..
Le Roitelet rubis, *De Buffon*, v. 373.—*Latham*, ii. 511.—LEV. MUS.

320.   RUBY-
CROWNED..

W. With a rich ruby-colored ſpot towards the hind part of the head : reſt of the head, upper part of the neck, body, and coverts of the wings, of an olive-colour : coverts croſſed by two white lines : primaries and tail duſky, edged with yellow : from

bill

Size. bill to tail a light yellow. LENGTH four inches: extent five: weight four drams.

Place. Inhabits *North America*, from *Hudson's Bay* to *Pensylvania*; probably through the whole continent. A most delicate bird, to be found in the rude climate of the bay.

321. GOLDEN-
 CRESTED.

Br. Zool. i. N° 153.—*Catesby*, App. 13.
M. Regulus. Kongsfogel, *Faun. Suec.* N° 262.—*Latham*, ii. 508.
Le Roitelet, *De Buffon*, v. 363.—*Pl. Enl.* 651. 3.—LEV. MUS.—BL. MUS.

W. With a black crown, divided lengthways with a rich scarlet line, which it shews or conceals at pleasure. In other respects, the colors and marks resemble the former. The least of all *European* birds. LENGTH only three inches and a half.

Place. Is found in *New York*; and inhabits the red cedars. Is met with in *Europe* as high as *Drontheim* *. Crosses annually from the *Orknies* to the *Shetland* isles; where it breeds, and returns again before winter: a long flight, of sixty miles, for so small a bird. Rare in *Russia*. Frequent in *Sibiria*, about the *Jenesei*.

322. WREN.

Br. Zool. i. N° 154.
M. Troglodytes, *Faun. Suec.* N° 261.—*Latham*, ii. 506.
Le Troglodyte, *De Buffon*, v. 352.—*Pl. Enl.* 651. 2.—LEV. MUS.—BL. MUS.

W. With head and back brown, obscurely barred with dusky: coverts of wings, quil-feathers, and tail, elegantly barred with black and ferruginous: whole under side of a dirty white, mottled with pale brown.

Twice the size of the *European* Wren; yet appears to be of the same kind. Is one of the exceptions to the remark made, that the

* *Av. Nidr. Catal. MS.*

animals

animals of the fame fpecies in the new are leffer than thofe of the old world.

Appears in the province of *New York* in *May*, and lays in *June*. Builds its neft in holes of trees, with fibres of roots and fticks, lining it with hairs and feathers. Lays from feven to nine eggs, white, thinly fpotted with red. Has the fame actions with the *European* Wren: fings, but with a different note. Retires fouth in *Auguft*.

The *European* kind reaches to the *Feroe* ifles; where it enters the cottages, to peck the dried meat of the inhabitants *. Found alfo in *Norway*; but not far north. Rare in *Sweden* and *Ruffia*. Unknown in *Sibiria*.

Little Sparrow? *Catefby*, i. 35.
Hedge Sparrow, *Lawfon*, 144.—*Latham*, ii. 420.

W. With the body entirely brown.
Lefs than the *European* Hedge Sparrow. Mr. *Catefby* fays, that it partakes much of the nature of that fpecies. Mr. *Law-fon* fays, that the Hedge Sparrow of *Carolina* differs fcarcely from the *Englifh*; only that he never heard it fing. They are not numerous; are ufually feen fingle, hopping under bufhes: feed on infects: and are commonly feen near houfes in *Carolina* and *Virginia*, where they continue the whole year.

* *Brunnich*, N° 284.

A. NIGHTINGALE, *Br. Zool.* i. N° 145.
Nàchtergahl, *Faun. Suec.* N° 345.—*Latham*, ii. 410.
Le Roſſignol, *De Buffon*, v. 81.—*Pl. Enl.* 615.—LEV. MUS.—BL. MUS.

W. With head and neck tawny, daſhed with olive: throat, breaſt, and belly, gloſſy aſh-color: tail deep tawny.

PLACE.   Inhabits the groves of *Oland, Gothland, Upſal*, and *Schonen*; but not farther north. Appears about the middle of *May:* retires about the time of hay-harveſt *. Found in the temperate parts of *Ruſſia*; and in *Sibiria*, as far as *Tomſk* only; not as yet in the eaſtern parts. None in *Scotland*. Extends over every temperate part of *Europe*; to *Syria* †, *Perſia* ‡, and the *Holy Land* §; and to the banks of the *Nile*.

B. REDSTART, *Br. Zool.* i. N° 146.
M. Phœnicurus Rodſtjert, *Faun. Suec.* N° 257.—*Latham*, ii. 421.
Le Roſſignol de muraille, *De Buffon*, v. 170.—*Pl. Enl.* 351.—LEV. MUS.—BL. MUS.

W. With white front: crown and back deep blueiſh grey: cheek and throat black: breaſt, rump, and ſides, red: two middle feathers of tail brown; the reſt red. FEMALE, head and back aſh-color: chin white.

PLACE.   Inhabits *Europe*, as high as *Drontheim*. In all parts of *Ruſſia* and *Sibiria*: in the laſt, the colors are extremely vivid. Extends to *Kamtſchatka*, and even to the *Arctic* circle.

* *Amœn. Acad.* iv. 597.   † *Ruſſell*, as quoted by Mr. *Latham*.   ‡ *Fryer's Trav.* 248.   § *Haſſelquiſt*.

GREY

C. GREY REDSTART.
M. Erithacus, *Faun. Suec.* N° 258.
Le Rouge-queue, *De Buffon*, v. 180.—*Latham*, ii. 423.

W. With a hoary crown : back and wings cinereous : whole un-
der fide of the body and tail ferruginous.
Inhabits *Sweden.* Lives in trees. Lays nine blueifh grey eggs. *PLACE.*
Seen alfo near the *Volga.*

D. Red-breaft, *Br. Zool.* i. N° 147.
Rotgel, *Faun. Suec.* 260.—*Latham*, ii. 442.
Le Rouge-gorge, *De Buffon*, v. 196.—*Pl. Enl.* 361.—LEV. MUS.

W. With front, chin, and breaft, of a deep orange red : upper
part of the body, wings, and tail, olivaceous.
Inhabits *Europe* as far as *Drontheim.* Scarce in *Ruffia.* Is feen *PLACE.*
above the *Kama* ; but never in *Sibiria.* Its familiarity with mankind has occafioned it, · in many countries, to receive a fond name :
thus the *Danes* call it *Tommi-Liden* ; the *Norwegians, Peter Ronfmad* ;
the *Germans, Thomas Gierdet* ; and we, *Robin Red-breaft* *.

E. BLUE-THROAT. M. Suecica, N° 259.
Bloukropfl, *Kram. Auft.* 375.—*Latham*, ii. 444.
La Gorge-bleue, *De Buffon*, v. 206.—*Pl. Enl.* 361.—LEV. MUS.

W. With a tawny breaft, marked with a fky-blue crefcent : over
each eye a white line : head and back brown : tail dufky,
ferruginous towards the bafe, and tipt with yellow : belly whitifh :
the vent yellowifh.
Inhabits *Weft Bothnia* and *Lapland.* Lives among the alders and *PLACE.*
willows, and is fuppofed not to migrate from that fevere climate †.
Is found in all the northern parts of *Ruffia* and *Sibiria.* Sings
finely.

* Mr. *Latham.*   † *Amæn. Acad.* iv. 597.

3 H   A bird,

A bird, differing from this only by a blue line below each eye, is figured by Mr. *Edwards*, tab. 28, and drawn from one ſhot on the rock of *Gibraltar*.

F. BLACK-CAP, *Br. Zool.* i. Nº 148.
    M. Atricapilla, *Faun. Suec.* Nº 256.—*Latham*, ii. 415.
    La Fauvette à tête noire, *De Buffon*, v. 125.—*Pl. Enl.* 580.—LEV. MUS.—BL. MUS.

W. With a black crown: hind part of neck pale aſh: back, and coverts of wings, greyiſh olive: breaſt and belly light aſh. Crown of the FEMALE dull ruſt-color.

PLACE.   Found in *Sweden*; chiefly in *Schonen.* Not in the *Ruſſian* catalogue.

G. PRTTY-CHAPS, *Br. Zool.* i. Nº 149.
    M. Hippolais, *Faun. Suec.* Nº 248.—*Latham*, ii. 413.
    La Fauvette, *De Buffon*, v. 117.—*Pl. Enl.* 579.—LEV. MUS.

W. With inſide of the mouth red: head, back, and wings, olivaceous aſh: inner coverts yellow: breaſt white, tinged with yellow: belly ſilvery: tail duſky.

PLACE.   Found as far as *Sweden*.

H. HEDGE, *Br. Zool.* i. Nº 150.
    M. Modularis Jarnſparf, *Faun. Suec.* Nº 245.—*Latham*, ii. 419.
    Le Traîne Buiſſon, ou Mouchet, ou la Fauvette d'hiver, *De Buffon*, v. 151.—*Pl. Enl.* 615.—LEV. MUS.

W. With a deep brown head, mixed with aſh: throat and breaſt of a dull ſlate-color: belly dirty white: ſides, thighs, and vent, of a tawny brown: tail duſky.

PLACE.   Inhabits *Sweden*, its fartheſt northern reſidence. Lays four or five fine pale blue eggs.

I. Bog-rush.
M. Schænobænus, *Faun. Suec.* N° 246.—*Latham*, ii. 418.
La Rouffette, ou la Fauvette des bois, *De Buffon*, v. 139.

W. With head, back, and rump, of a teftaceous brown; the two
firft fpotted : the wings teftaceous on their outmoft fides :
throat and belly of the fame color : tail dufky.   Size of a Wren.
Inhabits among the bog-rufhes of *Schonen* in *Sweden*.   Place.

K. Fig-eater.
M. Ficedula, *Faun. Suec.* N° 251.—*Latham*, ii. 432,
Le Bec-figue, *De Buffon*, v. 187.—*Pl. Enl.* 668.

W. With head and upper part of the body and wings dufky,
mixed with chefnut : breaft of a cinereous white ; that of
the female white : tail of the male black ; of the female inclined
to chefnut : legs of the male chefnut, of the female black.
Inhabits (but rarely) the gardens and cultivated parts of *Sweden*.   Place.

L. Grasshopper, *Br. Zool.* i. N° 382.
Alauda trivialis, *Lin. Syf.* 288.—*Latham*, ii. 429.—Lev. Mus.

W. With head and upper part of the body of an olive brown,
fpotted with black : primaries dufky, edged with olive brown :
breaft and belly dirty white : tail very long, and cuneiform, compofed
of twelve fharp-pointed brown feathers.
Inhabits *Sweden*.   Is frequent in *Sibiria*.   Scarce in *Ruffia*.   Has   Place.
the note of a Grafshopper.

M. Sedge, *Br. Zool.* i. N° 155.
M. Salicaria, *Faun. Suec.* N° 249.—*Latham*, ii. 430.
La Fauvette de rofeaux, *De Buffon*, v. 142.—Lev. Mus.

W. With a brown head, ftreaked with dufky : over each eye a line
of white, bounded above by another of black : throat white :
breaft and belly white, tinged with yellow : back reddifh brown,

fpotted

fpotted with black : rump tawny : tail brown ; circular when fpread.

PLACE.   Inhabits *Sweden.*   Is frequent in *Ruffia* and *Sibiria*, in willow thickets near rivers, even to the *Arctic* circle.

N. Scotch, *Br. Zool.* i. N° 152.
M. Acredula, *Faun. Suec.* N° 263.—*Latham*, ii. 513.

W. With front and under fide of the body of a fine pale yellow : back and wings green, dafhed with afh-color : tail forked and brown.   Size of a Wren.

PLACE.   Inhabits *Sweden,* about *Upfal.*   Found alfo in *Ruffia* and *Sibiria.*

O. Long-billed.

W. With a very long flender bill : forehead, cheeks, and chin, pale ruft-color : upper part of body and tail brown, tinged with olive : under part of the body of the fame color, but lighter : middle of the belly white.   Leffer than a Hedge Sparrow.

PLACE.   Inhabits *Kamtfchatka.*

** **   WITH PARTICOLORED TAILS.**

P. Wheat-ear, *Br. Zool.* i. N° 157.
M. Oenanthe. Stenfquetta, *Faun. Suec.* N° 254.—*Latham*, ii. 465.
Le Motteux, ou Cul blanc, *De Buffon,* v. 237.—*Pl. Enl.* 554.—Lev. Mus.

W. With head and back grey, tinged with red : from the bill to the hind part of the head, acrofs each eye, is a broad bar of black ; above that a line of yellow : breaft and belly white, tinged with yellow : rui..p and lower half of the tail white ; the end black. Female wants the black bar acrofs the eyes : lefs white on the tail, and the colors in general are duller.

A fpecies

A fpecies which extends from the fultry climate of *Bengal* * to the frozen region of *Greenland*. Is migratory, at left in the temperate and frigid zones. Goes even in fummer as high as beyond the *Arctic* circle, in *Europe* and *Afia*, wherever the country is rocky. In *Greenland*, is converfant among rills of water. Feeds on infects and worms, efpecially thofe o. places of interment; is therefore detefted by the natives †. Breeds in that country in *June*. Is found in *Iceland* and the *Feroe* iflands. On its firft appearance in *Sweden*, the peafants expect to be freed from the fevere nocturnal frofts ‡. Its winter retreat unknown.

PLACE.

Q. STAPAZINA.
M. Stapazina, *Lin. Syft.* 331.—*Latham*, ii. 468.
Le Motteux, ou Cul blanc roufsâtre, *De Buffon*, v. 454.—LEV. MUS.

W. With head, neck, and breaft, of a reddifh brown: throat and belly white: acrofs the eyes a brown bar: rump white: tail like that of the former.

Is frequent, with the preceding, in *Ruffia* and *Sibiria*; and extends to *Kamtfchatka*. Often found in the warmer parts of *Europe*.

PLACE.

R. WHIN-CHAT, *Br. Zool.* i. N° 158.
Le Tarier, *De Buffon*, v. 224.—*Pl. Enl.* 678.
M. Rubetra, *Faun. Suec.* N° 255.—*Latham*, ii. 245.—LEV. MUS.

W. With head and back of rufty brown, fpotted with black: over each eye a white line; under that a broad bed of black: breaft reddifh yellow: two middle feathers of the tail black; the reft white at their bottoms, black at their ends. The FEMALE has on the cheeks a bed of brown inftead of black, and the other colors lefs vivid.

---

* *Edw. Birds*, i. Preface, xii.   † *Faun. Groenl.* N° 84.   ‡ *Amœn. Acad.* iv. 597.

Found

PLACE.　　Found not farther north than *Sweden*. Is found in the temperate parts of *Ruffia,* as far as the *Urallian* chain; but has not reached *Sibiria*.

S. WHITE-THROAT.
M. Sylvia. Skogfneter mefar, *Faun. Suec.* N° 250.—*Latham,* ii. 428.
La Grifette, ou Fauvette grife, *De Buffon,* v. 132.—*Pl. Enl.* 579. 3.—Lev. Mus.

W. With head of a brownifh afh: back tinged with red: leffer
coverts of wings pale brown; greater dufky, edged with
tawny brown: wings and tail dufky, with reddifh brown margins:
exterior fide, and part of the interior fides, of the outmoft feather
of the tail white.

PLACE.　　Not farther north than *Sweden*. Scattered over all *Ruffia* and
Sibiria.

T. AWATCHA.

W. With crown, upper part of neck and body, deep brown:
primaries edged with white: lower part of the five outmoft
feathers of the tail deep orange; ends brown; two middle feathers wholly brown: throat and breaft white; the fides of the firft,
and all the laft, fpotted with black: from upper mandible to each
eye, an oblique white line: fides pale ruft-color: middle of the
belly white.

PLACE.　　Inhabits *Kamtfchatka.*

U. KRUKA.
M. Curruca. Kruka, *Faun. Suec.* N°. 247.—*Latham,* ii. 417.

W. With head, wings, and upper part of body, brownifh afh; lower
part white: tail dufky; but each outmoft feather ftriped
down with a line of white.

PLACE.　　Inhabits *Sweden,* and all parts of *Ruffia*; but not *Sibiria*. Its eggs
afh-colored, fpotted with ruft. Not our Hedge Sparrow, which
Linnæus makes fynonymous with it.

XXX. ΓIT-

XXX. T I T M O U S E. *Gen. Birds* LIV.

Crested Titmoufe, *Catefby*, i. 57.—*Latham*, ii. 544.
La Mefange huppée de la Caroline, *De Buffon*, v. 451.—*Briffon*, iii. 561.
Parus bicolor, *Lin. Syft.* 340.—LEV. MU^c —BL. MUS.

T. With the forehead, head, and upper part of the neck and
body, of a deep grey: under fide white, tinged with red;
deepeft under the wings: feathers on the head long, which it erects
occafionally into a pointed creft, like a toupet: legs of a lead-
color. FEMALE differs not in color.

PLACE.

Inhabits the forefts of *Virginia* and *Carolina* the whole year, and
feed on infects. Shuns houfes. Found alfo in *Greenland* *. Flies
fwift; and emits a weak note.

Yellow-rump, *Catefby*, i. 58.—*Latham*, ii. 546.
La Mefange à croupion jaune, *De Buffon*, v. 453.
Parus Virginianus, *Lin. Syft.* 342.—*Briffon*, iii. 575.

T. With the head, whole body, wings, and tail, brown, tinged
with green: rump yellow.

PLACE.

Inhabits *Carolina.* Frequents trees, and feeds on infects.

Finch Creeper, *Catefby*, i. 64.—*Latham*, ii. 558.
Parus Americanus, *Lin. Syft.* 341.—BL. MUS.

T. With a blueifh head: white fpot above, and another beneath
each eye: upper part of the back of a yellowifh green; reft
of the back, tail, and wings, of a dufky blue; the laft croffed with
two bars of white: throat yellow, bounded beneath by a black

* *Faun. Groenl.* 123.

3

band,

band, extending to the hind part of the neck; which is of the same color: breaft yellow: belly white: fides tinged with red: legs dull yellow.   FEMALE dufky.

PLACE.

Inhabits *Carolina* all the year.   Creeps up and down the bodies of trees, and picks infects out of the bark.

327. COLEMOUSE.

Br. Zool. i. N° 164.
Parus ater, *Faun. Suec.* 268.—*Latham,* ii. 540.
La petite Charbonniere, *De Buffon,* v. 400.—LEV. MUS.

T. With a black head, marked on the hind part with a white fpot: back and rump of a cinereous green; brighteft on the laft: coverts of the wings of a dufky green; the loweft order tipt with white.

PLACE.

Shot during fummer in *Newfoundland*.   Is found in *Sibiria,* even beyond the *Lena*; and winters in that climate.

328. CANADA.

Mefange à tête noire du Canada, *De Buffon,* v. 408.—*Briffon,* iii. 553.
Parus Atricapillus, *Lin. Syft.* 341.—*Latham,* ii. 542.

T. With the head and chin black: fides of the neck, cheeks, and all the under part of the body, white: upper fide of the neck, back, and rump, of a deep afh-color: coverts of the wings, and primaries, brown; the firft edged with grey; the exterior fides of the laft with a lighter grey; the inner with white: the two middle feathers of the tail cinereous; the others brown on the inner fide, and afh-colored on the outmoft, edged with light grey.

PLACE.

Inhabits *Canada* and *Hudfon's Bay,* and as high as lat. 64. 30, on the weftern fide of *North America*.   Is a moft hardy bird; and continues about *Albany* Fort the whole year; but moft numerous in cold weather, probably compelled by want of food.   Feeds on worms and infects: makes a twittering noife; from which the natives call it *Kifs-kifs-kefhifh* *.

* *Phil. Tranf.* lxii. 407.

I cannot

I cannot add a bird of this kind from *Louisiana* as a new species, as it differs in nothing, except having the black spot on the chin larger, and the colors deeper. The FEMALE has a tinge of red amongst the cinereous, and on the head *.

Parus Hudsonicus, *Forster.—Ph. Transf.* lxii. 408. 430.—*Latham,* ii. 557.

T. With the head of a rusty brown: a white line beneath each eye: black throat: feathers on the back long, brown tipt with olive: feathers on the breast and belly black, tipt with white: sides under the wings ferruginous: wings brown: edges of the primaries cinereous: tail rounded; brown, edged with cinereous: legs black. Male and Female resemble each other. LENGTH five inches and an eighth. EXTENT seven. WEIGHT half an ounce.

Continues, even about *Severn* river, the whole year. Frequents the juniper-bushes, on buds of which it feeds. Lays five eggs. In winter collects in small flocks, flying from tree to tree. The natives call them *Peche-ke-ke-shish*.

PLACE.

A. GREAT TITMOUSE, *Br. Zool.* i. N° 162.—*Latham,* ii. 536.
Le Charbonniere, ou grosse Mesange, *De Buffon,* v. 392.—*Pl. Enl.* 3.
Talg-oxe, *Faun. Suec.* 265.—LEV. MUS.—BL. MUS.

T. With white cheeks: bill, head, and throat, black: belly yellowish green, divided lengthways with a bed of black: rump blueish grey: coverts of wings blue: primaries edged with blue: tail

* *De Buffon,* v. 407.—*Pl. Enl.* 502.

dusky,

dufky; exterior fides of the outmoft feathers white; of the others blueifh: legs lead-color. Size of a Chaffinch.

PLACE.   Inhabits *Norway*, *Sweden*, *Ruffia*, and *Sibiria*, even in the winter.

B. Strömian, *Strom. Sond.* i. 240.—*Brunnich*, p. 73.—*Latham*, ii. 537.

T. With bill black above, yellow below: neck and upper part of the body yellowifh green: throat yellow: breaft yellow, fpotted with bay: belly blue, yellowifh near the vent: tail bifurcated, of the fame color with the back; the two middle feathers greenifh; the two outmoft edged with white: legs black.

PLACE,   Difcovered by Mr. *Ström*, in *Sondmor*.

C. Azure Titmouse.
Parus Cyaneis, *Nov. Com. Petrop.* xiv. 498. tab. xiii. fig. 1.—588. tab. xxiii. fig. 1.
Parus Indicus, *Aldr.*—*Raii. Syn. Av.* 74.—*Latham*, i. 538.

T. With a very fhort and thick bill: crown and hind part of the neck of a hoary whitenefs; the lower part of the laft bounded by a tranfverfe band of dark blue: cheeks white, croffed by a deep blue line, extending beyond the eyes: back light blue: rump whitifh: under fide of the neck, breaft, and belly, of a fnowy whitenefs, with a fingle dufky fpot on the breaft: wings varied with rich blue, dufky, and white: tail rather long; of a dufky blue, tipt with white: legs dufky blue.

Size of the *Englifh* Blue Titmoufe. The plumage of this elegant fpecies is extremely loofe, foft, and of moft exquifitely fine texture, and fo liable to be raifed, that when the bird is fitting, but efpecially when it is afleep, it appears like a ball of feathers.

PLACE,   It inhabits, in great abundance, the rorthern woods of *Sibiria* and *Ruffia*, and about *Synbirfk*, in the government of *Kafan*. It is a migratory bird, and appears in winter converfant about the houfes in *Peterfburgh*. It twitters like the common Sparrow, but with a fofter and fweeter note.

BLUE,

D. Blue, *Br. Zool.* i. N° 163.
Blamées, *Faun. Suec.* N° 267.—*Latham*, ii. 543.
La Mefange bleue, *De Buffon*, v. 413.—*Pl. Enl.* 3. 2.—Lev. Mus.—Bl. Mus.

T. With a rich blue crown, wings, and tail: a black line over each eye: cheeks and forehead white: back yellowish green: breast and belly yellow.

Inhabits as high as *Sondmor* * Found in fouthern *Ruffia*, but not in *Sibiria*.   Place.

E. Marsh, *Br. Zool.* i. N° 165.
Entita, Tomlinge, *Faun. Suec.* N° 269.—*Latham*, ii. 541.
La Nonuette cendrée, *De Buffon*, v. 403.—*Pl. Enl.* 3. 3.—Lev. Mus.—Bl. Mus.

T. With head wholly black: under fide of the body white: back cinereous. Like the *Colemoufe*, N° it wants the white spot on the hind part of the head: its tail is longer, and the bulk larger.

Is found as far as *Sondmor* †. Inhabits all parts of *Ruffia* and *Sibiria*, even as far as *Kamtfchatka;* and endures the hardeft frofts.   Place.

F. Crested.
Parus criftatus. Tofsmyffa. Tofstita, *Faun. Suec.* N° 266.
La Mefange huppée, *De Buffon*, v. 447.—*Pl. Enl.* 502.—*Latham*, ii. 545.

T. With a large upright creft: chin black: reft of the plumage a mixture of black, afh-color, and white.

Is found in *Sweden*, and in the weft and temperate parts of *Ruffia;* but does not reach *Sibiria*.   Place.

---

* *Strom.* 239.   † *Ibid.*

G. LONG-TAILED, *Br. Zool.* i. N° 166.

Lanius caudatus Ahltita, *Faun. Suec.* N° 83.—*Latham,* ii. 551.

La Mefange à longue queue, *De Buffon,* v. 436.—*Pl. Enl,* 502. 3.—LEV. MUS.—
BL. MUS.

T. With crown white, mixed with dark grey : head furrounded by a bed of black, beginning at the bafe of the bill : from the hind part of the head to the rump a line of black ; feathers on each fide of that line, and thofe on the breaft, a fine purplifh red : tail very long and cuneiform ; black, with the interior edges of the three outmoft feathers white.

PLACE.

Inhabits *Sweden.* Frequent, even in winter, in thickets and woods, all over *Ruffia* and *Sibiria.* Its elegant neft defcribed in the *Br. Zool.* i. p. 395.

H. BEARDED, *Br. Zool.* i. N° 167.—*Latham,* ii. 552.

La Mouftache, *De Buffon,* v. 418.—*Pl. Enl.* 618.—LEV. MUS.—BL. MUS.

T. With a fine grey head : beneath each eye a deep black triangular tuft of feathers : back, fides, and thighs, orange-colored : fecondaries black, edged with orange : middle of the breaft bloom-colored : tail long, cuneiform, and ferruginous. FEMALE wants the black tufts : crown of a dirty brown : outmoft feathers of the tail black ; the ends white.

PLACE.

Found but rarely in *Schonen* in *Sweden.* Is very common about the *Cafpian* and *Palus Mæotis,* and among the rufhes of the rivers which fall into them ; but in no high latitudes in *Afia.* None in *Sibiria.*

XXXI. SWAL-

XXXI. SWALLOW. *Gen. Birds*, LV.

Br. Zool. Nº 168.—*Latham*, ii. 560.
Hirundo ruftica. Ladu Swala, *Faun. Suec.* Nº 270.
L'Hirondelle de cheminée, *De Buffon*, vi. 591.—*Pl. Enl.* 543.—Lev. Mus.—
  BL. Mus.

330. Chimney.

SW. With the head, upper part of the body, and coverts of the wings, black, gloffed with rich purplifh blue: forehead red: under fide ferruginous. That of *Europe* white; in the Male tinged with red: tail black; every feather, unlefs the two middle, marked with a white fpot near the end.

Differs in nothing from the *Englifh* chimney Swallow, but in the rednefs of the under fidé.

Thefe birds inhabit, during fummer, *Newfoundland*, and other parts of *North America*. Build on lofty rocks and precipices, efpecially fuch as yield fhelter by overhanging their bafe. Others, fince the arrival of the *Europeans*, affect the haunts of mankind, and make their nefts in barns, ftables, and out-houfes: in fome parts they are, on that account, called Barn Swallows. The *Swedes* give them the fame name, *Ladu Swala*, becaufe in their country they alfo neftle in barns.

Place.

They appear in the *Jerfies* the beginning of *April*, wet, fays Mr. *Kalm*, from the fea or lakes, at the bottom of which they had paffed torpid the whole winter—I fhould rather imagine, from the cafual fhowers they met with in their long flight from their winter quarters: and that they do take fuch, Mr. *Kalm* himfelf is witnefs to, by meeting with them on their paffage at fea, nine hundred and twenty miles from any land *.

In the province of *New York* they appear in *May*. Make the fame fort of neft with the *European*. Lay in *June*. Difappear in *Auguft*, or early in *September*.

* Voy. i. 24.—See alfo *Br. Zool.* i. p. 344, &c.

Is

Is found in *Europe* as far north as *Drontheim*, and sometimes frequents the *Feroe* isles.

In Sibiria. This species is very common all over *Sibiria*; but those which are found beyond the *Jenesei*, and in all the north-east part of that country, have their lower part rust-colored, like the *American* variety; for they cannot be deemed a distinct species.

331. MARTIN.

Br. Zool. i. N° 169.—*Latham*, ii. 564.
Hirundo urbica. Hus-Swala, *Faun. Suec.* N° 271.
L'Hirondelle au Croupion blanc, ou de Fenêtre, *De Buffon*, vi. 614.—
Pl. Enl. 542.—Bl. Mus.

SW. With a white rump, breast, and belly : head and back black, glossed with blue : wings and tail black : feet covered with white down.

Place. In *Europe* is seen as high as *Drontheim*.

Inhabits, during summer, *Newfoundland* and *New York*. It was also found by the navigators on the western coast in the month of *October* : it was inferior in size to those found in *Europe*. A specimen, with a black rump, was sent from *Hudson's Bay* *; doubtful whether a variety or distinct species. They build there under the windows of the few houses, or against the steep banks of rivers.

In Sibiria. Is very common in *Sibiria* and *Kamtschatka*.

332. SAND.

Br. Zool. i. N° 170.—*Latham*, ii. 568.
Hirundo riparia. Strand-Swala. Back-Swala, *Faun. Suec.* N° 273
L'Hirondelle de rivage, *De Buffon*, vi. 632.—*Pl. Enl.* 543. 2.—Bl. Mus.

SW. With the head and upper part of the body of a mouse-color: wings and tail dusky : under side white : throat crossed by a mouse-colored ring : feet smooth and black.

* *Ph Transf.* lxii. 408.

Arrives

Arrives in *June* in *New York*. Builds in deep holes of banks, over lakes and rivers; and departs in *August* or the beginning of *September*. It is frequent in *Sibiria* and *Kamtschatka*. Is found in *Europe* as far north as *Sondmor* *

Purple Martin, *Catesby*, i. 51,
Great American Martin (fem?) *Edw.* 120.
Hirundo purpurea. H. Subis, *Lin. Syst.* 344.—*Latham*, ii. 574. Nº 21.—575. Nºs 23, 24.
Le Martinet coleur de pourpre, *De Buffon*, vi. 676.
L'Hirondelle de la Baie de Hudson, *Ib.* 677.
L'Hirondelle de la Louisiane, *Ib.* 674.—*Pl. Enl.* 722.—LEV. MUS.—BL. MUS.

SW. With its whole plumage black, glossed most richly with variable blue and deep purple: wings and tail of a duller color: legs and feet naked, large, and strong; three toes only standing forward, not all four, as in the *European* kind. In SIZE far superior to the *English* Swift; but the wings in proportion shorter.

The colors of the FEMALE are less glossy on the upper part of the body; below of a dirty white: in some, the ridge of the wings is white, and the breast grey. Such is the specimen engraven by Mr. *Edwards*; which I suspect to be a young bird, and not to differ in species, although it may in sex, from that of Mr. *Catesby*; for I have had opportunity of examining both male and female from *New York*. I must also unite the *Louisiane* of the Count *De Buffon*, to this species.

Inhabits *North America*, from *Hudson's Bay* to *South Carolina* and *Louisiana*. Appears in *New York* in *April*. Leaves the province the latter end of *August*. By the self-interest of mankind, they are welcome guests, and provided with lodgings, in form of earthen pots or boxes, placed on the outsides of the houses, against their arrival, and sometimes with empty calabashes hung on the tops of poles †. In these they make their nests, and lay four or five eggs. In return for these benefits, they are the guardian of the poultry; driving

* *Strom.* 249,   † *Lawson,* 144.

away,

away, and purfuing with great noife, Crows, Hawks, and all kinds of vermin. On the approach of any thing noxious, they fet up a loud note; which the chickens confider as an alarm, and inftantly run under fhelter.

334. SWIFT.

> Br. Zool. i. Nº 171.—Latham, ii. 584.
> Swift, or Diveling, Lawfon, 145.
> Hirundo apus. Ring-Swala, Faun. Suec. 272.
> Le Martinet noir, De Buffon, 643.—Pl. Enl. 542.—Bl. Mus.

SW. With a very fmall bill: white chin: all the plumage befides dufky: all the toes ftanding forward.

PLACE.

According to Mr. Lawfon, inhabits Carolina. Found in vaft abundance beyond lake Baikal, on the loftieft rocks; chiefly about the river Onon, where a variety with a white rump is very common. Extends in Europe as high as Drontheim.

335. ACULEATED.

> American Swallow, Catefby, i. 8.
> Chimney Swallow, Kalm. ii. 146.
> Hirundo pelafgia, Lin. Syft. 345.—Latham, ii. 583.
> Le Hirondelle brune acutipenne, De Buffon, vi. 699. — Pl. Enl. 726.
> —Lev. Mus.—Bl. Mus.

SW. With the bill fhort, broad, and black: head, upper part of the neck, and wings, dufky: breaft cinereous: back, tail, and belly brown: tail even at the end; extremities of each fhaft ñaked and fharp-pointed: wings extend far beyond the tail: legs longer than common to this tribe, and naked a little below the knee. LENGTH five inches and a half.

PLACE.

Inhabits many parts of North America. Arrives in New York and Penfylvania in May; fomtimes early, fometimes late in the month. Builds in chimnies, forming a moft curious neft, with bits of fmall fticks, cemented by peach-tree gum. It is open at top, and forms about a third of a circle. Lays four or five eggs in June, and quits the

country

country in *August.* They often ftick clofe to the chimney-wall by
their feet, and fupport themfelves by applying their fharp tail to
the fides. They make all day a great thundering noife, by flying
up and down the funnel.

It is remarkable, that three fpecies of the *American* Swallows, in
general feek the protection of houfes for their places of building
their nefts, ovation, and nutrition; yet it is very certain, that before the arrival of the *Europeans* they muft have had recourfe to
rocks or hollow trees for thofe purpofes; for the miferable hovels
of the *Indians* had neither eaves for the ufes of the 331ft and
330th fpecies, nor chimnies for that of the bird in queftion.
The two firft muft therefore have fixed their neft againft the face
of fome precipice, as fome of the Houfe Swallows do at prefent in
America, and this fpecies does about the fteep rocks about *Irkutfk*
in *Sibiria.* The inftinct that directs part of this genus to fly to
the protection of mankind, as foon as opportunity, unknown to
preceding broods, offered, is as wonderful as it is inexplicable.

In Sibiria.

The Comte *De Buffon* mentions another of this fpecies *, which
is found in *Louifiana.* It differs only in the fuperior length of the
wings, from the bird I defcribe: I therefore can confider it but as
a mere variety.

Louisiane.
A Variety.

* vi. 700.

XXXII. GOATSUCKER. *Gen. Birds,* LVI.

336. SHORT-
WINGED.

Goatsucker of Carolina, *Catesby,* i. 8.
East India Bat, or Musqueto Hawk, *Lawson,* 144.
L'Engoulevent de la Caroline, *De Buffon,* vi. 532.—*Latham,* ii. 592.

G. With the head, back, breast, and coverts of the wings, elegantly mottled with black and bright rust-color, and spotted with large ragged black marks: the scapulars of the same color, here and there spotted with white: on the lower part of the
back is a mixture of ash-color: primaries and secondaries most beautifully varied with narrow bars of black and ferruginous: the four
middle feathers of the tail barred and mottled with the same colors;
as are the external webs of the three outmost on each side; but the
inner webs of a snowy whiteness. Wings, when closed, reach little

SIZE.   farther than half the length of the tail. LENGTH twelve inches:
EXTENT twenty-four.

PLACE.   I received this species from Doctor GARDEN of *Charlestown, South
Carolina*; where it is called, from one of its notes, *Chuck, Chuck
Will's widow*; and in the northern provinces, *Whip poor Will,* from the
resemblance which another of its notes bears to those words. This,
Mr. *Kalm* says, is the fancy of the *Europeans*; for the real sound is
likest to *Whipperiwhip,* with a strong accent on the first and last
syllable *. It begins its note about the time that the cherry-trees
begin to blossom, or near the 22d of *April,* in the *Jersies*; probably
sooner in the southern provinces. Mr. *Blackburne* observed them

EGGS.   first, in the province of *New York,* in *May.* Adds, that they lay two
eggs on the bare ground; and that they will scarcely quit them on
the nearest approach. They disappear in *August.*

FOOD.   Their food is entirely insects, which they catch night and morning, at the time in which they emit their song. They never settle

* *Kalm,* ii. 152.

9   on

on high trees; but on bushes, rails, or the steps of houses, which they frequent, as insects swarm more, near to habitations, than other places. They give their note sitting: if they see an insect pass, they fly up, catch it, and then settle again, and renew their song. Oft-times numbers perch near one another, make a vast noise, repeating their song as if in emulation. They continue their call till it is quite dark: their note ceases during night; but commences at the dawn, and is continued till the sun rises, when they again desist for the whole day * I must add, that, besides these notes, it has that strange sound resembling the turning of a great spinning-wheel; probably common to the whole genus †.

They are extremely rare towards the sea-side; but swarm towards the mountains. Doctor *Garden* never got but this one. Mr. *Clayton* confirms their scarcity in the maritime parts of the provinces; and favors us with the following account of them.

" I never heard but one in the maritime parts; though my abode
" has been always there; but near the mountains, within a few mi-
" nutes after sun-set, they begin, and make so shrill and loud a noise,
" which the echoes from the rocks and sides of the mountains increase
" to such a degree, that the first time I lodged there I could hardly
" get any sleep. The shooting them in the night is very difficult;
" they never appearing in the day. Their cry is pretty much like
" the sound of the pronunciation of the words *Whip poor Will*, with
" a kind of a *chucking* between every other, or every two or three
" cries; and they lay the accent upon the last word *Will*, and left of
" all upon the middle one.

" The *Indians* say, these birds were never known till a great
" massacre was made of their country folks by the *English*, and that
" they are the departed spirits of the massacred *indians*. Abundance
" of people here look upon them as birds of ill omen, and are very

* *Kalm,* ii. 153.　　† *Br. Zool.* i. p. 352, 4to—417, 8vo.

3 K 2　　　　　　　　" melancholy

" melancholy if óne lights on their houfe or near their door, and fets
" up its cry (as they will fometimes upon the very threfhold); for
" they verily believe one of the family will die very foon after *."

Whip poor Will, or leffer Goatfucker, *Edw.* 63. — *Catefby*, App. 16.—
    Latham, ii. 595.
Caprimulgus minor Americanus, *Lin. Syft.* 346.—Lev. Mus.—Bl. Mus.

G. With the head and body dufky, mottled with white and pale
ruft-color : primaries black, marked near the middle with a
white bar: under the throat is a white crefcent, with the ends pointing
upwards : breaft barred with dirty white and dufky : tail black,
marked regularly on each web with fpots, mottled with black and
white : near the ends of each feather is a large white fpot, the ends
quite black : wings, when clofed, extend beyond the end of the tail.
LENGTH nine inches and a half : EXTENT about twenty-three.

PLACE.

Inhabits the fame provinces with the former, and feems to have
the fame manners and notes; for, according to Doctor *Garden*, each
are known in different places, by the name of *Whip poor Will*. It is
found as far north as *Henly* Houfe, a fettlement for about a hundred
miles up *Albany* river in *Hudfon's Bay*, where it is called the *Mufqueto*
Hawk.

* *Catefby*, App. 16.

XVIII

P. Mazell sculp.

Aculeated Swallow, N.°335. Longwinged Goatsucker, N.°337.

A. EUROPEAN.

Caprimulgus Europeus. Nattſkafwa. Quallknarran, *Faun. Suec.* Nᵒ 274. — *Latham*, ii. 593.

L'Engoulevent, *De Buffon*, vi. 512.—*Br. Zool.* i. Nᵒ —LEV. MUS.—BL. MUS.

G. With head and back elegantly ſtreaked with narrow lines of black and grey, and with a few long oblong ſtrokes of black and ruſt : belly barred with black and grey : wings black ; each web finely marked with ruſty ſpots : near the ends of the three firſt primaries, a large oval white ſpot : tail duſky, with regular ſpots, mottled with ruſt and black ; ends of the two firſt feathers white. FEMALE wants the ſpots on the wings and tail.

It is found in *Europe* as far north as *Sondmor*, and is common all over *Sibiria* and *Kamtſchatka* ; and lives not only in foreſts, but in open countries, where it finds rocks or high banks for ſhelter.

PLACE..
IN KAMTS-
CHATKA.

DIV.

DIV. II.

WATER FOWLS.

SECT. I. CLOVEN-FOOTED.

D I V. II.   Water-Fowls.

SECT. I. Cloven-Footed.

XXXIII. SPOON-BILL.   *Gen. Birds*, LIX.

338. ROSEATE.

Ajaja, *Maregrave*, 204.—*Wil. Orn.*—*Raii Av.*—Platalea ajaja, *Lin. Syst.* 231.
—*Latham*, iii.
La Spatule d'Amerique, *De Buffon*, vii. 456.—*Pl. Enl.* 165.—*Du Pratz*, ii. 84.

PLACE.

SP.B. With the fore part of the head and throat naked and whitifh: the whole plumage white, tinged with a beautiful rofe-color, deepeft about the wings and coverts of the tail, where it nearly approaches crimfon. SIZE of a Goofe. Is an eatable fowl. Is converfant in *Louifiana*, about the fhores and rivers; and lives on water-infects and fmall fifh. Is found alfo in *Mexico**, *Guiana* †, *Brafil*, and in *Jamaica*, and the greater *Antilles*. The plumage acquires its beauty in proportion to the age of the bird ‡; fo probably is whitifh when young. It foon grows tame.

* *Fernandez*, 49.   † *Barrere*, 125.   ‡ The fame.

A. SPOON-BILL, *Br. Zool.* ii. App. N° ix.—La Spatule, *De Buffon,* vii. 448. tab. xxiv.
—*Pl. Enl.* 405.—*Latham,* iii.

Platalea Leucorodia, Pelekan, *Faun. Suec.* N° 160.—LEV. MUS. ?

SP. B. Wholly white, with a pendent creft: legs and bill black:
at the angles of the bill, on each cheek, a bright orange fpot.
From the end of the bill to end of the claws, forty inches. Ex-
TENT fifty-two.

Inhabits the *Feroe* ifles *; and on the continent is fometimes found
in fummer as high as *Weft Bothnia* and *Lapland* †. Inhabits alfo the
temperate parts of *Ruffia* and *Sibiria,* both in flocks and folitary,
frequenting the vaft lakes of the country. Is feen even beyond
lake *Baikal.* Winters in the fouth. Builds its neft on high trees,
and is very clamorous in the breeding feafon. Lays four eggs. Feeds
on fifh, which it is faid to take from the diving tribe of birds,
frightening them from their prey by clattering its bill ‡. It devours frogs and fnakes; and will even feed on vegetables.

PLACE.

---

* *Worm. Muf.* 310.   † *Faun. Suec.* N° 160.   ‡ *Worm. Muf.* 310.

XXXIV. HERON. *Gen. Birds*, LXIII.

Hooping Crane, *Catefby*, i. 75.—*Edw.* 132.—*Latham*, iii.
Ardea Americana, *Lin. Syft.* 234.
La Grue blanche, *De Buffon*, vii. 308.—*Pl. Enl.* 889.

H. With a yellowifh brown bill, ferrated near the end: crown covered with a red fkin, thinly befet with black briftles: from the bill, beneath each eye, extends a fimilar ftripe: on the hind part of the head a triangular black fpot: quil feathers, and a few of the greater coverts, black: fecondaries, and the whole plumage, of a pure white: webs of the tertials elegantly loofe and unconnected, and, falling over the primaries, almoft conceal them : legs and feet black and fcaly. LENGTH from the bill to the tip of the claws five feet feven inches.

PLACE.   Inhabits all parts of *North America*, from *Florida* to *Hudfon's Bay*. Is migratory: appears early in the fpring about the *Alatamaha*, and other rivers near *St. Auguftine*, and then quits the country in great numbers, and flies north in order to breed in fecurity. They appear in fummer in *Hudfon's Bay*, and return fouthward with their young on approach of winter. They make a remarkable hooping noife: this makes me imagine thefe to have been· the birds, whofe clamor Captain *Philip Amidas* (the firft *Englifhman* who ever fet foot on *North America*) fo graphically defcribes, on his landing on the ifle of *Wokokou*, off the coaft of *North Carolina:* " WHEN," fays he, " fuch a flock of Cranes (the moft part white) arofe under us, with " fuch a cry, redoubled by many ecchoes, as if an armie of men " had fhowted all together." This was in the month of *July* *; which proves, that in thofe early days this fpecies bred in the then defert parts of the fouthern provinces, till driven away by population, as was the cafe with the common Crane in *England*; which abounded in our undrained fens, till cultivation forced them entirely to quit our kingdom.

* *Smith's Hift. Virgin.* &c. 2.

Brown

Brown and afh-colored Crane, *Edw.* 133.—Grus Canadenfis, *Lin. Syft.* 234.— Toquil Coyotl, *Fernandez,* 44 —*Latham,* iii. La Grue brune, *De Buffon,* vii. 310.—LEV. MUS.

H. With a dufky bill, near four inches long : crown red and naked : cheeks and throat white : hind part of the head and whole neck cinereous ; reft of the plumage of the fame color, tinged with pale ruft : primaries black, fhafts white ; the row of feathers incumbent on them light afh : tertials brown with elegant loofe webs, incurvated, and extending beyond the ends of the primaries : tail cinereous : legs black. LENGTH three feet three. EXTENT three, five. WEIGHT feven pounds and a half.

This fpecies is found in *Mexico* ; but migrates into the north to breed. About the middle of *February* they are feen in their flight over the *Jerfies,* fteering northerly ; and in the fpring fome make a fhort halt there *. They arrive in *May* about *Severn* river in *Hudfon's Bay.* Frequent lakes and ponds. Feed on fifh and infects. Hatch two young ; and retire fouthward in *autumn* †. I muft obferve, that they formerly made a halt in the *Hurons* country, at the feafon in which the *Indians* fet their *maiz* ; and again on their return from the north, when the harveft was ready, in order to feed on the grain. The *Indians,* at thofe times, were ufed to fhoot them with arrows headed with ftone ; for *Theodat* ‡, my authority, made his remarks in that country in the beginning of the laft century.

Largeft crefted Heron, *Catefby,* App. 10.—Ardea Herodias, *Lin. Syft.* Le grand Heron d'Amerique, *De Buffon,* vii. 385.—*Latham,* iii.

H. With a bill eight inches long : on the hind part of the neck a long creft of flender herring-bone feathers, of a brown color, to be erected at pleafure : the head, neck, and whole of the body, brown,

---

* *Kalm,* ii. 72.   † *Ph. Tranf.* lxii. 409.   ‡ As quoted by *De Buffon.*

paleft

paleſt on the under part, and ſpotted : primaries black : legs brown.
HEIGHT, when erect, four feet and a half.

PLACE.

Inhabits *Virginia*. Feeds on fiſh, frogs, and lizards.

342. RED-SHOUL-
DERED.

Aſh-colored Heron, *Edw.* 135.—Ardea Hudſonias, *Lin. Syſt.* 238.—*Latham*, iii.
Le Heron de la Baie d'Hudſon, *De Buffon*, vii. 386.—LEV. MUS.

H. With a white forehead : black creſt : hind part of the neck of
a reddiſh brown ; fore part white, ſpotted with black : feathers on the breaſt long and narrow : belly black and white, bounded with black: ſides grey : primaries and tail duſky : coverts and
ſecondaries cinereous : ſhoulders and thighs of an orange red : bill
yellowiſh : legs duſky. In SIZE ſuperior to the *Engliſh* Heron.

PLACE.

Inhabits *Hudſon's Bay*, frequenting, during ſummer, the inland
lakes.

FEMALE ?

Head ſmooth, deep cinereous : neck paler : throat white : breaſt
and belly white, ſtriped downwards with black : back, tail, and coverts of wings, light aſh : primaries black : ſhoulders and thighs of
a dirty yellow : legs duſky.

PLACE.

Sent to Mrs. *Blackburn* from *New York*, under the name of the *Hen
Heron*. It probably is the female of the laſt. Its LENGTH was three
feet to the tail : to the end of the toes four feet nine.

343. COMMON
HERON.

Br. Zool. ii. N° 173 —Ardea cinerea. Hagen, *Faun. Suec.* N° 165.—*Latham*, iii.
Le Heron commun, *De Buffon*, vii. 34.—*Pl. Enl.* 787. 755.—LEV. MUS.—BL.
MUS.

H. With a white crown : long pendent black creſt : white neck,
ſtreaked before with black : coverts of the wings, ſcapulars,
and tail, grey : belly white : primaries duſky. Creſt on the FEMALE
very ſhort. LENGTH three feet three.

PLACE.

Is frequent in *Carolina* * ; and I think a ſpecimen was ſent to Mrs.
Blackburn from *New York*, where they breed in flocks as they do

* *Cateſby*, App. xxxvi.—*Lawſon, Hiſt. Carol.* 148.

in

in *England*. If I miftake not the kind, they come to *New York* in *May*, and retire in *October*. They are found in *Ruffia* and *Sibiria*, but not very far north. *Crantz* fays, that they have been feen in the fouth of *Greenland*; but were never obferved by *Fabricius* * : but it certainly inhabits *Romfdal* and *Nordmer* †, in the fevere climate of the diocefe of *Drontheim*.

It may be here remarked, that this, and the whole tribe of what *Linnæus* calls *Grallæ*, or the *Cloven-footed Water Fowl*, quit *Sweden*; and of courfe the more northern countries, at approach of winter; nor is a fingle fpecies feen till the return of fpring ‡.

White Heron, *Br. Zool.* i. Nº 175.—Ardea Alba, *Lin. Syft.* 239.—*Faun. Suec.* Nº 166.—*Latham*, iii.

Le Heron blanc, *De Buffon*, vii. 365.—*Pl. Enl.* 886.—Lev. Mus.—Bl. Mus.

344. GREAT WHITE.

H. With a very flender yellow bill: plumage entirely of a milk white: legs black. LENGTH to the toes four feet and a half.

Inhabits *America*, from *Jamaica* and *Mexico*, to *New Eng-land*. It migrates, being feen in *New York* from *June* to *October* only. Is found, but rarely, in *Sweden*. Inhabits the *Ruffian* domi-nions, about the *Cafpian* and *Black Seas*, the lakes of *Great Tartary*, and the river *Irtifh*, and fometimes extends north as high as lat. 53. Captain COOK obferved this fpecies in *New Zealand* ‖.

PLACE.

Garzetta, *Alar. Av.* lib. iii. 161.—*Will. Orn.* 280.—*Raii Syn. Av.* 99.— *Catefby*, i. 77.—*Latham*, iii.

La Garzette blanche, *De Buffon*, vii. 371.—Lev. Mus.—Bl. Mus.

345. LITTLE WHITE.

H. With the bill and legs black: whole plumage white: on the head a fhort creft. LENGTH two feet.

This fpecies is found in *New York*. Is met with again in *New Zea-land* and *Otaheite*.

PLACE.

* *Faun. Greenl.* 106.   † *Leems*, 242.   ‡ *Amæn. Acad.* iv. 538.
‖ *Voy. towards S. Pole*, i. 87.

The

The little white *Carolina* Heron of *Catesby*, with a red bill and green legs, seems only a variety of this.

346. GREAT EGRET. La grande Aigrette, *De Buffon*, vii. 377.—*Pl. Enl.* 925 —*Latham*, iii.

H With a long slender crest: bill and legs black: whole plumage of a silvery whiteness: the feathers on the back inexpressibly elegant, long, silky, narrow, and with unwebbed plumes, hanging over the wings and tail; the same kind are pendent from the breast. Of double the Size of the *European* species. LENGTH of which, from bill to the tip of the tail, is two feet *.

PLACE. Inhabits *Louisiana* and *Guiana*. Does not frequent the shores; but the vast morasses and overflown tracts, where it nestles on the little isles formed by the inundations. The *Guiritinga* of the *Brasilians* † is probably the same species. It extends to the *Falkland* isles; for *Bougainville* observed these Egrets, which he first thought were common Herons. They fed towards night, and made a barking noise ‡.

The feathers of the Great Egret would prove a valuable article of commerce, being very much sought after for the ornamental part of dress.

347. LITTLE EGRET. Br. *Zool.* ii. App. N° vii.—Ardea Garzetta, *Lin. Syst.* 237.—*Latham*, iii. L'Aigrette, *De Buffon*, vii. 372. tab. xx.—*Pl. Enl.* 901.—LEV. MUS.

H. With yellow irides: a crest with some short and two long pendent feathers: whole plumage of a delicate silvery white: feathers on the breast and scapulars very delicate, loose, and unwebbed: legs a blackish green. WEIGHT about one pound. LENGTH to the tip of the tail two feet.

Br. *Zool.* ii. App. N° vii.   † *Marcgrave*, 209.   ‡ *Voy. round the World*, Engl. ed. 67.

7

Is

Is frequent in *New York* and *Long Island:* about the *Black* and *Caspian* feas; but feldom farther north. Are found in *France,* and the fouth of *Europe.* Migrates into *Austria* in fpring and autumn *. Is frequent in *Senegal, Madagascar, Isle de Bourbon,* and *Siam* †.

L'Aigrette rouffe, *De Buffon,* vii. 378.—*Pl. Enl.* 902.—*Latham* iii.

H. With the body of a blackifh grey: the filky long feathers of the neck and back of a rufty red. Length about two feet.
Inhabits *Louifiana.*

Small Bittern, *Catesby,* i. 80.—Ardea virefcens, *Lin. Syst.* 238.—*Latham,* iii. Le Crabier vert, *De Buffon,* vii. 404.—Lev. Mus.—Bl. Mus.

H. With a green head, and large green creft: bill dufky above, yellow beneath: throat white: neck a bright bay, ftreaked before with white: coverts of the wings dufky green, edged with white: tail and primaries dufky: feathers on the back cinereous, long, narrow, and filky: belly of a cinereous red: legs yellowifh. The colors of the Female lefs brilliant: wings fpotted with ruft-color. It wants the long filky feathers. Length eighteen inches.
Inhabits from *New York* to *South Carolina.* Ufually fits, with its long neck contracted, on trees hanging over rivers. Feeds on fmall fifh, frogs, and crabs. From the laft, the *French* call feveral of thefe Herons *Crabiers.*

They are fuppofed to migrate, even from *Carolina,* at approach of winter.

* *Kram Austr.* 346.   † *De Buffon, Ois.* vii. 375. 376.

350. LOUISIANE.   Le Crabier roux à tête & queue vertes, *De Buffon*, vii. 407.—*Pl. Enl.* 909 —
Latham, iii.

H. With the crown and tail of a dull green: the neck and belly red, tinged with brown: coverts of the wings dufky green, edged with tawny: the back covered with long flender feathers, faintly dafhed with purple.

PLACE.   Inhabits *Louifiana.*

351. BLUE.   Blue Bittern, *Catefby*, i. 76.—Le Crabier bleu, *De Buffon*, vii. 398.—
Ardea Cærulea, *Lin. Syft.* 238.—*Latham*, iii.—LEV. MUS.

H. With a blue bill, dufky at the point: head and neck of a changeable purple; the firft adorned with a beautiful creft of long flender feathers: the remainder of the plumage entirely of a fine deep blue: from the breaft depend feveral long feathers: the back is covered with others a foot in length, hanging four inches beyond the tail; they are filky, and of the fame fine texture with thofe of the creft: the legs are green.   WEIGHT fifteen ounces.

PLACE.   It appears, but not in numbers, in *Carolina*, and that only in the fpring of the year.   Its winter refidence feems to be *Jamaica* *.

352. YELLOW-CROWNED.   Crefted Bittern, *Catefby*, i. 79 —Ardea Violacea, *. Lin. Syft.* 238.—*Latham*, iii.
Le Crabier gris de fer, *De Buffon*, vii. 399.

H. With a black, ftrong, and thick bill: crown of a pale yellow: from the hind part iffue three or four long flender white feathers, erigible at pleafure; fome are fix inches long: a broad white ftripe runs from the corner of the lower mandible as far as the ears: the reft of the cheeks and head are of a blueifh black: head, breaft, belly, and coverts of wings, of a dufky blue: the primaries brown, tinged with blue: the back ftriped with black, mixed with

---

* *Sloane's Hift. Jamaica*, ii. 315.

white :

white : from the upper part arife tufts of elegant flender filky feathers, falling beyond the tail : the legs and feet yellow. WEIGHT
one pound and a half.

This fpecies appears in *Carolina* in the rainy feafons : but their
native places are the *Bahama* iflands, where they breed in amazing
numbers, amidft the bufhes in the rocks. They are called by the
iflanders *Crab-catchers*, as they chiefly live on thofe cruftaceous animals. They are of great ufe to the inhabitants ; who take the
young birds before they can fly, and find them delicious eating.
They fwarm fo on fome of the rocky ifles, that two men, in a few
hours, will fill a fmall boat with them, taking them when perched on
the rocks or bufhes ; for they will make no attempt to efcape, notwithftanding they are full grown.

PLACE.

H. With a black ftrong bill : crown dufky : cheeks and chin
whitifh : neck of a pale cinereous brown, ftreaked before with
white : back, wings, and tail, cinereous, clouded round each feather
with dufky : feathers on the fides of the back long and broad,
hanging over the ends of the wings : belly white : legs yellowifh.
LENGTH two feet one inch.

353. ASH-COLOR-
ED.

Inhabits *New York.* Arrives there in *May :* breeds, and leaves the
country in *October.*—BL. MUS.—*Latham,* iii.

PLACE.

H. With a bill about two inches long : crown, back, and tail, of
an uniform dufky color : hind part of the neck and cheeks
rufty and black : chin and throat white : fore part of the neck
marked with ftreaks of white and black : coverts of the wings,
with ftreaks of black and yellowifh white : ridge of the wing white :
primaries dufky.—LEV. MUS.—*Latham,* iii.

354. STREAKED.

Another, in the fame *Mufeum* and fame cafe (probably differing
only in fex) has, from the lower mandible, a white line bounding
the lower part of each cheek : the greater coverts of the wings and

3 M

fcapulars

scapulars dufky, each feather tipt with white. In other refpects it agrees with the former : the legs of each are greenifh : the form of their bodies flender and elegant. LENGTH, from bill to the tip of the tail, about feventeen inches.

PLACE.   Sent to Sir *Afhton Lever* from *North America*.

355. GARDENIAN.   Le Pouacre de Cayenne ? *Pl. Enl.* 939.—*Latham*, iii.

H. With a dufky ftrong bill : head, neck, breaft, and belly, whitifh, elegantly ftreaked downwards with fhort fine lines of black ; the crown and hind part the darkeft : upper part of the back ftreaked with white ; the lower dufky and plain : the whole wing of the fame color : the leffer coverts marked with fmall yellowifh fpots ; the greater coverts marked with a white fpot at the end of each feather, forming, acrofs the wings, two rows : the primaries edged with dull white ; the ends tipt with the fame : tail dufky : legs of a deep dirty yellow. LENGTH about twenty-two inches.

PLACE.   Doctor GARDEN, of *South Carolina*, favoured me with this bird. From the characteriftic lines of white fpots in the wings, I do not doubt but that the Brown Bittern of *Catefby*, i. 78, is the fame * with this : notwithftanding, it would hardly be known, had he not preferved the fpots in his very bad figure of it. He fays it frequents ponds and rivers in the interior part of the country remote from the fea.—LEV. MUS.

356. NIGHT.   Ardea Nycticorax, *Lin. Syft.* 135.—*Will. Orn.* 279.—*Latham*, iii.
Le Bihoreau. Le B. de Cayenne, *De Buffon*, vii. 435. 439. tab. xxii.—*Pl. Enl.* 758. 759. 899.—LEV. MUS.—BL. MUS.

H. With a black bill, crown, back, and fcapulars ; the laft broad and long : forehead, cheeks, neck, and under fide of the body, white : wings and tail of a very pale afh-color : the hind part of the

* Alfo l'Etoile of *De Buffon*, vii. 428.

head

head is moſt ſpecifically diſtinguiſhed by three very ſlender white feathers, five inches long, forming a pendent creſt: legs of a yellow-iſh green. The LENGTH, to the tip of the tail, one foot ſeven inches.

Inhabits *New York*; and a variety is found as low as *Cayenne*. Is PLACE. common to *Europe*. Is frequent in the ſouthern parts of the *Ruſſian* dominions; but does not extend farther than lat. 53. It muſt not at this time be ſought for in the wood near *Sevenhuys* in *Holland*, ſo noted in the days of Mr. *Willughby* for the vaſt *rendezvous* of Shags, Herons, Spoon-bills, and theſe birds, beſides Ravens, Wood-pigeons, and Turtles *, it being now cut down. When Mr. *Willughby* viſited the place it was rented, for the birds and graſs, for three thouſand gilders a year.

This bird is not the *Nyƈticorax* of the Antients; which was ſome rapacious fowl, probably of the Owl kind. It is the *Nacht-rab*, or Night-Raven of the *Germans*; ſo called from its noƈturnal cry, re-ſembling the ſtraining of a perſon to vomit.

Br. Zool. ii. N° 174.—Ardea Stellaris. Rordrum, *Faun. Suec.* N° 164.—*Latham*, iii. 357. BITTERN. Bittern from Hudſon's Bay, *Edw.* 136.—Le Butor, *De Buffon*, vii. 411. 430.—*Pl. Enl.* 789.—LEV. MUS—BL. MUS.

H. With the upper mandible duſky; lower yellow: feathers on the crown black and long; on the cheeks tawny; on the throat white: hind part of the neck browniſh red; fore part white, beau-tifully marked with ſhort ſtripes of red, bounded on each ſide with one of black: feathers on the breaſt very long: the belly of the colors of the fore part of the neck: back, coverts of wings, and the tail, are ferruginous, traverſed with duſky lines: primaries black: legs yellowiſh green. Rather inferior in SIZE to the *European* Bit-tern; but ſo like, as not to merit ſeparation.

* *Ray's Travels*, i. 33.

3 M 2 It

PLACE.

It inhabits from *Hudson's Bay* to *Carolina* *. In the former, it appears the latter end of *May:* lives among fwamps and willows: lays two eggs. Like the *European* fpecies, is very indolent; and, when difturbed, takes but a fhort flight †.

That of the old continent is found in *Ruffia*; and, in *Afia*, in *Sibiria*, as far north as the river *Lena*, and is continued confiderably to the north. Inhabits *Sweden* ‡; but, with all the other Herons, difappears at approach of winter ‖.

The fecond fpecies of Bittern, mentioned by *Lawfon*, p. 148. as being leffer than the former, with a great topping, of a deep brown color, and a yellowifh white throat and breaft, is at prefent unknown to us.

358. RUSTY-CROWNED.

H. With yellow irides: very fmall creft: the bill feven inches long, flender, and of the fame color with the former: forehead dufky: throat white: creft and hind part of the neck of a deep ferruginous color: the fore part of the neck marked with four rows of black fpots: the feathers towards the breaft long: a dark line paffes from the breaft upwards to the back of the neck: the upper part of the body, and coverts of the wings, deep ferruginous, marked with a few large black fpots: primaries dufky: tail fhort, and of a lead color: belly and breaft of a dirty white, ftriped with black: legs of a dirty yellow. The creft on the head is very fmall, and the feathers lie univerfally fmooth. SIZE of the *European* Bittern.

PLACE.

Inhabits *North America*; the province unknown. Defcribed from a live bird at *Amfterdam*.—LEV. MUS?—*Latham*, iii.

---

* *Lawfon*, 148.   † *Ph. Tranf.* lxii. 410.   ‡ *Faun. Suec.* Nº 164.
‖ *Amæn. Acad.* iv. 588.

Little Bittern, *Br. Zool.* ii. App. Nº x. tab. viii.—*Pl. Enl.* 323.—*Latham*, iii.   359. LITTLE.
Ardea Minuta, *Lin. Syst.* 240.—LEV. MUS.—BL. MUS.

H. With a smooth head: crown black: hind part of the neck and cheeks ferruginous: coverts on the ridge of the wing, and ends of the greater, of a bright bay; the rest of the coverts of a very pale clay color: primaries and secondaries dusky, with ferruginous tips: lower side of the neck and belly of a yellowish white: breast crossed with a band of black: tail black: legs of a dusky green. LENGTH, to the end of the tail, fifteen inches. The body narrow: neck very long.

Inhabits from *New York* to *South Carolina*, and many parts of *Europe*. Extends to, and perhaps winters in *Jamaica*. Its eggs are of a sea-green color.   PLACE.

A. COMMON CRANE, *Br. Zool.* ii. App. Nº vi.—Ardea Grus, Trana, *Faun. Suec.* Nº 161.
 —*Latham*, iii.   —La Grue, *De Buffon*, vii. 286. tab. xiv.—*Pl. Enl.* 769.
 —LEV. MUS.

H. With a bald crown: fore part of the neck black: primaries black: a large tuft of elegant unwebbed curling feathers springing from one pinion of each wing: those, and all the rest of the plumage, cinereous. LENGTH six feet. WEIGHT about ten pounds.

Cranes arrive in *Sweden* in great flocks in the spring season; pair,   PLACE.
and disperse over the whole country; and usually resort to breed to the very same places which they had used for many years past *.

* *Amœn. Acad.* iv. 588.

No

No augural attention is paid to them there; yet *Hefiod* directs the *Grecian* farmer " to think of ploughing whenever he hears the an- " nual clamor of the Cranes in the clouds *."

PLACE.

Inhabits all *Ruffia* and *Sibiria*, even as far eaft as the river *Anadyr*; and migrates even to the *Arctic* circle. None feen in *Kamtfchatka*, except on the very fouthern promontory, which they probably make a refting-place, on their re-migration; *Kamtfchatka* being deftitute of ferpents and frogs, on which they feed in countries where corn is unknown. They lay two blueifh eggs on the rufhy ground: the young are hatched late; and as foon as they can fly attend their parents in their fouthern migration:

Poturæ te, *Nile*, GRUES.

For *Egypt* is generally fuppofed to be the great winter quarters of thefe birds. Previous to their retreat, they affemble in amazing numbers, choofe their leader, foar to a confiderable height, and then, with continued clamor, proceed to their defigned place. *Milton,* when he touches on this wonderful inftinct of nature, defcribes their progrefs with equal truth and elegance.

> Part loofely wing the region: part more wife,
> In common, rang'd in figure (>) wedge their way,
> Intelligent of feafons, and fet forth
> Their aery caravan, high over feas
> Flying, and over lands with mutual wing
> Eafing their flight. So fteers the prudent CRANE
> Her annual voyage, borne on winds;
> The air flotes as they pafs, fann'd with unnumber'd plumes.

* Εργων και Ημερων. II. v. 66.

B. SIBIRIAN

B. SIBIRIAN CRANE. Grus Leucogeranos, *Pallas Itin.* ii. 714.—*Latham*, iii.

H. With a red bill like the former, ferrated near the end : face naked beyond the eyes : coverts and primaries black : all the reft of the bird of a fnowy whitenefs: legs red. Its HEIGHT is four feet and a half.

Inhabits the vaft moraffes of *Sibiria*, and every part where lakes abound ; and penetrates far north into the boggy forefts about the *Ifchim*, *Irtifh*, and *Oby*. Makes its neft among the inacceffible reeds, with layers of plants. Lays two great grey eggs, ftreaked with numerous dufky lines. Makes a clamorous noife, and that frequently, efpecially during its flight. Feeds on fmall fifh, frogs, and lizards. Winters ufually about the *Cafpian* fea. Obferved to migrate in fpring northward along the courfe of the *Wolga*, always in pairs *.

PLACE.

C. WHITE STORK. Ardea Ciconia. Storck, *Faun. Suec.* Nº 162.—La Cigogne, *De Buffon*, vii. 253.—*Pl. Enl.* 866.—*Latham*, iii. —LEV. MUS.

H. With red bill and legs : primaries black : the reft of the plumage white : fkin of the color of blood. Larger than the common Heron.

Inhabits moft parts of *Europe*, except *England*. In the weft of *Ruffia*, is not found beyond 50 degrees north, nor to the eaft of *Mofcow*. It appears in *Sweden* in *April*; retires in *Auguft* † : does not reach *Norway*, unlefs tempeft-driven.

PLACE.

This fpecies is femi-domeftic : haunts towns and cities ; and in many places ftalks unconcerned about the ftreets, in fearch of offals and other food. Removes the noxious filth, and clears the fields of ferpents and reptiles. They are, on that account, protected in *Holland*; held in high veneration by the *Mahomedans*; and fu greatly

* *Extracts*, ii. 146.　　　† *Amœn. Acad.* iv. 588.

refpected

respected were they in old times by the *Theffalians*, that to kill one of thefe birds was a crime expiable only by death *.

The Storks obferve great exactnefs in the time of their autumnal departure from *Europe* to more favorable climates. They pafs a fecond fummer in *Egypt*, and the marfhes of *Barbary* † : in the firft they pair, and lay again, and educate a fecond brood ‡. Before each of their migrations they rendezvous in amazing numbers; are for a while much in motion among themfelves; and after making feveral fhort flights, as if to try their wings, all of a fudden take flight with great filence, and with fuch fpeed, as in a moment to attain fo great a height as to be inftantaneoufly out of fight. The beautiful and faithful defcription which the NATURALIST'S POET ‖ gives of this annual event, ought not by any means to be omitted.

> Where the *Rhine* lofes his majeftic force
> In *Belgian* plains, won from the raging deep
> By diligence amazing, and the ftrong
> Unconquerable hand of Liberty,
> THE STORK-ASSEMBLY meets; for many a day
> Confulting deep and various, ere they take
> Their arduous voyage thro' the liquid fky.
> And now, their route defign'd, their leaders chofe,
> Their tribes adjufted, clean'd their vigorous wings;
> And many a circle, many a fhort effay,
> Wheel'd round and round, in congregation full
> The figur'd flight afcends, and riding high
> The aerial billows, mixes with the clouds.

D. BLACK STORK. Ardea nigra. Odenfwala, *Faun. Suec.* N° 163.—*Latham*, iii. La Cigogne noire, *De Buffon*, vii. 271.—*Pl. Enl.* 399.

H. With the bill, legs, and fkin, red : head, neck, body, and wings, black, gloffed with blue : breaft and belly white. About the SIZE of the former.

* *Pliny*, lib. x. c. 23.   † *Shaw's Trav.* 428.   ‡ *Belon Oyf.* 201.
‖ THOMSON.

Inhabits

Inhabits many parts of *Europe*. It is not uncommon in the temperate parts of *Ruſſia* and *Siliria*, as far as the *Lena*, where lakes and moraſſes abound. Migrates to warmer countries in autumn. Is a ſolitary ſpecies. Preys on fiſh, which it not only wades for, but, after hovering over the waters, will ſuddenly plunge on its prey. It alſo eats beetles and other inſects. Perches on trees: and builds its neſt in the depths of foreſts.

Theſe birds paſs over *Sweden* in the ſpring in vaſt flocks, flying towards the extreme north. They ſometimes reſt in the moors at night; but it is reckoned a wonder, if any one is found to make its neſt in the country. They return ſouthward in autumn; but, in both their paſſages, ſoar ſo high as to appear ſmall as ſparrows *.

---

* *Amœn. Acad.* iv. 589.

XXXV.

XXXV. I B I S. *Gen. Birds,* LXV.

360. WOOD.

Wood Pelecan, *Catefby,* i. 81.—*Latham,* iii.
Curicaca, *Marcgrave,* 191.—*De Buffon,* vii. 276.—*Pl. Enl.* 868.—*Briffon,* v. 335.
Tantalus Loculator, *Lin. Syft.* 240.

I. With a bill near ten inches long; near feven in girth at the bafe; ftrait till near the end, where it bends downwards: fore part of the head and face covered with a bare dufky blue fkin: hind part of the head, and the whole neck, of a pale yellowifh brown: under the chin is a pouch capable of containing half a pint: the greater primaries, and fome of the greater coverts, are black gloffed with green; the reft of the wing, back, and belly, white: tail fquare, fhort, and black: legs very long, black, and femi-palmated. The body of the SIZE of a Goofe.

PLACE.

Appears in *Carolina,* at the latter end of fummer, during the great rains, when they frequent the overflown *favannas* in vaft flocks; but retire in *November.* They perch erect on tall cyprefs and other trees, and reft their monftrous bills on their breafts for their greater eafe. They are very ftupid and void of fear, and eafily fhot. Fly flowly. Their food is herbs, fruits and feeds, fifh, and water infects; notwithftanding which they are excellent eating.

The refidence of thefe birds, the reft of the year, is *Brafil, Guiana*,* and perhaps other parts of *South America.*

361. SCARLET.

Guara, *Marcgrave,* 203.—*De Buffon,* vii. 35.—*Pl. Enl.* 81.
Red Curlew, *Catefby,* i. 84.—*Latham,* iii.
Tantalus Ruber, *Lin. Syft.* 241.—LEV. MUS.—BL. MUS.

I. With a flender incurvated bill, and naked fkin on the face, both of a pale red color: the whole plumage of the richeft fcarlet, only the ends of the wings are black: legs pale red. In SIZE fomewhat larger than the *Englifh* CURLEW.

* *Des Marchais,* iii. 326.

Frequent

Frequent the coasts of the *Bahama* islands. Are common in *East Florida*: in *Georgia* are frequent in the months of *July* and *August*, after which they retire southward. A few are seen in the south of *Carolina*, and the parts of *America* within the tropics, and very seldom to the north. These birds perch, like the former, upon trees, and make a most resplendent appearance. They lay their eggs in the tall grass. When first hatched, the young are of a dusky color: their first change is to ash-color; then to white; and, in their second year, to columbine; and with age acquire their brilliant red. In *Guiana* it is often domesticated, so as never to leave the poultry yard. It shews great courage in attacking the fowls, and will even oppose itself to the cat. The flesh is esteemed excellent. Its rich plumage is used by the *Brasilians* for various ornaments.

Brown Curlew, *Catesby*, i. 83.—*De Buffon*, vii. 42.—*Latham*, iii. Tantalus Fuscus, *Lin. Syst.* 242.

362. BROWN.

I. With the bill six inches and a half long, resembling the former: bill, face, and legs, red: neck, upper part of the back, and tail, of a cinereous brown: lower part of the back, breast, and belly, white.

White Curlew, *Catesby*, i. 82.—*De Buffon*, vii. 41.—*Pl. Enl.* 915.—*Latham*, iii. Tantalus Albus, *Lin. Syst.* 242.

363. WHITE.

I. With the face, bill, and legs, like the former: the whole plumage of a snowy whiteness, except the ends of the four first primaries, which are green. The flesh and fat is of saffron-color.

These birds arrive in *South Carolina*, with the *Wood Ibis* and the *Brown Curlew*, in great numbers, in *September*, and frequent the low watery tracts: continue there about six weeks, and then retire south to breed. This species goes away with egg. There was a suspicion, that the BROWN and the WHITE differed only in sex; but experiment proved the contrary. The white kind are twenty

PLACE.

3 N 2

times

times more numerous than the others; the flesh of the latter alfo differs, being of a dark color.

All thefe fpecies frequent the fides of rivers, and feed on fmall fifh, cruftaceous animals, and infects.

A. Bay Ibis. Tantalus Falcinellus, *Lin. Syft.* 241.—*Muller,* N° 178.—*Latham,* iii. —Lev. Mus.

I. With a black face: violet-colored wings and tail: blue legs.

Inferted here on the authority of Mr. *Muller*; but this fpecies has hitherto been known only to fouthern *Europe,* and about the *Cafpian* and *Black Seas.*

XXXVI.

XXXVI. CURLEW. *Gen. Birds*, LXVI.

Br. Zool. ii. Nº 177.—Eſkimaux Curlew, *Faun. Am.*—Latham, iii. —Lev. Mus.

364. ESKIMAUX.

C. With a duſky bill, near three inches long: crown of a deep brown, divided lengthways by a white ſtripe: cheeks, neck, and breaſt, of a very pale brown, marked with ſmall duſky ſtreaks, pointing down the back: ſcapulars and coverts ſpotted with black and pale reddiſh brown: primaries duſky; ſhafts white; inner webs marked with red oval ſpots: tail barred with black and light brown: legs blue. Larger than the *Engliſh* WHIMBREL, of which it is a variety; and differs only in having its back brown inſtead of white.

PLACE.

Were ſeen in flocks innumerable, on the hills about *Chateaux Bay,* on the *Labrador* coaſt, from *Auguſt* the 9th to *September* 6th, when they all diſappeared, being on the way from their northern breeding-place. They kept on the open grounds, fed on the *empetrum nigrum*, and were very fat and delicious. They arrive in *Hudſon's Bay* in *April* or the beginning of *May:* pair and breed to the north of *Albany Fort*, among the woods: return in *Auguſt* to the marſhes; and all diſappear in *September* *.

* *Ph. Tranſ.* lxii. 411. where it is called the *Eſkimaux* Curlew.

A. CURLEW,

A. CURLEW, *Br. Zool.* ii. N° 176.—Scolopax arquata, *Faun. Suec.* N° 168.—*Latham*, iii.
Le Courlis, *De Buffon*, viii. 19.—*Pl. Enl.* 818.—LEV. MUS.—BL. MUS.

C. With an incurvated bill, feven inches long : head, and upper
part of the wings, pale brown, fpotted with black : back white :
tail white, barred with black : legs blueifh. WEIGHT from twenty-
two to thirty-feven ounces.

PLACE. Inhabits *Europe*, as high as *Lapmark* * and *Iceland* † ; and is found
on the vaft plains of *Ruffia* and *Sibiria*, quite to *Kamtfchatka.*

B. WHIMBREL, *Br. Zool.* ii. N° 177.—Scolopax Phæopus. Windfpole. Spof. *Faun.*
Suec. N° 169.—*Latham*, iii.
Le Courlieu, ou petit Courlis, *De Buffon*, viii. 27.—*Pl. Enl.* 842.—LEV. MUS.

C. With a bill near three inches long : head marked lengthways
by a whitifh line, bounded on each fide by one of black :
neck, coverts of wings, and upper part of the back, pale brown,
fpotted with black : lower part of the back and the belly white : tail
light brown, barred with black : legs blueifh grey. WEIGHT twelve
ounces.

PLACE. Inhabits the fame places with the former.

<hr>

* *Leems*, 249.   † *Brunnich*, p. 49.

XXXVII.

XIX

Eskimaux Curlew, Nº 364.   *Little Woodcock, Nº 365.*

P. Mazell sculp.

XXXVII. SNIPE. *Gen. Birds,* LXVII.

SN. With the upper mandible of the bill two inches and a half long; the lower much fhorter: forehead cinereous; hind part black, with four tranfverfe yellowifh bars: from bill to the eye a dufky line: chin white: under fide of the neck, breaft belly, and thighs, of a dull yellow, paleft on the belly: hind part of the neck black, edged with yellowifh red: back, and leffer coverts, of the fame colors; reft of the coverts marked with zigzags of black and dull red: primaries dufky: inner coverts ruft-colored: tail black, tipped with brown: legs fhort, pale brown. LENGTH, from tip of the bill to the end of the tail, eleven inches and a half.

365. LITTLE WOODCOCK.

This fpecies has entirely the form of the *European* Woodcock; but differs in fize and color. They appear in the province of *New York* in the latter end of *April,* or beginning of *May.* They lay, the latter end of the laft month or beginning of *June,* from eight to ten eggs? and ufually in fwampy places. Mr. *Lawfon* * found them in *Carolina* in *September.* He prefers them, in point of delicacy, to the *European* kind.—*Latham,* iii. —LEV. MUS.—BL. MUS.

PLACE.

Br. Zool. ii. N° 187.—Scolopax gallinago, Horfgjok, *Faun. Suec.* N° 173.— *Latham,* iii.
La Becaffine, *De Buffon,* vii. 483.—*Pl. Enl.* 883.—LEV. MUS.—BL. MUS.

366. COMMON SNIPE.

SN. With head divided lengthways with two black lines, and three of reddifh brown: throat white: neck mottled with brown and teftaceous: ridge of the wing dufky: greater coverts and primaries dufky, tipt with white: belly white: lower half of the tail black; upper orange, with two dufky ftripes: toes divided to the origin. The *American* is leffer than the *Britifh.*

* P. 44.

Inhabits

PLACE.　　Inhabits *Europe* as far as *Iceland*: extends even to *Greenland**; and common in all parts of *Ruſſia* and *Sibiria*. Is found all over *North America*, and ſwarms in *South Carolina*. The lines on the head are leſs diſtinct in the *American* kind: the bill is alſo ſhorter than that of the *Engliſh*.

367. JACK.　　　*Br. Zool.* ii. N° 189.—Petite Becaſſine, ou le Sourde, *De Buffon*, vii. 490. —*Latham*, iii.　—LEV. MUS.—BL. MUS.

SN. With crown of the head black, tinged with ruſt: a yellow ſtroke over each eye: neck mottled with white, brown, and teſtaceous: rump of a gloſſy purpliſh blue: tail brown, edged with tawny. WEIGHT under two ounces.

PLACE.　　Theſe two inhabit *Europe*, *North America*, and *Sibiria*.

368. RED-BREAST-ED SNIPE.　　SN. With a bill like the common kind, two inches one-eighth long: head, neck, and ſcapulars, varied with black, aſh-color, and red: under ſide of the neck and breaſt ferruginous, thinly ſpotted with black: coverts and ſecondaries dark cinereous; the laſt tipt with white: back and rump white, concealed by the ſcapulars: tail barred with duſky and white: legs dark green: middle and outmoſt toe connected by a ſmall web. SIZE of the *Engliſh* Snipe.

PLACE.　　Inhabits the coaſt of *New York*.—*Latham*, iii.　—BL. MUS.

369. BROWN.　　SN. With a bill like the former: from that to each eye a white bar: head, neck, and ſcapulars, of a fine uniform cinereous brown, in a very few places marked with black: coverts and primaries dark brown; ſhaft of the firſt primary white: ſecondaries light brown, edged with white: back white: rump and tail barred

* *Faun. Greenl.* N° 71.

9

with black and white : breaft mottled with brown and white : belly
white : legs and toes like the former.   Its SIZE the fame.—BL. MUS.
   Inhabits the fame place.

S N. With the bill very flender, long, and black : the crown, and
   upper part of the back, dufky, ftreaked with red : cheeks cinereous, ftreaked with black : neck and breaft cinereous, mixed with
ruft-color, and marked obfcurely with dark fpots : belly white :
thighs fpotted with black : leffer coverts of wings afh-colored; greater
dufky, edged with brown : primaries and fecondaries dufky ; the
laft tipt with white : lower part of the back white, fpotted with
black : tail barred with black and white ; tips reddifh : legs
greenifh : the toes bordered by a narrow plain membrane.   SIZE of
the *Englifh* Snipe.
   Obferved in *Chateaux Bay*, on the coaft of *Labrador*, in *September*.
Are perpetually nodding their heads.—*Latham*, iii.

Edw. 137.—Scolopax Fedoa, *Lin. Syft.* 244.—La Barge rouffe de Baie de
   Hudfon, *De Buffon*, vii. 507.—*Latham*, iii.—LEV. MUS.

S N. With bill fix inches long : throat white : from the bill to the
   eye extends a dufky line : head and neck mottled with dufky
and light brown : breaft barred with black : belly plain brown :
back and coverts varied with rufty brown and dufky : the primaries and fecondaries ferruginous on their exterior webs : tail barred
with light brown and black : legs very long, black, and naked
very high above the knees.
   Inhabits *Hudfon's Bay* and *Connecticut*.—BL. MUS.

372. **RED.**

Red Godwit, *Br. Zool.* ii. N° 181.—*Edw.* 138.—Scolopax Lapponica, *Faun. Suec.* N° 174.—*Ph. Tranf.* lxii. 411.—*Latham*, iii. La Barge rouffe, *De Buffon*, vii. 304.—*Pl. Enl.* 900.—Lev. Mus.

S N. With a bill three inches three quarters long, reflecting a little upward; yellow near the bafe; dufky towards the end: head, neck, breaft, and upper part of the back, ferruginous, and ftreaked with black, excepting the neck, which is plain: lower part of the back white: leffer coverts of the wings cinereous brown: exterior webs of the primaries black; the lower parts white: the lower part of the tail white; the upper black; the tips white: the legs dufky. LENGTH, to the tip of the tail, one foot fix inches. EXTENT two feet four. WEIGHT twelve ounces. Varies much in colors, according to age.

PLACE.

Is common to the north of *Europe* and of *America*. Very numerous, in fummer time, in the fens of *Hudfon's Bay*; where they breed, and then retire. Appear about the *Cafpian* fea in the fpring; but never in *Sibiria*, nor in the north of *Afia*.

373. **COMMON.**

Godwit, *Br. Zool.* ii. N° 179.—*Catefby*, App.—*Latham*, iii. La Barge Aboyeufe, *De Buffon*, vii. 501.—*Pl. Enl.* 876.—Lev. Mus.—Bl. Mus.

S N. With the bill turning a little up; four inches long; pale purple at the bafe; dufky at the end: head, neck, back, and coverts of the wings, of a very light brown, with a brown fpot in the middle of each feather: primaries dufky; inner webs of a reddifh brown: tail barred with black and white: belly and vent white: legs dufky; in fome of a greyifh blue. In SIZE fomewhat inferior to the laft.

PLACE.

Common to the north of *Europe*, *Afia*, and *America*.

Br. Zool. ii. N° 186 —*Phil. Tranf.* lxii. 410.—*Latham*, iii.—BL. Mus.

SN. With a very flender bill, a little more than two inches long: orbits, chin, and throat, white: from bill to eye a line of white: cheeks and under fide of the neck white, with fhort ftreaks of a dufky color: crown and upper part of the neck brown, with white ftreaks: coverts of the wings, primaries, fecondaries, and fcapulars, black, with elegant triangular fpots of white: tail barred with black and white: breaft and belly white: legs long; and in the live bird of a rich yellow; fometimes red. LENGTH, to the tip of the tail, near fixteen inches.

PLACE.

Arrives in *Hudfon's Bay* in fpring. Feeds on fmall fhell-fifh and worms; and frequents the banks of rivers. Called there, by the natives, from its noife, *Sa-fa-fkew*; by the *Englifh, Yellow legs*. Retires in autumn: Makes a ftop in the province of *New York*, on its return, but does not winter there. This, and feveral other fpecies of Snipes and Sandpipers, are called, in *North America, Humilities.*

Leffer Godwit, *Br. Zool.* ii. N° 188.—La Barge commune, *De Buffon*, vii. 500. —*Pl. Enl.* 874.—*Latham*, iii.
Scolopax Limofa, *Faun. Suec.* N° 172.
Jadreka, *Olaf. Iceland*, ii. 201. tab. xlviii.

SN. With a bill near four inches long: head and neck cinereous: cheek and chin white: back of an uniform brown: wings mark- ed with a white line: rump and vent feathers white: middle fea- thers of the tail black; in the reft the white predominates more and more, to the outmoft: legs dufky. WEIGHT nine ounces. LENGTH, to the tail, feventeen inches.

PLACE.

Inhabits *Iceland, Greenland* *, and *Sweden*. Migrates in flocks in the fouth of *Ruffia*.

* *Faun. Groenl.* N° 72.

SN.

376. STONE.   SN. With a black bill: head, neck, and breaft, fpotted with black and white: back, fcapulars, and greater coverts, of the fame colors: primaries dufky: rump and tail barred with black and white: belly white: legs long and yellow. Double the SIZE of a Snipe.

PLACE.   Obferved in autumn feeding on the fands on the lower part of *Chateaux Bay*, continually nodding their heads. Are called there *Stone Curlews*.

377. REDSHANK.   *Br. Zool.* ii. N° 184.—Scolopax Totanus, *Faun. Suec.* N° 167.—*Latham*, iii. Le Chevalier aux pieds rouges, *De Buffon*, vii. 513.—*Pl. Enl.* 845.—LEV. MUS.—BL. MUS.

SN. With bill red at the bafe, black at the end: head, hind part of neck, and fcapulars, of a dufky afh-color, faintly marked with black: back white, fprinkled with black: under fide of neck white, ftreaked with dufky: breaft and belly white: tail barred with black and white: legs of a bright orange-color.

PLACE.   Inhabits *North America*, in common with the north of *Europe*, as high as *Finmark* *; and is found in *Sibiria*. Grows lefs common towards the north of that country. Is fometimes found, in *Hudfon's Bay*, quite white †.

378. YELLOW-SHANKS.   SN. With a flender black bill, an inch and a half long, a little bent at the end: head, hind part of the neck, back, and greater coverts of the wings, dirty white, fpotted with black: leffer coverts plain brown: primaries dufky: breaft and fore part of the neck fpotted with black and white: belly and coverts of the tail pure white: tail barred with brown and white: legs yellow. LENGTH, from tip of the bill to the tail, eleven inches.

PLACE.   Appears in the province of *New York* in autumn.—BL. MUS.

Leems, 253.   † *Edw.* 139.

Br.

379. GREEN-
SHANK.

Br. Zool. ii. N° 183.—La Barge varié, *De Buffon,* vii. 503.—*Latham,* iii.
Scolopax Glottis, *Faun. Suec.* N° 171.—LEV. MUS.

S^N. With a bill two inches and a half long, very flender, and a
little recurvated: head, and upper part of neck, cinereous,
with dufky lines: over each eye a white line: coverts of wings, fcapulars, and upper part of the back, of a brownifh afh-color: breaft,
belly, and lower part of the back, white: primaries dufky: tail white,
finely marked with waved dufky bars: legs green. LENGTH fourteen inches. WEIGHT only fix ounces.

PLACE.

Inhabits the province of *New York*; and in every latitude of
Ruffia and *Sibiria,* in plenty. Is found in *Europe* as high as *Sondmor* *

380. SEMIPAL-
MATED.

S^N. With a bill two inches long, and dufky: head and neck ftreaked with black and white: breaft white, with round brown fpots:
belly and fides white; the laft marked with tranfverfe bars of brown:
back and coverts of the wings cinereous, with great fagittal fpots
of black: primaries dufky, with a tranfverfe white bar: fecunda-
ries white: the middle feathers of the tail cinereous, barred with
black; outmoft white: legs dufky: toes femipalmated. LENGTH
fourteen inches.

PLACE.

Inhabits *New York.*—BL. MUS.

381. BLACK.

S^N. With red bill and legs: the plumage moft intenfely
black.

PLACE.

Obferved by *Steller* in the iflands towards *America.*

* *Strom.* 235.

A. EUROPEAN

A. **European Woodcock**, *Br. Zool.* ii. N° 178.—Le Becaſſe, *De Buffon*, vii. 462.— *Pl. Enl.* 885.—*Latham*, iii.
Scolopax ruſticola. Morkulla, *Faun. Suec.* N° 170.—Lev. Mus.—Bl. Mus.

SN. With a reddiſh cinereous front: hind part of the head barred with reddiſh brown: upper part of the body and wings barred with ruſt-color, black, and grey: breaſt and belly dirty white, barred with duſky lines. Weight twelve ounces.

PLACE. Inhabits, during ſummer, *Scandinavia*, *Lapland*, and *Iceland*. Migrates ſoutherly at approach of winter. Common in *Ruſſia* and *Sibiria*, but only in the time of migration; and breeds in the northern marſhes and *Arctic* flats.

B. **Great Snipe**, *Br. Zool.* ii. N° 188.—*Latham*, iii.
Scolopax Media, *Friſch.* tab. 228.—Lev. Mus.

SN. With head divided lengthways by a teſtaceous line, bounded on each ſide by one of black; above and beneath each eye another: neck and breaſt yellowiſh white, marked with ſemicircular ſpots of black: ſides undulated with black: back and coverts teſtaceous, ſpotted with black, and edged with white: primaries duſky: tail ruſt-color; external feathers barred with black. Weight eight ounces.

PLACE. Inhabits the *Arctic* regions of *Sibiria*. Found in *England* and *Germany*. The link between the Woodcock and Snipe.

2   C. Dusky

C. Dusky Snipe.   Scolopax Fufca, *Lin. Syft.* 243.—*Briffon*, v. 276. tab. xxiii.
La Barge brun, *De Buffon*, vii. 508.—*Pl. Enl.* 875.—*Latham*, iii.

SN. With the upper part dufky black, with the edges of the feathers whitifh : lower part of the body dark cinereous : two middle feathers of the tail dufky, ftreaked croffways with white : the fide
feathers brown, ftreaked in the fame manner : legs black.   Length
twelve inches ; of the bill two and a quarter.
   Breeds within the *Arctic* circle.   Migrates into *Ruffia* and *Sibiria*.   PLACE.

D. Finmark Snipe.   Scolopax Gallinaria, *Muller*, N° 183.—*Latham*, iii.

SN. With the bill tuberculated like that of the common Snipe :
head entirely grey : legs yellow.   In other refpects, has great
agreement with the common fort.
   Inhabits *Finmark*.   PLACE.

XXXVIII.

XXXVIII. SANDPIPER. *Gen. Birds,* LXVIII.

382. HEBRIDAL.   *Br. Zool.* ii. N° 200.—Tringa interpres, Tolk, *Faun. Suec.* N° 178.
Turnstone, *Catesby,* i. 72.—Tringa Morinellus, *Lin. Syst.* 249.—*Latham,* iii.
Le Tourne-pierre, *De Buffon,* viii. 130 —*Pl. Enl.* 130.—Lev. Mus.—Bl. Mus.

S. With black bill, red at the base : white forehead, throat, belly, and vent: crown white, spotted with black : sides of the head white : a black line passes above the forehead ; is continued under each eye to the corner of the mouth ; drops on each side of the neck to the middle, which is surrounded with a black collar : breast black : coverts cinereous : upper part of the back, scapulars, and tertials, ferruginous, spotted with black : lower part of back white, traversed with a black bar : tail black, tipped with white : legs short ; of a bright orange. WEIGHT three ounces and a half. LENGTH eight inches and a half.

PLACE.   Inhabits *Hudson's Bay, Greenland*,* and the *Arctic* flats of *Sibiria,* where it breeds ; wandering southerly in autumn. It lays four eggs. It migrates in *America* as low as *Florida.* In spring it possesses itself of the isles of the *Baltic,* till it quits them in autumn †.

383. STRIATED.   Tringa Striata, *Lin. Syst.* 248.—*Faun. Groenl.* N° 71.—*Faun. Dan.* N° 194.—
Latham, iii.

S. With the base of the bill and legs yellow : upper part of the body undulated with dusky and cinereous : front of the neck dusky : breast and belly white : primaries and secondaries black ; the last tipt with white : tertials white, with a stripe of black : tail black : the feathers on the side cinereous, edged with white. SIZE of a Stare.

PLACE.   Inhabits *Sweden, Norway,* and *Iceland.* Is found, but not frequently, in *Russia* and *Sibiria* ; and is conversant, even during winter, in the

* *Faun. Groenl.* N° 74.   † *Amœn. Acad.* iv. 590.

frosty

frofty climate of *Greenland*; but retires to breed into the bottom of the bays: flies very fwiftly along the furface of the water, catching the infects on the furface. Never touches the water with its feet or body; but dexteroufly avoids the rifing of the higheft waves. Twitters with the note of the Swallow *.

Br. Zool. ii. Nº 193.—*Le Canut, De Buffon,* viii. 142.—*Latham,* iii. Tringa Canutus, *Faun. Suec.* Nº 183.—LEV. MUS.

384 KNOT.

S. With a black bill: between the bafe and eyes a white line: crown and upper part of the body of a dufky brown: wings of the fame color, croffed with a line of white: the breaft and under fide of the neck white, fpotted with black: rump white, with large black fpots: legs fhort, of a blueifh grey: toes divided to the origin.

I have feen this fpecies from the province of *New York.* Obferved by Dr. *Pallas* only about lake *Baikal.*

PLACE.

Br. Zool. ii. Nº 196.—Tringa Macularia, *Lin. Syft.* 249.—*Edw.* 277.— *Latham,* iii. La Grive d'eau, *De Buffon,* viii. 140 —LEV. MUS.—BL. MUS.

385. SPOTTED.

S. With a white line above each eye: crown, upper part of neck and body, and coverts of the wings, olive brown, with triangular black fpots: under fide, from neck to tail, white, with brown fpots: middle feathers of the tail brown; thofe on the fide white, with dufky lines: legs of a dirty flefh-color. FEMALE has no fpots on the lower part of the body. SIZE of the Striated.

Inhabits *North America.* Arrives in *Penfylvania* in *April,* and ftays there all the fummer. Vifits *Hudfon's Bay* in *May:* breeds there, and retires in *September.*

PLACE.

---

* *Faun. Groenl.* Nº 73.

Br.

386. Ash-colored.

Br. Zool. ii. Nº 194.—Tringa Cinerea, *Brunnich*, Nº 179.—*Latham*, iii.
—Bl. Mus.

S. With a dusky cinereous head, spotted with black : neck cinereous, marked with dusky streaks : back and coverts of wings finely varied with concentric semicircles of black, ash-color, and white : coverts of the tail barred with black and white : tail cinereous, edged with white : breast and belly pure white ; the first spotted with black : legs dusky green : toes bordered with a narrow membrane, finely scolloped. LENGTH ten inches. WEIGHT five ounces.

PLACE.

Seen in great numbers on *Seal* Islands near *Chateaux Bay*. Continues the whole summer in *Hudson's Bay*, and breeds there. Breeds in *Denmark?* Has been shot on the *Flintshire* shores in the winter season.

387. New York.

S. With the under side of neck and body white : the breast spotted with brown : lesser coverts dusky, edged with white : back and greater coverts dusky in the middle ; the edges cinereous : the secondaries of the same colors : coverts of the tail barred with black and white : sides beneath the wings streaked with brown : tail cinereous.

PLACE.

Inhabits the province of *New York.*—BL. Mus.

388. Common.

Br. Zool. ii. Nº 204.—La Guignette, *De Buffon*, vii. 540.—*Pl. Enl.* 850.—*Latham*, iii.

Tringa Hypoleucus Snappa, *Faun. Suec.* Nº 182.—LEV. Mus.

S. Above of a deep brown, spotted with rugged marks of black : the plumage most glossy and silky : fore part of the neck and breast white, with a few black spots : belly white : primaries and secondaries dusky ; the last with their bottoms and ends white : the outmost feathers of the tail spotted with white and brown ; the middle brown, tipt with white : legs yellowish. WEIGHT about two ounces. Differs little from the *European* kind, but in the colors of the legs.

PLACE.

Inhabits *Chateaux Bay*, and the northern latitudes of *Sibiria*, as far as *Kamtschatka*.

B.

Br. Zool. ii. N° 201.—Le Becaſſeau, *De Buffon,* vii. 535.—*Pl. Enl.* 843.   389. GREEN.
Tringa Ocrophus, *Faun. Suec.* N° 180.—Tr. Littorea ? *Faun. Suec.* N° 185.—
    Latham, iii.   —LEV. MUS.

S. With head and upper part of the neck cinereous brown, ſtreaked with white : lower part of neck mottled with brown and
white : back, ſcapulars, and coverts of wings, of a duſky green,
gloſſy as ſilk, and elegantly ſpeckled with white : primaries duſky :
rump, breaſt, and belly, white : tail white, the feathers marked
with different numbers of ſpots : legs a cinereous green. About a
third larger than the COMMON.

I have ſeen this ſpecies in Mr. *Kuckan*'s collection, which he made   PLACE.
in *North America.* It is alſo found in *Iceland ;* and is very common
in all the watery places of *Ruſſia* and *Sibiria.* The *Tringa Littorea*
is ſaid to migrate from *Sweden* to *England,* at approach of winter *.

Br. Zool. ii. N° 206.—Tringa Cinclus, *Lin. Syſt.* 251.—*Latham,* iii.   390. PURRE.
    L'Alouette de mer, *De Buffon,* vii. 548.—*Pl. Enl.* 851.—LEV. MUS.—BL. MUS.

S. With a ſlender black bill : head and upper part of neck aſh-
colored, ſtreaked with black : from bill to eyes a white line :
under ſide of the neck white, mottled with brown : back and coverts of wings a browniſh aſh-color : greater coverts duſky, tipt with
white : breaſt and belly white : two middle feathers of the tail
duſky ; the others aſh-color, edged with white : legs of a duſky green.
WEIGHT an ounce and a half. LENGTH ſeven inches and a half.

Inhabits the coaſt of *New York,* and extends as low as *Jamaica*   PLACE.
and *Cayenne.* Not mentioned among the *Scandinavian* birds. Common in all latitudes of *Ruſſia* and *Sibiria.*

* *Amœn. Acad.* iv. 590.

391. DUNLIN.   *Br. Zool.* ii. N° 205.—La Brunnette, *De Buffon,* vii. 493.—*Latham,* iii.
Tringa Alpina, *Faun. Succ.* N° 181.—*Faun. Groenl.* N° 77.
Loar Thrælt, *Olaf. Iceland.* i. N° 677. tab. xli.—LEV. MUS.—BL. MUS.

S. With head, hind part of the neck, and back, ferruginous, marked with great black fpots : lower part of neck white, ftreaked
with black : coverts of the wings afh-colored : breaft and belly
white, marked with a black crefcent : tail afh-colored ; the two
middle feathers longeft and darkeft : legs black.   In SIZE fomewhat larger than the laft.

PLACE.   Inhabits *Greenland, Iceland, Scandinavia,* the alps of *Sibiria,* and,
in its migration, the coafts of the *Cafpian* fea.

392. RED.   *Br. Zool.* ii. N° 202.—Aberdeen Sandpiper, N° 203 —*Latham,* iii.
Tringa Icelandica, *Lin. Syft.* Add.

S. With the bill black and flender, a little bending : head, upper
part of the neck, and beginning of the back, dufky, marked
with red : lower fide of the neck and breaft cinereous, mixed with
ruft-color, and obfcurely fpotted with black : leffer coverts of the
wing cinereous ; primaries dufky ; fecondaries tipt with white :
two middle feathers of the tail are dufky, and extend a little beyond the others ; the reft are cinereous : legs long and black.

PLACE.   The birds which I have feen of this kind came from the coafts
of *New York, Labrador,* and *Natka Sound.*   They are alfo found in
Iceland *.   Probably are the fame with the *Scolopax Subarquata,* which,
during fummer, frequent the fhores of the *Cafpian* fea, lake *Baikal,* and
efpecially the mouth of the *Don* †.   I believe them to be the young,.

* *Brunnich,* N° 180.
† *Nov. Com. Petrop.* xix. 471. tab. xix.—The *Tringa Ruficollis,* PALLAS *Iter.* iii.
100, is another red-necked fpecies, found about the fame falt lakes.

or

or the females, of that which is defcribed in the *Br. Zool.* under the name of the Red. The laft differs in nothing, but in having the whole under fide of a full ferruginous color. LENGTH from eight to ten inches. Varies greatly in fize.

SIZE.

Br. Zool. ii. N° 191.—Grey Plover, *Wil. Orn.* 309.—*Latham*, iii.
Tringa Squatarola, *Faun. Suec.* N° 186.
Le Vanneau Pluvier, *De Buffon*, viii. 68.—*Pl. Enl.* 854.—LEV. MUS.—BL. MUS.

393. GREY.

S. With a ftrong black bill: head, back, and coverts of wings, black, edged with grey, tinged with green, and fome white: cheeks and throat white, ftreaked with black: primaries dufky, white on their interior lower fides: belly and thighs white: rump white: tail barred with black and white: legs of a dirty green: back toe very fmall. WEIGHT feven ounces. LENGTH, to the end of the tail, twelve inches.

According to *Lawfon*, frequents the vallies near the mountains of *Carolina*. Are feen flying in great flocks; but feldom alight *. Very common in *Sibiria*; and appear in autumn in flocks, after breeding in the extreme north.

PLACE.

Br. Zool. ii. N° 198.—*Latham*, iii.
Tringa Gambetta, *Faun. Suec.* N° 177.—Tr. Variegata, *Brunnich*, N° 181.

394. GAMBET.

S. With head, back, and breaft, of a cinereous brown, fpotted with dull yellow: coverts of the wings, and fcapulars, cinereous, edged with yellow: primaries dufky: belly white: tail dufky, bordered with yellow: legs yellow. SIZE of the Greenfhank.

Taken in the frozen fea, between *Afia* and *America*, lat. 69½, long. 191½. Inhabits alfo *Scandinavia* and *Iceland* †; in the laft it is called *Stelkr*, from its note.

PLACE.

---

* *Hift. Carol.* 140. *Catefby*, App   † *Paulfen's Lift.*

Le

395. ARMED.   Le Vanneau armé de la Louisiane, *Brisson*, iv. 114. tab. viii.—*De Buffon*, viii.. 65.—*Pl. Enl.* 835.—*Latham*. iii.

S. With an orange bill, depressed in the middle : on each side of the base is a thin naked skin of a light orange-color, which rises above the forehead, extends beyond each eye, and falls, in form of a pointed wattle, far below the chin : crown dusky : hind part of the neck, back, rump, scapulars, and coverts of the wings, of a dusky grey : from the chin to the vent white, tinged with tawny : primaries mostly black : the fore part of the wing is armed with a sharp spur, a most offensive weapon : the tail short, whitish, crossed near the end with a black bar, and tipt with white : legs long, and of a deep red. LENGTH, from bill to tail, above ten inches. EXTENT about two feet two.

PLACE.   Inhabits *Louisiana*.

396. SWISS.   Tringa Helvetica, *Lin. Syst.* 250.—Le Vanneau Suisse, *De Buffon*, viii. 60.— *Pl. Enl.* 853.
Vanellus Helveticus, *Brisson*, v. 106. tab. x.—*Latham*, iii.   —LEV. MUS.— BL. MUS.

S. With a strong bill an inch long, depressed in the middle : front and sides of the head white ; hind part spotted with black and white : cheeks, under side of the neck, breast, and belly, black : thighs and vent white : primaries black : back, coverts of wings, and the tail, white, barred with black : legs black : back toe very small. SIZE of a Lapwing.

PLACE.   Inhabits the coast of *Connecticut* and *Hudson's Bay*. Visits the last in the spring. Feeds on berries, insects, and worms. Retires in *September*. Breeds also in the *Arctic* flats of *Sibiria* ; and, in the time of migration, appears in all parts of the south of *Russia* and *Sibiria*. Is found in *France* and *Swisserland*.

Br

Br. Zool. ii. N° 207.—Tringa pufilla, *Lin. Syſt.* 252.—Oddinſtiane Iſlandis, *Latham*, iii.

S. With crown black : upper part of the plumage brown, edged with black and pale ruſty brown : belly and breaſt white : tail duſky : legs black. SIZE of a Hedge Sparrow.

Inhabits the north of *Europe, Iceland,* and *Newfoundland.* Obferved alfo in *Natka Sound.* It is met with as far fouth as *St. Domingo* *. Probably migrates there in the winter.

A. RUFF, *Br. Zool.* ii. N° 192.—Le Combattant, ou Paon de mer, *De Buffon*, vii. 521.— *Pl. Enl.* 305, 306.—*Latham*, iii.
Tringa pugnax, Bruſhane, *Faun. Suec.* N° 175.—*Leems Lapm.* 246.—LEV. MUS.— BL. MUS.

S. With a long ruff of feathers on the fore part and ſides of the neck ; and a long tuft on the hind part of the head : legs yellow. REEVES, or the females, are of a pale brown : back ſlightly fpotted with black : breaſt and belly white : neck ſmooth.

Inhabits the north of *Europe* in ſummer, as far as *Iceland,* and is very common in the northern marſhes of *Ruſſia* and *Sibiria.*

* *Briſſon,* v. 222.

B. FRECKLED.

B. Freckled.—Calidris nævia, *Briffon*, v. 229. tab. xxi. fig. i.—*Latham*, iii.

S. Above of a dufky cinereous, fpotted with red and black ; the laft gloffed with violet : lower part of the body of a reddifh white, varied with dufky and chefnut-colored fpots : two middle feathers of the tail afh-colored, edged with white ; the reft dark cinereous : the outmoft feather on each fide marked lengthways, on the exterior fide, with a white line : legs greenifh. LENGTH near nine inches.

PLACE. Is found in the north of *Ruffia* and *Sibiria* ; and alfo in *France*.

C. Selninger, *Muller*, N° 206.—Tringa Maritima, *Brunnich*, 182.—*Leems*, 254. —*Latham*, iii.

S. Above varied with grey and black : the middle of the back tinged with violet : fore part of the neck dufky : lower part of the body white : tail dufky ; four outmoft feathers on each fide fhorter than the reft, and edged with white : legs yellow. SIZE of a Stare.

PLACE. Inhabits *Norway* and *Iceland*. Lives about the fea-fhores, and is always emitting a piping note.

D. Lapwing, *Br. Zool.* ii. N° 190.—Le Vanneau, *De Buffon*, viii. 48.—*Pl. Enl.* 242. —*Latham* iii.

Tringa Vanellus Wipa, Cowipa, Blæcka, *Faun. Suec.* N° 176.—Lev. Mus.— Bl. Mus.

S. With a black bill : crown and breaft black : head adorned with a very long flender creft, horizontal, but turning up at the end : cheeks and fides of the neck, and belly, white : back and fcapulars gloffy green ; the laft varied with purple : primaries and fecondaries black, marked with white : coverts of tail, and vent, orange : outmoft feathers of the tail white, marked with one dufky fpot : the upper half of the reft black ; the lower of a pure white. WEIGHT eight

eight ounces. Length thirteen inches and a half. Extent two feet and a half.

The Lapwing is common in moſt parts of *Europe*. Extends to the *Feroe* iſles, and even to *Iceland* *. Is very frequent in *Ruſſia* ; but becomes very rare beyond the *Urallian* chain ; yet a few have been obſerved about the rivers *Ob* and *Angara*, and beyond lake *Baikal* ; but never farther to the eaſt. They extend ſouthward as far as *Perſia* and *Egypt* †, where they winter ; but, unable to bear the vaſt heats of the ſummer, migrate to the countries about *Woroneſch* and *Aſtracan* ‡. Appears in *Lombardy*, in *April* ; re-tires in *September*. Continues in *England*, and I believe in *France*, the whole year ; but conſtantly ſhifts its quarters in ſearch of food, worms and inſects. In *France*, multitudes are taken for the table in clap-nets, into which they are allured by the playing of a mirror.

E. **Waved.** Tringa Undata, *Brunnich*, N° 188.—*Latham*, iii.

S. Of a duſky color, marked witn undulated lines of white and yellowiſh clay-color: ſhaft of the firſt primary white: tips of the ſecondaries, and their coverts, white: tail aſh-colored, tipt with black.

Inhabits *Denmark* and *Norway*.

F. **Shore.** Tringa Littorea, *Faun. Suec.* N° 183.—*Brunnich*, N° 177.—*Latham*, iii. Le Chevalier varié, *De Buffon*, vii. 517.—*Pl. Enl.* 300.

S. With a duſky neck, ſtriped obliquely with white: back and co-verts of the wings duſky, with ſmall ruſty ſpots, lighteſt on the laſt: primaries and ſecondaries duſky, the laſt tipt with white ; ſhaft of firſt primary white: breaſt and belly white: tail croſſed with wav-ed bars of duſky and white: legs duſky. Size of a Turtle Dove.

Inhabits the marſhes of *Sweden* ; and is found in *Denmark*.

* *Brunnich*. † *Haſſelquiſt*, 288. ‡ *Extracts*, i. 107. ii. 147.

3 Q G. Wood.

G. Wood. Tringa Glareola, *Faun. Suec.* N° 184.—*Latham*, iii.

S. With the back dufky, fpeckled with white : primaries and fecondaries dufky ; the laſt tipt with white : breaſt and belly
whitiſh : tail barred with black and white ; the outmoſt feathers lighteſt : rump white : legs of a dirty green. Size of a Stare.

PLACE.   Inhabits the moiſt woods of *Sweden*.

H. Uniform. Keildu-fuin, Iſlandis, *Muller*, N° 205.—*Latham*, iii.

S. With a ſhort black bill, and of an uniform light aſh-color on
all its plumage.

PLACE.   Inhabits *Iceland*.

XXXIX.

XXXIX. PLOVER. *Gen. Birds,* LXIX.

Spotted Plover, *Edw.* 140.—Le Pluvier doré à gorge noire, *De Buffon,* viii. 85.   398. ALWARGRIM.
Charadrius Apricarius. Alwargrim, *Faun. Suec.* N° 189.—*Latham,* iii.
—LEV. MUS.—BL. MUS.

PL. With a fhort bill, like that of the *Swifs* Sandpiper: the forehead white, from which a white line falls from each corner, along
the fides of the neck, and unites at the breaft; the whole intervening
fpace black, as are the breaft, belly, thighs, and vent; only on the
laft are a few white fpots: crown, hind part of neck, back, and
coverts of wings and tail, dufky, moft elegantly fpotted with bright
orange: the primaries dufky, edged with afh-color: fecondaries and
tail barred with dufky and brown: legs black. SIZE of a Golden
Plover.

Inhabits all the north of *Europe, Iceland, Greenland,* and *Hudfon's*   PLACE.
Bay, and all the *Arctic* part of *Sibiria;* and defcends fouthward in its
migrations. Is called in *Hudfon's Bay,* the *Hawk's eye,* on account
of its brilliancy. It appears in *Greenland* in the fpring, about the
fouthern lakes, and feeds on worms and berries of the Heath *. Arrives in *New York* in *May:* breeds there, and difappears in the latter
end of *October,* after collecting in vaft flocks. Is reckoned moft
delicious eating.

Br. Zool. ii. N° 208.—Le Pluvier doré, *De Buffon,* viii. 81.—*Pl. Enl.* 904.   399. GOLDEN.
Charadrius Pluvialis, Akerhoua, *Faun. Suec.* N° 190.—*Latham* iii.   —LEV.
MUS.—BL. MUS.

PL. With head, upper part of the neck, back, and coverts of
wings, dufky, elegantly fpotted with yellowifh green: breaft
brown, ftreaked with greenifh lines: belly white: middle feathers
of the tail barred with black and yellowifh green; the others with

* Faun. Grœnl. N° 79.

3 Q 2   black

black and brown : legs black. LENGTH, to the tip of the tail, eleven inches.

PLACE.

Inhabits *North America*, as low as *Carolina* *. Migrates to the *Labrador* coaſt, about a week after the *Eſkimaux* WHIMBRELS, in its way to *New York*; but not in ſuch numbers. Found in *Iceland* and the north of *Europe* ; but are ſcarcely ſeen in *Ruſſia*. Very frequent in *Sibiria*, where they breed in the north. They extend alſo to *Kamtſchatka*, and as far ſouth as the *Sandwich* iſles : in the laſt it is very ſmall. I have ſeen, which I ſuppoſe to be young, a variety with black breaſts : and ſometimes a very minute back toe.

400. NOISY.

Chattering Plover, *Cateſby*, i. 71.—Le Kildir, *De Buffon*, viii. 96.—*Latham*, iii. Charadrius Vociferus, *Lin. Syſt.* 253.—LEV. MUS.—BL. MUS.

PL. With a broad black bar, extending from the bill, beneath each eye, to the hind part of the head : orbits red : forehead, and ſpace before and beyond the eyes, white : fore part of the crown black ; the reſt brown : a white ring encircles the throat and neck ; beneath that another of black ; under that ſucceeds, on the breaſt, two ſemicircles, one of white, another of black ; all below is white : back and coverts of wings brown : primaries duſky : the feathers on the rump are orange, and extend over three parts of the tail ; the lower part of which is black. SIZE of a large Snipe.

PLACE.

Inhabits *New York*, *Virginia*, and *Carolina* ; where they reſide the whole year. Are the plague of the ſportſmen, by alarming the game by their ſcreams. They are called in *Virginia*, *Kill-deer*, from the reſemblance their note bears to that word. Migrate to *New York* in the ſpring : lay three or four eggs : ſtay late.

* *Lawſon*, 140.—*Cateſby*, App.

Br. Zool. ii. N° 211.—Le Pluvier à collier, *De Buffon*, viii. 90.—*Pl. Enl.* 921.   401. RINGED
Charadrius Hiaticula, Strandpipare, *Faun. Suec.* N° 187.—*Latham*, iii.
—LEV. MUS.—BL. MUS.

PL. With a fhort bill ; upper half orange-color ; the end black :
    crown light brown : forehead, and a fmall fpace behind each
eye, white : from the bill, beneath each eye, extends a broad bar
of black : the neck is encircled with a white ring ; and beneath that
is another of black : coverts of wings of a pale brown : primaries
dufky : tail brown, tipt with lighter : legs yellow.

Almoft all which I have feen from the northern parts of *North*   VARIES.
America, have had the black marks extremely faint, and almoft loft.
The climate had almoft deftroyed the fpecific marks ; yet, in the
bill and habit, preferved fufficient to make the kind very eafily afcertained. The predominant colors were white, and very light afh-
color. WEIGHT near two ounces. LENGTH, to tip of the tail, fe-   SIZE.
ven inches and a half. Thofe of the weftern coafts of *North America*
are much fmaller.

Inhabits *America*, down to *Jamaica* * and the *Brafils* ; in the laft   PLACE.
it is called *Matuitui* †, where it frequents fea-fhores and eftuaries.
Is found in fummer in *Greenland :* migrates from thence in autumn.
Is common in every part of *Ruffia* and *Sibiria*. Was found by the
navigators as low as *Owyhe*, one of the *Sandwich* ifles, and as light-
colored as thofe of the higheft latitudes.

PL. With the bill an inch long ; black towards the end ; red towards   402. BLACK-
    the bafe : forehead black : crown black, furrounded with a circle   CROWNED.
of white : throat white : neck and breaft of a very light afh-colored brown, divided from the belly by a dufky tranfverfe ftroke :
belly and vent white : back, fcapulars, and coverts of the wings, ci-

---

* *Sloane.*   † *Marcgrave*, 199.

nereous brown : primaries dufky ; white towards their bottoms : tail white towards the bafe ; black towards the end ; and tipt with white : legs very long, naked an inch above the knees, and of a blood-red : toes very fhort.   LENGTH, to the end of the tail, about ten inches.

PLACE.   Inhabits the province of *New York*. Has much the habit of the *European* DOTTREL.—LEV. MUS.

403. SANDER-
LING.

Br. Zool. ii. Nº 212.—Le Sanderling, *De Buffon*, vii. 532.
Charadrius Calidris, *Lin. Syft.* 255.—*Latham*, iii.   —LEV. MUS.—BL. MUS.

PL. With a flender, black, weak bill, bending a little at the end : head and hind part of the neck cinereous, ftreaked with dufky lines : back and fcapulars of a brownifh grey, edged with dirty white : coverts and primaries dufky : belly white : feathers of the tail fharp-pointed and cinereous : legs black. WEIGHT near an ounce and three quarters. LENGTH eight inches.

PLACE.   Inhabits *North America*.   Abounds about *Seal Iflands*, on the *La-brador* coaft.   I do not find it among the birds of northern *Europe* ; nor in *Afia*, nearer than lake *Baikal*.

404. RUDDY.

PL. With a black ftrait bill, an inch long : head, neck, breaft, fcapulars, and coverts of wings and tail, of a ruddy color, fpot-ted with black, and powdered with white ; in the fcapulars and coverts of wings the black prevails : the outmoft web of the four firft quil feathers brown ; the internal white, tipt with brown : the upper part of the others white ; the lower brown : the two middle fea-thers of the tail brown, edged with ruft ; the others of a dirty white : legs black : toes divided to their origin.

PLACE.   Inhabits *Hudfon's Bay*.—Mr. *Hutchins*.

Br. Zool. ii. Nº 209.—*Fl. Scot.* i. Nº 157.—*Latham,* iii.
Charadrius Himantopus, *Lin. Syst.* 255.—*Hasselquist,* 253.
L'Echasse, *De Buffon,* viii. 114.—*Pl. Enl.* 878.—Lev. Mus.

405. LONG-LEG-GED.

PL. With the crown, upper part of the neck, back, and wings, dusky; the last crossed with a white line: tail of a greyish white: forehead and whole under side of the neck and body white: legs the most disproportionably long and weak of any known bird; the *French,* for that reason, call it, very justly, *l'Echasse,* or the bird that goes upon stilts; they are of a blood-red, four inches and a half long, and the part above the knees three and a half. LENGTH, to the end of the tail, thirteen inches.

Inhabits from *Connecticut* to the islands of the *West Indies* *. Is not a bird of northern *Europe.* Frequent in the southern desert of independent *Tartary:* very common about the salt lakes; and often on the shores of the *Caspian* sea. Is again found on the *Indian* shores near *Madras* †.

PLACE.

A. DOTTREL, *Br. Zool.* ii. Nº 210.—Charadrius Morinellus *Lahul,* Lappis, *Faun. Suec.* Nº 188.—*Leems Lapmark,* 260.—*Latham,* iii.
Le Guignard, *De Buffon,* viii. 87.—*Pl. Enl.* 832.—Lev. Mus.

PL. With bill and crown black: from the bill, over each eye, a white line: breast and belly dull orange; the first crossed with a white line: vent white: back, coverts of wings, and tail, olivaceous, edged

* *Sloane.* † *Raii Syn. Av.* 193.

with dull yellow: tail dufky olive; ends of the outmoft feathers white. Colors of the female duller. WEIGHT four ounces. LENGTH ten inches.

PLACE.　　Inhabits *Europe*, even as high as *Lapmark*. Firft appears in *Drontheim*; then feeks the *Lapland* alps. Returns in fmaller numbers. Appears in *May* at *Upfal*, in its paffage northward. Breeds in all the north of *Ruffia* and *Sibiria*; but appears in the temperate latitudes only in their migrations.

B. ALEXANDRINE.—Charadrius Alexandrinus, *Lin. Syft.* 253.—*Brunnich,* App. p. 77. —*Haffelquift Itin.* 256.—*Latham,* iii.

PL. With a black bill: a white line over each eye, and collar round the neck: head, upper part of body, and coverts of wings, light cinereous brown: primaries dufky; from the fifth to eighth marked with an oblong white fpot on the exterior margin: fecondaries dufky, tipt with white: middle feathers of the tail black; outmoft white: under fide of the body white: legs dufky blue. SIZE between a Lark and a Thrufh.

PLACE.　　Found in the diocefe of *Drontheim, Norway*. Common about the falt lakes between the rivers *Argun* and *Onon*; but not obferved in any other part of *Ruffia* or *Sibiria*. Inhabits alfo the canal which conveys water from the *Nile* to *Alexandria* *.

* *Haffelquift Itin.* 256.

XL.

XL. OYSTER-CATCHER. *Gen. Birds.*

Br. Zool. ii. N° 213.—*Catesby*, i. 85.—L'Huitrier. La pie de mer, *De Buffon*,   406. PIED.
viii. 119.—*Pl. Enl.* 929.—*Latham*, iii.
Hæmatopus oftralegus Strandfkjura, *Faun. Suec.* N° 192.—*Brunnich*, N° 189.—
—LEV. MUS.—BL. MUS.

O. With a long depreffed bill, cuneated at the end, and of a rich orange-color : beneath the throat fometimes a white bar : the whole neck befides, with head, back, and coverts of wings, of a fine black : wings dufky, croffed with a bar of white : under fide of the body white : lower part of the tail white ; end black : legs ftrong and thick, of a dirty flefh-color. WEIGHT fixteen ounces. LENGTH feventeen inches.

Inhabits *North America,* from *New York* to the *Bahama* Iflands; and   PLACE. again is found in *Sharks Bay,* on the weft coaft of *New Holland* *, with fome variation of color. It is met with about *Curaçoa* in the *Weft Indies* †, and wholly black, with a red bill and cinereous legs.

Found as far as *Lapmark* ‡. Inhabits all *Ruffia* and *Sibiria.* Breeds on the great *Arctic* flats : and extends to *Kamtfchatka.*

---

* *Dampier,* iii. 85.   † *Feuillée, Obferv.* ed. 1725. p. 289.   ‡ *Leems* *Lapmark,* 252.

3 R   XLI.

XLI. RAIL. *Gen. Birds.* LXXIII.

407. CLAPPER.

R With the crown, and whole upper part of neck, back, and wings, of an olive brown, edged with pale ash-color: primaries dark, edged with tawny: tail of the same color: cheeks cinereous: throat white: under side of the neck and breast brown, tinged with yellow: space beyond the thighs barred with dark cinereous and white: legs brown. LENGTH fourteen inches; of the bill two.

PLACE.

Inhabits *New York.* Called there the *Meadow Clapper.* It arrives there in *May,* lays in *June,* and disappears in *October.*—LEV. MUS.—BL. MUS.

408. VIRGINIAN.

Rallus Virginianus, *Lin. Syst.* 263.—*Latham,* iii.
American Water Rail, *Edw.* 279.—LEV. MUS.—BL. MUS.

R. With a dusky bill, red at the base of the lower mandible: crown dusky: cheeks cinereous: from the bill to each eye a white line: throat whitish: upper part of the neck and back dusky, bordered with brown: ridge of the wing white: coverts ferruginous: primaries and tail dusky: under side of the neck and breast of a brownish orange: lower belly, sides, and thighs, dusky, barred with white: vent black, white, and orange: legs of a dirty flesh-color. In size and shape like the *English* Rail *, of which it seems a mere variety; ours having a deep ash-colored breast instead of a red one.

A VARIETY OF THE ENGLISH RAIL.

PLACE.

Inhabits *Pensylvania.* The common kind is found in the *Feroe* islands, *Norway,* as far as *Sondmor, Sweden, Russia,* and the west of *Sibiria.*

* Le Rale d'Eau, *De Buffon,* viii. 154.—*Pl. Enl.* 749.

XLII. GAL-

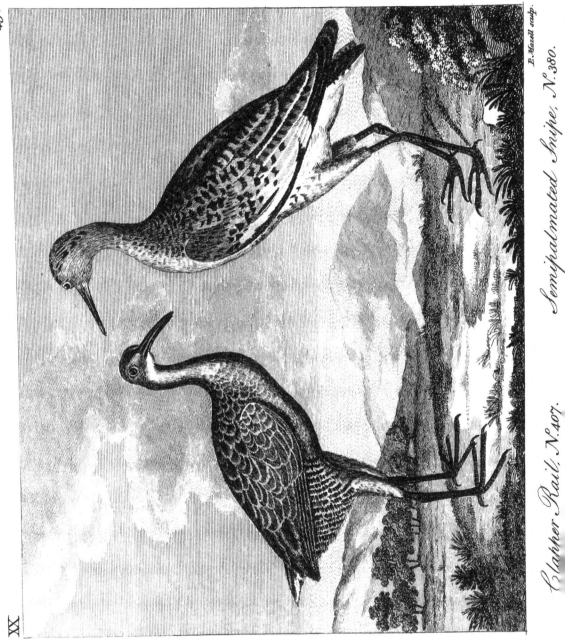

Clapper Rail. N.º 407.　　　Semipalmated Snipe. N.º 380.

XLII. GALLINULE. *Gen. Birds.* LXXV.

Little American Water Hen, *Edw.* 144.—*Latham*, iii.   409. Soree.
Soree, *Catesby*, i. 70.—Le Rale de Virginie, *De Buffon*, viii. 165.—Lev. Mus.

G. With a yellow bill: crown, hind part of the neck, back, tail, and wings, of a rufty brown, fpotted with black: coverts of the wing alone plain and more ferruginous: forehead, throat, and great part of the front of the neck, of a deep black: cheeks, fides of the neck, and breaft, of a fine blueifh afh-colour: belly and fides of a dirty white; the laft barred downwards with black: legs of a dull green. Rather larger than a Lark.

Thefe birds migrate in great numbers into *Virginia* the latter end   Place. of *September*, and continue there about fix weeks. During that fpace they are found in vaft multicudes in the marfhes, feeding on wild oats. On their firft arrival they are exceedingly lean; but foon grow fo fat as to be unable to fly. In that ftate they lie upon the reeds; and the *Indians* go in canoes and kill them with their paddles, or run them down. It is faid that they have taken a hundred dozen in a morning. They are moft delicious eating; and, during the feafon, are found on the tables of moft of the planters, for breakfaft, dinner, and fupper *.

G. With the crown and hind part of the neck dark olivaceous   410. Yellow-brown, fpotted with white: back plain brown: fcapulars   breasted. edged with yellowifh white: breaft dirty yellow: belly white: legs brown. In Size leffer than an *Englifh* Quail.

Inhabits the province of *New York.*—Bl. Mus.   Place.

---

* *Burnaby's Travels*, octavo ed. 42.

3 R 2   Br.

411. COMMON.

Br. Zool. ii. N° 217.—*Brunnich,* N° 191.—La Poule d'Eau, *De Buffon,* viii. 171.
—*Pl. Enl.* 877.—*Latham,* iii.

Fulica chloropus, *Lin. Syst.* 258.—LEV. MUS.—BL. MUS.

G. With the head and upper part of the neck, body, and coverts of the wings, of a fine deep olive green: primaries and tail dusky: breast and belly cinereous: vent white: legs green. WEIGHT of the male fifteen ounces: length fourteen inches.

Inhabits *New York,* and as low as *Carolina:* does not cross the *Baltic:* rare in *Denmark:* inhabits *Russia,* and the west of *Sibiria,* but not the east.

A. CRAKE, *Br. Zool.* i. N° 216.—Rallus crex. Angsnarpa; Kornkraka, *Faun. Suec.* N° 194.
Le Rale de Terre, Genet, ou Roi des Cailles, *De Buffon,* viii. 146.—*Pl. Enl.* 750.—
Latham, iii.   —LEV. MUS.—BL. MUS.

G. With the crown, hind part of the neck, and back, black, edged with bay: coverts of wings plain bay: tail of a deep bay: belly dirty white: legs cinereous. When lean weighs six, when fat eight ounces.

PLACE

Inhabits *Europe,* even as far north as *Drontheim.* Notwithstanding it is so short-winged a bird, and a bad flier, yet it is found in summer in the *Schetland* isles; not uncommon in the temperate parts of *Russia* and *Sibiria,* but none in the north, or towards the shores. Where Quails are common, in those countries this bird abounds; and the contrary where Quails are scarce. The Crakes depart at the same time with the Cranes. The *Tartars* observe how ill adapted the first are for a long flight; therefore believe that every Crane takes a Crake on its back, and so assists the migration*.

* *Gmelin.*

DIV.

DIV. II.

SECT. II. PINNATED FEET.

D I V. II.

SECT. II.　PINNATED FEET.

XLIII. PHALAROPE. *Gen. Birds,* LXXVI.

412. GREY.

Br. Zool. ii. N° 218.—Edw. 308.—Latham, iii.
Tringa lobata, *Faun. Suec.* N° 179.—*Muller,* N° 195.—*Brunnich,* N° 171.
Le Phalarope à feftons dentelés, *De Buffon,* viii. 226.—*Pl. Enl.* 766.—LEV. MUS.

PH. With a black bill, flatted and dilated near the extremity: eyes placed far back: white forehead: crown dufky: upper part of the neck light grey: back, rump, and fcapulars, deep dove-color, marked with dufky fpots: edges of the fcapulars yellow: coverts and primaries dufky; the firft edged with white: breaft and belly white: tail dufky, edged with afh-color: legs black: toes fcolloped; the margins of the membranes finely ferrated. SIZE of a Purre.

PLACE.

Inhabits *Scandinavia, Iceland,* and *Greenland:* in the laft, lives on the frozen fide, near the great lakes: quits the country before winter: is feen on the full feas in *April* and *September,* in the courfe of its migration. Swims flowly: cannot dive. Twitters like a Swallow. The feathers being very foft, the *Greenlanders* ufe it to wipe their rheumy eyes. Is not found in *Ruffia,* but is frequent in all *Sibiria,* about the lakes and rivers, efpecially in autumn; probably in its migration from the *Arctic* flats: it was alfo met with among the ice, between *Afia* and *America.*

413 RED.

Br. Zool. ii. N° 219.—Edw. 142, 143.—Latham, iii.
Tringa fulicaria, *Faun. Suec.* N° 179.—*Brunnich,* N° 172.—*Muller,* N° 196.
　—*Faun. Groenl.* N° 76.
Le Phalarope rouge, *De Buffon,* viii. 225.

PH. With bill in form of the laft: crown, hind part of the neck, and upper part of the breaft, of a dufky afh-color: fides of the neck ferruginous: throat, belly, and vent, white: wings black: greater

coverts

coverts and fecondaries tipt with white: back and fcapulars dufky; the laft edged with bright ferruginous: rump white, barred with cinereous: tail dufky: toes like the former. The whole under fide of the neck, the breaft, and belly, of the fuppofed FEMALE are ferruginous. *Linnæus* calls the male *Tringa Hyperborea*, and feparates them *.

PLACE.

Found in *Hudfon's Bay* and *Scandinavia*; common about the *Cafpian* fea, and lakes and rivers adjacent, during fpring; but does not extend to the farther part of *Sibiria*; yet was found by the navigators between *Afia* and *America*. They go in pairs. Swim in the ponds of the fens; and are perpetually dipping their bills in the water in fearch of infects.

Edw. 46.—*Latham,* iii.

414. BROWN.

PH. With a flender black bill, a little bending at the end: crown black: cheeks and neck of a light afh-color, tingedwith bloom-color: breaft and belly white: back, wings, and tail, dufky: greater primaries and greater coverts tipt with white: legs like the preceding.

PLACE.

Taken on board a fhip off the coaft of *Maryland*, the wind blowing from land. The form of the bill is a fpecific diftinction from the laft.

PH. With a flender black bill, dilated at the end: crown dufky and dull yellow: acrofs each eye a black line: cheeks and fore part of the neck a pale clay-color: breaft and belly white: back and tertials dufky, edged with dull yellow: coverts, primaries, and tail, cinereous; the laft edged like the tertials: legs yellowifh: toes bordered with a plain or unfcolloped membrane.

415. PLAIN.

Taken in the frozen fea, about *Lat.* 69 ½. *Long.* 191 ½.

PLACE.

* *Syft. Nat.* 249.

XLIV.

XLIV. C O O T. *Gen. Birds.* LXXVII.

COMMON.

Br. Zool. i. N° 220.—La Foulque, *De Buffon,* viii. 211.—*Pl. Enl.* 197.—*Latham,* iii.

Fulica atra, *Faun. Suec.* N° 193.—*Brunnich,* N° 196.—LEV. MUS.—BL. MUS.

C. With a white bill: head, neck, body, wings, and tail, of a full black: legs yellowish green. WEIGHT from twenty-four to twenty-eight ounces.

PLACE.

Inhabits the shores of *Sweden* and *Norway:* appears in spring, and very rarely visits the lakes or moors *. Frequent in *Russia,* and even to the east of *Sibiria.* I found it among the birds sent to Mrs. *Blackburn* from *North America.* The *Indians* about *Niagara* dress their skins, and use them for pouches. They are frequent in the rivers of *Carolina;* where they are called *Flusterers* †; I suppose, from the noise they make in flying along the surface of the water.

* *Aman. Acad.* iv. 591.   † *Lawson,* 149.

XLV.

XLV. GREBE. *Gen. Birds*, LXXVIII.

Eared or horned Dobchick, *Edw.* 49.—*Latham*, iii.

GR. with the head very full of feathers, and of a mallard green color: from each eye issues a long tuft of yellowish orange-colored feathers, almost meeting at the hind part of the head; beneath them is a large ruff of black feathers: fore part of the neck and breast of an orange red; the hind part and back dusky: coverts of the wings cinereous: primaries and tertials black: secondaries white: belly glossy and silvery: legs of a bluish ash-color before; flesh-colored behind. Of the SIZE of a Teal. Not the male of my Eared Dobchick, as Mr. *Edwards* supposes*; there being in that species no external difference of sexes, as I have had frequent opportunities of observing.

Inhabits *Hudson's Bay*: appears there in the fresh waters in *June*: lays its eggs among the aquatic plants. Retires towards the south in autumn: is called by the natives, *Sekeep*. It appears in *New York* in that season, and continues there till spring, when it returns to the north. For its vast quickness in diving is called, in *New York*, the *Water Witch*.

Colymbus podiceps, *Lin. Syst.* 223.—*Latham*, iii.
Pied-bill Dobchick, *Catesby*, i. 91.—LEV. MUS.—BL. MUS.

GR. with a strong arched bill, not unlike that of the common poultry; of an olive-color, crossed through the middle of both mandibles with a black bar: nostrils very wide: chin and throat of a glossy black, bounded with white: upper part of the neck and back dusky: cheeks and under part of the neck pale brown: breast silvery, mottled with ash-color: belly silvery: wings brown: ends of the

* See *tab.* 96. *Edw.* and my account of that species, *Br. Zool.* ii. N° 224.

3 S

secondaries

fecondaries white : toes furnifhed with broad membranes. The female wants the black bar on the bill. Length fourteen inches.

PLACE.　　Inhabits from *New York* to *South Carolina* : is called in the firft, the *Hen-beaked Wigeon*, or *Water Witch*. Arrives there late in the autumn, and goes away in *April*.

419. Louisiane.　Le Grebe de la Louifiane, *De Buffon*, viii. 240.—*Pl. Enl.* N° 943.—*Latham*, iii.

GR. The end of the bill flightly bent : middle of the breaft white, tinged with dufky : fides of the neck and body, quite to the rump, ruft-colored : from the bafe of the neck to the thighs marked with large tranfverfe black fpots : upper part of the body and wings deep brown : legs dufky. In Size rather lefs than the common Grebe.

PLACE.　　Inhabits *Louifiana*.

420. Dusky.　　*Br. Zool.* ii. N° 225.—*Edw.* 96. fig. 1.—Le petit Grebe, *De Buffon*, viii. 232.— *Pl. Enl.* 942.—*Latham*, iii.　　—Lev. Mus.—Bl. Mus.

GR. With the crown, neck, back, and primaries, dufky : ridge of the wings and fecondaries white : reft of the wings dufky : breaft and belly filvery, but clouded. Size of a Teal.

PLACE.　　Sent from *New York* with the *Horned Grebe*, as its female ; but is certainly a diftinct fpecies.

A. Great Crested Grebe, *Br. Zool.* ii. N° 223.—Le Grebe cornu, *De Buffon*, viii. 235.—*Faun. Suec.* N° 151.—*Latham*, iii.　　—Lev. Mus.—Bl. Mus.

GR. With the cheeks and throat furrounded with a long pendent ruff, of a bright tawny color : on the head a great dufky creft : hind part of the neck and back dufky : primaries of the fame color : fecon-

daries

daries white: breaſt and belly of a gloſſy ſilvery white: outſide of the legs duſky; inſide greeniſh. WEIGHT two pounds and a half. LENGTH twenty-one inches. EXTENT thirty.

Inhabits *Iceland*, northern *Europe*, and the temperate and northern parts of *Sibiria*, in every reedy lake. PLACE.

B. EARED, *Br. Zool.* ii. N° 225.—*Edw.* 96. fig. 2.—*Latham*, iii.
    Colymbus Auritus. Fiorna. Skrænlom, *Faun. Suec.* N° 152.—LEV. MUS.—BL. MUS.

GR. With crimſon irides: behind each eye a large tuft of ferruginous feathers. the head, upper part of the body, and primaries, duſky: ſecondaries white: whole under ſide white: feathers above the thighs ferruginous. LENGTH, to the rump, twelve inches.

Inhabits *Iceland* *, *Norway*, and *Sweden*, and alſo the lakes of *Lapland*, where it makes a floating neſt: quits thoſe countries in winter. Common in *Sibiria* and *Ruſſia*, in all latitudes. Inhabits *England* the whole year. I have ſeen numbers ſhot in *Lincolnſhire*. Could obſerve no external ſexual differences; ſo am certain they are not of the ſame ſpecies with the *Horned Grebe*, N° 417. PLACE.

C. RED-NECKED GREBE.

GR. With the crown, hind part of the neck, back, and wings, duſky brown: ſecondaries white? cheeks and throat white; the firſt marked with a few brown ſtreaks: under ſide of the neck bright ferruginous: belly white: legs duſky.

This ſpecies was ſent to me by the late Mr. *Fleiſcher* of *Copenhagen*, from either *Denmark* or *Norway*. The ſame ſpecies is found, but very rarely, towards the *Caſpian* ſea †. PLACE.

* *Olaffen.* † *Doctor* PALLAS.

DIV. II.

SECT. III. WEB-FOOTED.

D I V.   II.

S E C T.   III.   W E B - F O O T E D.

•W I T H   L O N G   L E G S.

XLVI. A V O S E T. *Gen. Birds*, LXXIX.

421. AMERICAN.

AV. With a slender black bill, slightly turning up : head, neck, and upper part of the breast, of a pale buff-color : rest of the lower part of the body white : back and primaries black : lesser coverts white ; greater black ; beneath which is a long transverse bar of white : legs very long, and dusky : feet semipalmated ; the webs bordering the sides of the toes for a considerable way.   Larger than the *English* AVOSET.

PLACE.

This species is preserved in the LEVERIAN MUSEUM.   It is a native of *North America*.   I imagine that it sometimes is found entirely white ; for the bird called by Mr. *Edwards* (tab. 139.) the *White Godwit* from *Hudson's Bay*, seems to be the same with this.

A. TEREK. Scolopax cinerea, *Nov. Com. Petrop*, xix. 473. tab. xix.—*Latham*, iii.

AV. With a bill one inch ten lines long, slightly recurvated : whole upper part of the plumage cinereous : the middle of each feather marked with dusky ; on the rump in form of bars : lower part

of

XXI

P. Brown del.

P. Mazell sculp.

American Avoset. N.º 421.

of body white : tail cinereous ; outmoſt feather on each ſide varied with white and aſh-color : legs ſhort, ſemipalmated. Size of the *Red Sandpiper*.

Frequents and breeds in the north * ; and haunts, during the ſummer, the *Caſpian* ſea. Migrates through *Ruſſia* and *Sibiria*.

Place.

B. Scooping, *Br. Zool.* ii. Nº 228.—Skarflacka. Alſit, *Faun. Suec.* Nº 191.—*Amœn. Acad.* iv. 591.—*L'Avocette, De Buffon,* viii. 466.—*Pl. Enl.* 353.—*Latham,* iii. —Lev. Mus.

A V. With a black, thin, flexible bill, three inches and a half long, bending upwards half its length : head, hind part of the neck, and part of the wings and ſcapulars, black : reſt of the neck, breaſt, belly, and tail, white. Weight thirteen ounces. Length, to the end of the tail, eighteen inches. Extent thirty.

Inhabits, in *Scandinavia*, only the iſle of *Oeland*, off *Sweden* ; where it rarely appears in the ſpring. Is properly a ſouthern bird. Very frequent, in the breeding ſeaſon, about *Foſſdike Waſh*, in *Lincolnſhire* : are then very eaſily ſhot, flying about one's head like the Lapwing, repeating ſhrilly, *twit, twit*. Lay two eggs, white, tinged with green, and ſpotted with black. Are frequent about the ſalt lakes of the *Tartarian* deſart, and about the *Caſpian* ſea.

Place.

* *Doctor* Pallas.

XLVII. FLAMMANT. *Gen. Birds,* LXXXI.

422. RED.　Flamingo, *Catesby,* i. 73.—Phænicopterus ruber, *Lin. Syst.* 230.—Le Flammant, ou le Phenicoptere, *De Buffon,* viii. 475.—*Pl. Enl.* 63.—*Latham,* iii. —LEV. MUS.

FL. With the upper mandible extremely convex, ridged, and sloping to a point; the under very deep, and convex on the upper part; the edges of both furnished with numerous teeth: space between it and the eyes naked: neck very long: tail short: primaries black: all the rest of the plumage of a fine crimson color; but the *European* birds, which are the only I have seen, are roseate: legs very long: toes webbed: hind toe very small. The attitude is erect: and its HEIGHT usually five feet.

PLACES IN AMERICA.　These birds inhabit *Louisiana* *, the *Bahama* islands, and those of the *West Indies*; and frequent only salt waters. They live in flocks, and are uncommonly tame, or rather stupid. A person who can stand concealed may shoot as many as he pleases; for they will not arise at the report of the gun; but the survivors will stand as if astonished: nor will they take warning at the sight of the slain; but continue on the spot till most of them are killed. Such is *Catesby*'s account. But the honest and intelligent *Dampier* gives a very different one; and says that they are extremely shy, and very difficult to be shot.

NESTS.　They build their nests in shallow ponds; and form, for that purpose, with mud, hillocks with a broad base, which appear about a foot and a half above the water: these taper to the top, in which the birds make a hollow for the eggs. They lay two, and cover them with their rump, their legs resting beneath the water. As soon as the young are hatched they run very fast, but cannot fly till they are full grown †. They are for a long time of a grey color, nor do they attain that of red till near two years.

* *Du Pratz,* ii. 81.　　　　† *Voyages,* i. 71.

They

They ftand upright, and in a row; fo at a diftance look like a file of foldiers.   They feed on the feeds of water-plants : not fifh.   Their flefh is good, notwithftanding it is lean, and looks black.   *Dampier* alfo bears witnefs to the delicacy of the tongues ; which, fays he, are large, and furnifhed with a knob of fat at the root, the fo much boafted morfel.   *Apicius,* a *Roman,* probably cotemporary with *Tiberius,* had the honor of firft introducing them to table : the fame perfon whom *Pliny* fo forcibly ftigmatifes with the title of *Nepotum altiffimus gurges* *.

The FLAMMANT inhabits feveral parts of the old world : fuch as fome of the coafts of the *Mediterranean* fea ; the *Cape De Verd* iflands ; and the *Cape of Good Hope.*   They are common on the fhores of the *Cafpian* fea, on the *Perfian* and *Turcomannian* coafts : repairing in flocks to the mouth of the river *Yemba* ; and fometimes to that of the *Yaik* ; and alfo to that of the *Volga,* below *Aftracan.*

PLACES IN EUROPE AND ASIA.

* *Lib.* x. c. 48.

• W I T H

WITH SHORT LEGS.

XLVIII. ALBATROSS. *Gen. Birds,* LXXXII.

423. WANDERING.   Diomedea Exulans, *Lin. Syft.* 214.—*Pl. Enl.* 237.—*Latham,* iii.
Albatrofs, *Edw.* 83.—*Pallas, Spicil. Zool. Fafc.* v. 28.
Tfchaiki of the Kamtfchatkaas, LEV. Mus —BL. Mus.

ALB. With a ftrong bill, finking a little in the middle; hooked at the end of the upper mandible, abrupt at that of the lower; noftrils covered with a ftrong guard, and opening forward; color red; tip dufky: the plumage, in fome, wholly dufky, with the color moft intenfe on the upper part; others again have their under fide entirely white: the tail is rounded: legs and feet of a dufky red: the webs dufky.

SIZE. ALBATROSSES differ greatly in fize. Whether they differ in fpecies I cannot determine. They weigh from twelve to twenty-eight pounds: and vary in extent of wings, from feven feet feven inches to ten feet feven.

PLACE. The white and the brown variety or fpecies appear annually in flocks of thoufands, about the end of *June,* and fpread over the whole *Ochotfchan* fea, the gulph of *Penfchinfhi,* and the *Kurile* iflands; but very rarely on the eaftern coaft of *Kamtfchatka.* They alfo arrived in great numbers about *Bering*'s Ifland, at the time when *Steller* was preparing to depart from his long confinement, after the fhipwreck of his illuftrious commander. He failed from thence on the 10th of *Auguft.* This coincides with the re-migration of thefe birds, who retire from the former places about the end of *July* or the middle of *Auguft.* Their arrival is the certain forerunner of fifh. It is probable that they purfue their prey northward, as they do not return the fame way. They fpread to the coafts of *America* *, and tend

* Seen the 4th of *July,* in lat. 56. 30, off the weftern coaft of *America.*—*Ellis's Voy.* i. 292.

from

from each continent to their breeding-places in the fouthern hemifphere, which they may arrive at by the feafon of fummer in that
adverfe part of the globe.

They feek the northern fhores, in purfuit of the vaft fhoals of fal-
mon which frequent thofe diftant places. They are the moft voracious of birds; and will fo fill themfelves with fifh, that fometimes
a large one will hang half out of their mouths till thofe in their
ftomach are digefted. They will be at times fo loaden with food
as to become incapable of flying; and even fo ftupified, as to be
readily hunted down by boats, or transfixed in the water by darts:
neither can they arife till they have vomited up their prey, which
they ftrive to do with all their might.

MANNERS.

The *Kamtfchatkans* are very folicitous about the capture of thefe
birds, not fo much for the fake of their flefh (which is very tough
and dry, and never eaten unlefs hunger compels) but on account of
the inteftines, which they blow into bladders, in order to form
floats for their nets. They angle for the Albatroffes as they do for
the fcaly race, baiting with a whole fifh, a large hook fixed to a long
cord. This they fling into the water, when there is an inftant con-
teft among thefe greedy birds, which fhall firft lay hold of it *

CAPTURE.

They have only the veftige of a tongue; which is one of the characters of the Gannet, Corvorants, and other voracious birds. Their
voice, like that of the Pelecan, refembles the braying of an afs.

The voyage which thefe birds undertake, from perhaps almoft the
extremity of the fouthern hemifphere to that of the northern, urged by
inftinct, to attend the migration of certain fifhes, is very amazing.
They are, indeed, feen in *April* and *May* off the *Cape* of *Good Hope* †,
fometimes foaring in the air with the gentle motion of a Kite, at a
ftupendous height; at others, nearer the water, watching the motions of the Flying Fifh, which they catch while thofe miferable be-
ings fpring out of their element to fhun the jaws of the *Coryphenes*.

MIGRATIONS.

* *Hift. Kamtfchatka*, Engl. edit. 155.   † *Dampier*, i. 531.—*Ofbeck*, i. 109.

I have not authority to say that some of them do not reside about the *Cape* the whole year: but I am acquainted with only two places in which they breed; one is the *Falkland* islands, the other the coast of *Patagonia* *. In the first, they begin to lay their eggs in *October*, the spring of those regions, and continue that function about a

NESTS AND EGG.   month. They build their nests with sedges, in form of a haycock, about three feet high, leaving a hollow in the summit for the egg; for they lay but one, which is four inches and a half long, white, with some small obscure spots at the thicker end. They are constantly watched while on their nests by multitudes of Hawks, who no sooner see the Albatros quit its nest, but they instantly dart down and carry off the egg. This obliges them to lay another, and prolong the season of incubation.

The remaining part of the summer they wander over all parts of the *Antarctic* seas; and were seen as low as lat. 67. 20. in the middle of our *January* †; and have been seen in several successive months to the northward, shunning the winter of their native hemisphere, and seeking warmth and food in the remote climate of *Kamtschatka*.

STELLER takes notice of some birds which the *Russians* rank under the name of *Gloughpichi*. He says they are found in great numbers on the isles between *Asia* and *America*; that they were of the size between an Eagle and a Goose, had a yellow crooked bill, and their plumage of the color of umber, spotted with white. He also saw numbers feeding on a dead whale. And in crossing the *Penchian* sea, he observed several flying: some white, others black ‡. All these I suspect to have been different sorts of ALBATROSSES, which may have wandered here; for the *Antarctic* voyagers observed at left three species in their approaches towards the southern pole ‖.

---

* This account is given by a distinguished officer in our navy, who had visited these islands.

    † *Cook's Voy. S. Pole,* i. 256.   ‡ *Descr. de la Kamtschatka,* 492.   ‖ *Cook's Voy. towards the S. Pole,* i. 43, 256, 258.

XLIX.

XLIX. A U K. *Gen. Birds*, LXXXIII.

Br. Zool. ii. N° 229.—*Edw.* N° 147.—*Latham*, iii.
Alca Impennis, *Faun. Suec.* N° 140.—*Brunnich*, N° 105.—*Muller*, N° 139.—
LEV. MUS.

224. GREAT.

A. With a ſtrong bill, bending at the end; black, and furrowed tranſverſely: between that and the eyes, a bean-ſhaped white ſpot: above, the whole plumage is of a gloſſy black; the ſecondaries alone tipt with white: breaſt and belly white: wings very ſhort, and uſeleſs for flight, covered with very ſhort feathers: legs black. To the end of the toes, three feet.

Inhabits (but not very frequently, the coaſts of *Norway*) the *Feroe* iſlands (in a certain number of years, *St. Kilda*) *Iceland*, *Greenland*, and *Newfoundland*. It lives chiefly on the ſea; but never wanders beyond the ſoundings. Feeds much on the *Lump-fiſh*, *Br. Zool.* ii. N° 57. and *Father Laſher*, *Br. Zool.* iii. N° 99. and other fiſh of that ſize. Builds on rocks remote from ſhore. Lays one egg, ſix inches long, irregularly marked with purpliſh lines on a white ground, or blotched near the thicker end with black or ferruginous ſpots. Hatches late. The young, in *Auguſt*, are only covered with grey down. Their food, at that period, is vegetable, the *Rhodiola Roſea*, and other plants, having been found in their ſtomachs. The *Greenlanders* uſe the gullet as a bladder to ſupport their darts *: and I think I have ſeen ſome habits of the *Eſkimaux* made of the ſkins.

PLACE.

Br. Zool. ii. N° 230.—Le Pingoin, *Pl. Enl.* 1004, 1005.—*Latham*, iii.
Alca Torda, *Faun. Suec.* N° 139.—LEV. MUS.—BL. MUS.

425. RAZOR-BILL.

A. With a white line from the bill to the eyes: bill thick, bent at the end, croſſed with tranſverſe groves; the largeſt white, and paſſing over each mandible: head, back, wings, and tail black:

* *Faun. Groenl.* p. 82.

9

ſecondaries

fecondaries tipt with white: breaft and belly white: legs black. WEIGHT near twenty-three ounces. LENGTH eighteen inches. EXTENT twenty-feven.

PLACE.   Inhabits the north of *Europe, Iceland, Greenland,* and the coaft of *Labrador* *. Extends in *Europe,* along the White fea, into the *Arctic Afiatic* fhores; and from thence to *Kamtfchatka* and the gulph of *Ochotfk,* wherever there are lofty rocks †. It is the only one of this fpecies which reaches the inland *Baltic.* It is found there on the *Carls-Ozar* ifles, near *Gothland,* and the ifle of *Bondon* off *Angerman-land.*

426. BLACK-BIL-   *Br. Zool.* ii. N° 231.—Alca Pica, *Lin. Syft.* 210.—*Latham,* iii.
LED.   Alca Unifuleata, *Brunnich,* N° 102.—*Muller,* N° 138.

A. With a black bill, marked with one furrow: whole upper fide of the body black; the under, from chin to tail, white. WEIGHT only eighteen ounces. LENGTH fifteen inches and a half. EXTENT twenty-five.

PLACE.   Inhabits the north of *Europe,* and the fame countries with the former. It extends farther fouth than any of the genus; being found on the coafts of *Candia,* the antient *Crete;* where it is called *Vutha-maria,* and *Calicatczu* ‡. In *Greenland* neither of them frequent the bays till the intenfe cold fets in; but live in the fea, reforting, in the

---

* Doctor PALLAS; to whom it was fent by a *Moravian* miffionary.

† In the *Britifh Zoology,* ii. N° 230, I made matter of wonder, the manner in which this bird placed its egg on the naked rock, with fo fecure a balance that it would not roll off. Mr. *Aikin* referred me to the following paffage in HARVEY *de Generatione Anim.* which moft clearly explains the caufe: " In the fame ifland" (the *Bafs),* fays he, " una mihi monftratur avis, quæ ovum duntaxat fingulare, five unicum, parit, " idemque fuper cujufdam lapidis acuti faftigium collocat (nullo nido, aut conquifitâ " ftrue fuppofitâ), idque tam firmiter, ut mater abire & redire, falvo ovo, poffit. Hoc " autem fi quis loco dimoveat, nullâ arte poftea ftabiliri poteft; quin inde devolutum " præceps in mare ruat. Locus nempè (ut dixi) cæmento albo incruftatur; ovumque, " cum nafcitur, lentâ & vifcofâ madet humiditate, quâ citò concrefcente, tanquam " *ferrumine* quodam fubftrato faxo agglutinatur."

‡ *Belon. Obf.* 12.

breeding

breeding feafon, to the cliffs in flocks, where each knows its ftation. Feed on the *cancer pedatus*, and other marine infects; and grow very fat. In winter, refort to the bays to feed; but at night return to fea. Vie with the *Eider-duck*, in point of utility to the *Greenlanders*. The fkins are ufed for cloathing: the raw fat is fucked as broth: the flefh, half putrid, is much admired: and the whole fowl, dreffed with the inteftines in it, efteemed a high delicacy.

They are taken in the fea with darts; or, chaced in canoes, are driven on fhore, and killed by the perfons who wait for them; or are taken in nets made of fplit whalebone. They are the chief food of the natives during *February* and *March* *.

The *Alca Balthica* of *Brunnich*, N° 115, is a variety of thefe birds, only wants the white line from the bill to the eyes.

Br. *Zool.* ii. N° 232.—*Edw.* 358.—Le Macareux, *Pl. Enl.* 275.—*Latham*, iii. Alca Arctica, *Faun. Suec.* N° 141.—LEV. Mus.—BL. Mus.

427. PUFFIN.

A. With a fhort bill, deep at the bafe, ridged, triangular, ending in a fharp point; bafe ftrongly rimmed; upper part blueifh grey; lower red; both furrowed tranfverfely: crown, and upper part of the body, wings, and tail, black: cheeks white, bounded by grey: breaft and belly white: legs orange-colored. WEIGHT twelve ounces. LENGTH twelve inches. EXTENT twenty-one.

Inhabits all the coafts of northern *Europe*, the icy fea, and all the way to *Kamtfchatka*; where they are larger and blacker than ufual, and their crown cinereous. Found in the *Feroe* ifles, where they are called *Lunda*; extends to *Iceland*, *Greenland*, and *Spitzbergen*. *Catefby* enumerates this fpecies, and fays that the GREAT AUK, and RAZOR-BILL, frequent the coafts of *Carolina* during winter †. In the fame feafon, numbers of thefe birds, and the RAZOR-BILLS, frequent the coaft of *Andalufia*; but difappear in the fpring. It is fuppofed that they continue fwimming from the northern parts in fearch of food; the fifh of the fofter latitudes not retiring to the great deeps out of their reach, as is the cafe with the fifh of more rigorous climates.

PLACE.

* *Crantz*, i. 48. † *Catefby*, App. xxxvi.

A. With

428. LABRADOR. A. With a bill about an inch and a quarter long, much carinated
at top, not very deep, a little convex; upper mandible dufky,
lower whitifh, marked with a black fpot, and angulated like that
of a Gull: crown, and upper part of the body, wings, and tail,
dufky: lower part white: legs red. SIZE of the former.

PLACE.   Inhabits the *Labrador* coaft?—BR. MUS.

429. LITTLE.   *Br. Zool.* ii. N° 233.—*Edw.* 91.—*Latham*, iii.
Alca Alle, Faun. Suec. N° 142.—*Brunnich*, N° 106.—Rot-ges, or Rottet, *Marten's
Spitzb.* 85.—LEV. MUS.

A. With a fhort, black, convex, and thick bill: whole upper part
black: cheeks and lower parts white: fcapulars ftreaked
downwards with white: legs dirty greenifh white: webs black: throat
of the male is black. In SIZE not fuperior to that of a Blackbird.

PLACE.   Inhabits the north of *Europe*, as far as *Spitzbergen*; but I believe
does not extend to *Afia*. Frequent in *Greenland*. Dives well. Is
always putting its bill to the water as if drinking. Grows very fat
in ftormy weather, when the waves bring fmall Crabs and little fifh
near the furface. Is called in *Newfoundland* the *Ice-bird*, being the
harbinger of ice *. Varies to quite white; and fometimes is found
with a reddifh breaft. Is called by the *Dutch, Rottet,* from its note.

430. ANTIENT. A. With a black bill, crown, and throat: on each fide of the
head a fhort whitifh creft: on the hind part of the neck are
numbers of white, long, loofe, and very narrow feathers, which give
it an aged look: wings, back, and tail, footy: breaft and belly
white. SIZE of the former.

PLACE.   Inhabits from the weft of *North America* to *Kamtfchatka* and the
Kurile iflands.—PALLAS, MS.—LEV. MUS.

* *Crantz,* i. 85.

A. With

A. With the bill black and ridged : crown, upper part of the neck, back, wings, and tail, dufky : under fide of the neck and breaft of a deep iron grey : belly of a dirty white : legs dufky. SIZE of the *Water Ouzel*.

431. PYGMY.

Seen in vaft multitudes about *Bird Ifland*, between *Afia* and *America*.

PLACE.

Alca Cirrhata, *Pallas, Spicil. Zool.* fafc. v. 7. tab. i.—*Latham*, iii. —LEV. MUS.

432. TUFTED.

A. With a ftrong thick bill, of a fub-triangular form, arched, hooked near the end ; the lower mandible truncated ; the upper mandible, near the bafe, rifing into a diftinct prominence : the bill of the male marked with three furrows ; of the female with two : its colors a fine red, yellow, and corneous : from the fides of the head are two long filky tufts of a yellow color, falling down the fides of the neck to the back : cheeks white : the reft of the plumage entirely black ; lighteft beneath : legs of a bright red. In SIZE fuperior to the *Puffin Auk*.

Inhabits only the fhores of *Kamtfchatka*, the *Kurile* iflands, and thofe intervening between *Kamtfchatka* and *America*.

PLACE.

In manners this fpecies greatly refembles the Puffin. Lives all day at fea, but at no great diftance from the rocks : comes on fhore at night : burrows a yard deep under ground, and makes there a neft, with feathers and fea plants : is monogamous, and lodges there the whole night with its mate. Lays one white egg, in the latter end of *May* or beginning of *June*. Bites fiercely when taken. Feeds on Crabs, Shrimps, and fhell-fifh, which it forces from the rocks with its ftrong bill.

MANNERS.

The females of *Kamtfchatka* copy from this bird the fafhion of hanging, from behind each ear, tufts made of flips of the white part of the fkins of the *Glutton*. Thefe are reckoned the moft valuable

3 U

prefent

prefent a lover can give his miftrefs, or a hufband to his wife, and the ftrongeft proof of affection.

Their magicians formerly ufed to recommend the bills of this bird, mixed with thofe of the Puffin, and the parti-colored hairs of Seals, as a powerful amulet. The diftant iflanders ftill bear the bills in their helmets and caps, and make their garments of the fkins; but the *Kamtfchatkans* at prefent make no ufe of any part but the eggs, which are a common food; the flefh being hard and infipid.

433. PERROQUET.   Alca Pfittacula, *Pallas, Spicil. Zool.* fafc. v. 15. tab. ii.—*Latham,* iii.

A. With an oval bill, or the upper and lower parts convex, and of a bright red color : from the remote corner of each eye is a very flender tuft of fine white feathers, hanging down the neck : the head and upper part of the body dufky ; the lower whitifh, varied with black edges : legs dirty yellow : webs dufky. About the SIZE of the Little Auk.

PLACE.   Inhabits *Kamtfchatka,* the ifles towards *Japan,* thofe towards *America,* and the weftern fhores of *America,* in great abundance. They fwim in flocks; but never, unlefs tempeft-driven, go far from the rocks, to which they refort towards night, and fhelter themfelves in the fiffures or holes, without keeping any certain neft. Are the moft ftupid of all birds, and caught by the natives in this ridiculous manner :—towards evening they put on their garment with great fleeves, pull out their arms, and leave the fleeves diftended, which the birds will creep into by flocks, and thus become an eafy prey.

The ftupidity of this fpecies has often been the falvation of mariners failing by night in thefe dangerous parts ; being often warned of the neighborhood of a dreadful rock, by thefe birds flying on board their veffel, miftaking it for their defigned lodging. They lay one egg, uncommonly great for their fize ; it is of a whitifh color, fpeckled with brown, dufky, or yellow ; and is efteemed for its delicacy.

Alca.

Alca Criftatella, *Pallas Spicil. Zool.* fafc. v. 18. tab. iii.—*Latham,* iii.

A. With a fcarlet bill : upper mandible convex, and end hooked; and near the angle of the mouth a fcarlet heart-fhaped fub-ftance : on the forehead rife fome upright feathers; and above that a fine creft of longer feathers, curling forwards : head and neck black : behind each eye hang a few narrow feathers : back black, marked with dufky ferruginous ftrokes : rump hoary : wings footy : under fide of the body of a dufky cinereous caft : legs livid : webs dufky. SIZE of the Miffel Thrufh.

Frequent on *Bird Ifland,* between *Afia* and *America.*

Alca Tetracula, *Pallas Spicil. Zool.* fafc. v. 23. tab. iv.—*Latham,* iii.

A. With a fmall arched bill, dufky and yellow : above the bafe, on the forehead, the feathers divide into two points : beyond each eye is a whitifh defcending line, in which are a few fetaceous feathers : general color of the bird dufky : belly hoary : on the hind part of the head and neck, and near the tail, a few rufty marks : tail very fhort, dufky, and fome of the feathers tipt with ferruginous : legs livid : webs intenfely black. LENGTH eleven inches. EXTENT eighteen.

Frequent about *Kamtfchatka,* the *Kurile* ifles, and thofe fituated towards *America.* Live in flocks on the rocks; but appear at fea folitary : moft ftupid, and clumfy. Can fcarcely fly; or ftand, except they reft againft the rocks, where they lodge in the fiffures the whole night, or in burrows, which they make with great facility. They fwim and dive admirably well. Are exceedingly bad food; but are eaten by the almoft-famifhed natives.

L. GUIL-

L. GUILLEMOT. *Gen. Birds,* LXXXIV.

436. FOOLISH.   *Br. Zool.* ii. N° 234 —Colymbus Troille, *Faun. Suec.* N° 149.—*Latham,* iii.
—LEV. MUS.—BL. MUS.

G. With the bill three inches long : head, neck, back, wings, and tail, of a deep moufe-color : fecondaries tipt with white : breaft and belly pure white : legs dufky. WEIGHT twenty ounces, LENGTH feventeen inches. EXTENT twenty-feven and a half.

PLACE.   Inhabits all parts of the north of *Europe,* to *Spitzbergen* ; the coaft of *Lapmark,* and along the white and icy fea, quite to *Kamtfchatka* and *North America.* Found in *Newfoundland.* Not mentioned among the birds of *Greenland.* Is a fpecies that winters on the coaft of *Italy* * ; poffibly thofe which quit *England* before that feafon.

437. BLACK.   *Br. Zool.* ii. N° 236 —Colymbus Grylle, *Faun. Suec.* N° 148.—*Latham,* iii.
—LEV. MUS.—BL. MUS.

G. With the bill an inch and a half long : infide of the mouth of a fine red : bill, head, neck, back, tail, and under fide of the body, black : wings dufky ; the coverts marked with a bed of white ; fecondaries tipt with white : legs fcarlet. LENGTH fourteen inches. EXTENT twenty-two.

PLACE.   Found in all the fame places with the former, except *Italy* ; doubtful as to *Newfoundland.* Inhabits *Hudfon's Bay* the whole year. The young are mottled with black and white, and fometimes are quite white †. Are excellent divers ; therefore called fometimes *Diving Pigeons.* Make a twittering noife. The *Greenlanders* eat the flefh, ufe the fkin for cloathing, and the legs as lures for fifh.

* *MS. Lift. of Birds of Italy,* fent to me by that eminent Ornithologift, M. SCOPOLI, from *Pavia.*

† Mr. *Hutchins* affures me, that the old birds do not vary, as has been imagined.

XXII

Red billed Grebe. N.º 418.   Marbled Guillemot. N.º 438.

P. Mazell sculp.

In *Kamtfchatka* is a variety, with a white oblique line iffuing from the white fpot on the wings.

G. With a black bill: crown dufky: throat, breaft, and belly, mottled with black and white: back and fides very gloffy, and marbled with black and ruft-color: wings dufky; greater co-verts edged with white: tail black: legs yellow: webs black. LENGTH nine inches.

Inhabits *Prince William's Sound*, on the weftern coaft of *North* *America*, and probably *Kamtfchatka*.—LEV. MUS.

LI. DIVER.

LI. DIVER. *Gen. Birds,* LXXXV.

439. NORTHERN. *Br. Zool.* ii. N° 237.—L'Imbrim, ou grand Plongeon de la Mer de Nord, *De Buffon,* viii. 258.—*Pl. Enl.* 952.
Colymbus Glacialis, *Lin. Syſt.* 221.—*Latham,* iii.　　　—LEV. MUS.—BL. MUS.

D. With head and neck black: throat, and hind part of the neck, marked with a ſemilunar ſpot of white, and with white ſtreaks pointing downwards: upper part of the body and wings black, varied with white ſpots: tail duſky: breaſt and belly white: legs black. LENGTH three feet five inches. EXTENT four feet eight. WEIGHT ſixteen pounds.

PLACE. Inhabits the north of *Europe,* and ſpreads along the *Arctic* coaſts, as far as the mouth of the *Ob* only. Is found about *Spitzbergen, Iceland, Hudſon's Bay,* and as low as *New York.* Makes its neſt, in the more northern regions, on the little iſles of freſh-water lakes. Every pair keeps a lake to itſelf. Sees well: flies very high, and, darting obliquely, falls ſecure into its neſt. Tries to ſave itſelf by diving, not flying. The young defend themſelves ſtoutly with their bills. Appears in *Greenland* in *April* or the beginning of *May.* Goes away in *September* or *October,* on the firſt fall of ſnow. The natives uſe the ſkins for cloathing; and the *Indians* about *Hudſon's Bay* adorn their heads with circlets of their feathers.

440. IMBER. *Br. Zool.* ii. N° 238.—Le Grand Plongeon, *De Buffon,* viii. 251.—*Pl. Enl.* 251.
Colymbus Immer, *Lin. Syſt.* 222.—*Latham,* iii.　　　—LEV. MUS.—BL. MUS.

D. With a duſky head: back, coverts of wings, and tail, duſky, elegantly edged with greyiſh white: primaries and tail black: breaſt and belly ſilvery: legs black: webs marked with white ſtripes. Larger than the laſt.

PLACE. Inhabits *New York* during winter. Extends to *Kamtſchatka;* but in no part of *Sibiria* or *Ruſſia.* Found in *Iceland,* and moſt parts of northern *Europe.*

Br.

Br. Zool. ii. N° 239.—Le petit Plongeon, *De Buffon*, viii. 254.—*Pl. Enl.* 99?.

Colymbus Stellatus, Soehane, *Brunnich*, N° 130.—*Latham*, iii.   —LEV. MUS.

    —BL. MUS.

441. SPECKLED.

D. With the bill turning a little upwards: head dark grey, spotted with white: hind part of the neck of an uniform grey: back, coverts of wings, primaries, and tail, dusky; the two first spotted with white: from chin to tail a fine silvery white. WEIGHT two pounds and a half. LENGTH two feet three. EXTENT three feet nine.

Thefe three species visit *New York* in the winter, and return very far north to breed. This is common about the *Baltic* and White Sea; but not observed in other parts of *Ruffia*; yet is a native of *Kamtfchatka**. It lays two eggs in the grass, on the borders of the lakes, not far from the sea. The eggs are exactly oval, of the size of those of a Goose, dusky, marked with a few black spots.

PLACE.

D. With a strong black bill, three inches long: head and neck light grey, striped regularly downwards with long narrow black lines: back and scapulars dusky and plain: primaries, tail, and legs, dusky: cheeks, and whole under side of the body, of a glossy white. WEIGHT between two and three pounds.

442. STRIPED.

Inhabits the inland lakes of *Hudson's Bay*, about a hundred miles southward of *York Fort*. Lays, in *June*, two eggs. Flies high, and passes backwards and forwards, making a great noise; which is said to portend rain. Detected by the natives, who look on this note as supernatural.—Mr. *Hutchins*.

PLACE.

* *Steller*, in *Nov. Com. Petrop.* iv. 424.

Br.

443.   RED-
THROATED.

Br. Zool. ii. N° 240 —Le Lumme, ou petit Plongeon de Mer de Nord, *De
    B. ffon*, viii. 261.—*Pl. Enl.* 308.
Colymbus Septentrionalis, *Lin. Syst.* 220.—*Latham*, iii.   —Lev. Mus.

D. With head and chin of an uniform grey : throat, and lower
     side of the neck, of a dull red : upper part grey, streaked
with black : upper part of body, wings, and tail, dusky : back and
coverts of wings spotted with white : breast and belly white : legs
dusky. WEIGHT three pounds. EXTENT three feet five. LENGTH,
to the tip of the tail, two feet.

PLACE.   Found in the north of *Europe* and *Iceland*, along the northern coast of
Russia, *Sibiria*, and *Kamtschatka* ; but does not haunt the inland lakes.
Inhabits the rivers of *Hudson's Bay* during summer. Prey much on
the fish entangled in the nets ; but are often caught themselves in
their rapid pursuit of the fish. Mr. *Hutchins* took fourteen out of a
single net in one tide. Numbers of every species of Diver are frequently taken in this manner about *Hudson's Bay*.

444.   BLACK-
THROATED.

Br. Zool. ii. N° 241.—Colymbus Arcticus, Lomm. *Faun. Suec.* N° 150.
Le Lumme, &c. *De Buffon*, viii. 261.—*Latham*, iii.   —Lev. Mus.

D. With bill, forehead, and cheeks, black : hind part of the neck
     cinereous : sides of the neck streaked downwards with black :
fore part varying with black, purple, and green : back and coverts
of wings dusky ; the first marked with square, the last with round
white spots : breast and belly white : tail black : legs partly red,
partly dusky.

PLACE.   This, and the preceding, inhabit *Hudson's Bay* *, the north of
Europe, and *Iceland*. Few in *Russia* : but frequent in the inland
lakes of *Sibiria*, especially those of the *Arctic* regions ; but in the
wandering season spread over all latitudes.

* *Edwards*, 147.

The

The *Norwegians* remark the fagacity of thefe birds, in prefaging the change of weather. When the fkies are big with rain, they fly wildly about, and make the moft horrible hoarfe noife, fearing that the fwelled waters fhould invade their neft; on the contrary, in fine weather, their note is different, and feemingly in an exulting ftrain. The *Norwegians* think it impious to deftroy, or even to dif- turb, this fpecies *.

The *Swedes* have lefs fuperftition : they drefs the fkins ; which, when prepared, fhew in the cleareft manner, on the infide, the quin- cuncial difpofition of feathers. They are exceedingly tough, and are ufed for gun-cafes and facings for winter-caps †.

* *Worm. Muf.* 304. † *Faun. Suec.*

LII. S K I M M E R. *Gen. Birds,* LXXXVI.

Cut-water, *Cateſby,* i. 90.—Le Bec en Ciſeaux, *De Buffon,* viii. 454. tab. 36.
—*Pl. Enl.* 357.
Rynchops nigra, *Lin. Syſt.* 229.—*Latham,* iii.   —LEV. MUS.—BL. MUS.

SK. With the bill greatly compreſſed, the edges ſharp ; lower
mandible four inches and a half long; upper only three ; baſe
red; the reſt black : forehead, chin, front of neck, breaſt, and
belly, white : head, and whole upper part of the body, black : wings
of the ſame color: lower part of the inner webs of the primaries
white : tail ſhort, and a little forked ; middle feathers duſky ; the
others white on their ſides : legs weak and red. LENGTH one foot
eight inches. EXTENT three feet ſeven.

PLACE.

Inhabits *America,* from *New York* to *Guiana* *. Skims nimbly
along the water, with its under mandible juſt beneath the ſurface, feeding on the inſects and ſmall fiſh as it proceeds. Frequents alſo oyſter-banks ; its bill being partly, like that of the Oyſ-
ter-catcher, adapted for preying on thoſe ſhell-fiſh. In Mr. *Ray's
Synopſis* † is a ſketch, ſent from *Madras,* of one of this ſpecies.

* *Barrere France Equin.* 135.   † 194. N° 5. tab. i. N° 5.

LIII.

LIII. T E R N. *Gen. Birds*, LXXXVII.

Catefby, i. 88.—Le Noddi, *De Buffon*, viii. 461.—*Pl. Enl.* 461.
Sterna Stolida, *Lin. Syft.* 227.—Lev. Mus.

446. Noddy.

T With a black long bill : crown white, gradually darkening to the hind part : whole plumage brown : wings and tail almoft dufky. WEIGHT four ounces.

Inhabit, in vaft numbers, the *Bahama* iflands, where they breed on the bare rocks. In the breeding feafon they, and numbers of other birds, are feen in great flights, flying near the furface of the water, continually dropping on the fmall fifh which are driven to the top, to fhun the perfecution of the greater. The whole air refounds with the noife of the birds, who feem in full exultation on their fuccefs ; which is expreffed in vaft variety of notes. A rippling and whitenefs in the water marks the courfe of the fhoals of fifh ; and above them the air is animated with their feathered enemies. Where the ftrongeft rippling is, there appear the thickeft fwarms of fowls. As foon as the time of nutrition is over, thefe birds difperfe over the ocean feparately ; and are feen at the diftance of hundreds of leagues from land ; but very feldom on the outfide of the tropics. Their ftupidity is notorious ; for they will fuffer themfelves to be taken by the hand, when they fettle, as they often do, on the yards of fhips.

Place.

447. Sooty.

T With a black bill, two inches long : forehead white : crown, hind part of the head and neck, back, and wings, of a footy blacknefs : cheeks, fore part of the neck, breaft, belly, and ridge of the wings, white : tail greatly forked ; tip of the exterior feathers white ; the reft of the tail dufky. Nearly the SIZE of the Common Gull.

Sent from NEW YORK to Sir *Afhton Lever*. Thefe birds are found in very remote climates. They fwarm in the ifle of *Afcenfion*, 8 fouth latitude. Emit a moft fharp and fhrill cry : are quite fearlefs ; and

Place.

3 X 2   fly

fly fo clofe to the few men who vifit that diftant ifle, as almoft to touch them. They lay only two or three eggs, which are of a furprizing fize, yellowifh, fpotted with brown and pale violet. The flocks which poffefs the different parts of the ifle, lay at different times. In fome quarters the young were found very large; in others only a fingle egg was found juft then layed. Mr. *Ofbeck* and Doctor *Forfter*, who were here in *April* and *May*, are filent about this fpecies: poffibly it was then on its migrations. It is to the Comte de *Buffon* * we owe this account; who received it from that obfervant nobleman the Marquis de *Querboënt*.

443. GREAT.

Br. Zool. ii. N° 254.—Sterna Hirundo, Tarna, *Faun. Suec.* N° 158.
Le Pierre garin, ou grande Hirondelle de mer, *De Buffon*, viii. 331.—*Pl. Enl.* 987.—*Latham*, iii.   —LEV. MUS.—BL. MUS.

T. With bill and feet of a fine crimfon color: forehead, throat, and whole under fide, of a pure white: crown black: upper part of the body, and coverts of the wings, an elegant pale grey: tail much forked, white, with the exterior edges of the three outmoft grey. WEIGHT four ounces one quarter. LENGTH fourteen inches. EXTENT thirty.

PLACE.

Inhabits *Europe*, as high as *Spitzbergen*; and alfo the northern parts of *North America*, as far as *Hudfon's Bay*. It appears in *New England* in *May*, and goes away in autumn: called there the *Mackerel Gull*. It is found on the *Arctic* coafts of *Sibiria* and *Kamtfchatka*. Retires even from *England* at approach of winter.

449. LESSER.

Br. Zool. ii. N° 255.—Sterna minuta, *Lin. Syft.* 228.
La petite Hirondelle de mer, *De Buffon*, viii. 337.—*Pl. Enl.* 996.—*Latham*, iii.
LEV. MUS.—BL. MUS.

T. With a yellow bill and legs; the firft tipt with black: forehead and cheeks white: from the back to the eyes is a black line: crown black: breaft and belly of the moft exquifite and gloffy whitenefs, unequalled by the fineft fatin: back and wings of a pale grey: tail

* *Oif.* viii. 345.

white;

white; lefs forked than that of the preceding. WEIGHT two ounces eight grains: LENGTH eight inches and a half: EXTENT nineteen and a half.

This fpecies is too tender to endure the high northern latitudes, nor even beyond the *Baltic*. Is met with in the fouth of *Ruffia*, and about the *Black* and *Cafpian* fea; and in *Sibiria* about the *Irtifh*. In *America* is feen, during fummer, about *New York*.   PLACE.

> Br. Zool. ii. Nº 256.—Sterna fiffipes, *Lin. Syft.* 228.
> Le Guifette noire, ou l'Epouvantail, *De Buffon*, viii. 341.—*Pl. Enl.* 333.—
>   Latham, iii.   —LEV. MUS.—BL. MUS.

450. BLACK.

T. With the bill, head, neck, breaft, and belly, black: vent white: wings and back of a deep afh-color: tail fhort; exterior fea-thers white; the others cinereous: legs a dirty red: webs deeply hollowed in the middle, fo as to form a crefcent.

This is the fpecies which I apprehend was fent to the Royal Society from *Hudfon's Bay*; and was feen in vaft flocks beyond lat. 41 north, long. 47 weft from *London*, by Mr. *Kahn* *, fomewhat fouth of the bank of *Newfoundland*. Is found in *Europe*, as far as *Iceland*. Very numerous in *Sibiria*, and about the falt lakes of the defarts of *Tartary*, which they animate by their note and active flight and motions.   PLACE.

A. KAMTSCHATKAN. PALLAS, *MS.*
La Guiffette ? *De Buffon*, viii. 339.—*Pl. Enl.* 924.

T. With the bill and crown black: forehead and fpace over the eyes white.

Obferved by *Steller* about *Kamtfchatka*. A bird feemingly of this fpecies was fhot on the *Severn* a few miles below *Shrewfbury*; and is   PLACE.

* *Voyage*, i. 23.

among

among the elegant drawings of my friend, *Joseph Plymley*, Esq; of *Longnor*.

B. CASPIAN T. Terna Cafpia, Mr. *Lepechin*, in *Nov. Com. Petrop.* xiv. 500. tab. xiii. —PALLAS, 483. tab. xxii.—*Latham*, iii.

T. With a fcarlet bill, three inches long : crown and hind part of the head cf an intenfe black, hoared with white : fpace round the eyes black ; under each a white crefcent : cheeks, lower fide of the neck, and whole under fide of the body, of a fnowy whitenefs : upper part cinereous and hoary : fix firft primaries darker, edged and tipt with black : tail deeply forked, and of a pure white : legs black. The fpecimen defcribed by Doctor PALLAS was of darker colors ; perhaps differed in age and fex. LENGTH near two feet. EXTENT three feet two inches. Seems, in the air, as big as a KITE.

PLACE.   Inhabits the *Cafpian* fea, about the mouth of the *Yaik*. Makes a laughing noife. Fifhes both in the fea and rivers. Remains long fufpended in the air, then dafhes on its prey ; and fkims the furface of the water like a Swallow. Lays, on the back of defart ifles, two eggs marked with dufky fpots. Wanders up the great river *Oby*, even towards the *frozen* ocean *.

* PALLAS *MS. Catalog.*

LIV. GULL. *Gen. Birds,* LXXXVIII.

Br. Zool. ii. N° 242.—Larus marinus, *Faun. Suec.* N° 155.
Le Goeland à manteau noir, *De Buffon,* viii. 405.—*Pl. Enl.* 990.—*Latham,* iii.
—Lev. Mus.—Bl. Mus.

G. With a ſtrong pale yellow bill; the lower mandible marked with a black ſpot, encircled with red: upper part of the back and wings black; primaries tipt with white: the reſt of the plumage of a ſnowy whiteneſs: legs pale fleſh-color. WEIGHT ſometimes five pounds. LENGTH twenty-nine inches. EXTENT five feet nine.

Inhabits northern *Europe,* as high as *Iceland, Lapmark,* and the *White Sea; Greenland,* and the coaſt of *North America* down to *New York* and *South Carolina,* where they are called *Old Wives.* Is obſerved, in *Greenland,* to attack other birds, eſpecially the *Eider Duck.* The *Eſkimaux* and *Greenlanders* make their garments of the ſkins of theſe, as well as other water fowl. This was a practice, in early times, with every people to whom manufactures were unknown. *Non avium plumæ in uſum veſtis conſeruntur* * ?

Br. Zool. ii. N° 246.—Larus fuſcus, *Faun. Suec.* N° 154.
Le Goeland à manteau gris brun, *De Buffon,* viii. 410.—*Latham,* iii.
—Lev. Mus.

G. With a yellow bill; lower mandible marked with a red ſpot: irides ſtraw-colored: head, neck, and tail, white: back and coverts of wings aſh-color: primaries duſky, with a white ſpot near their ends: legs of a pale fleſh-color; vary to yellow. WEIGHT about thirty ounces. LENGTH twenty-three inches. EXTENT four feet four.

Inhabits the north of *Europe, Iceland,* and *Greenland:* even in the laſt country a common ſpecies; and continues there the whole year.

* *Senecæ Epiſt.* Ep. xc.

9

Breeds

Breeds among broken rocks: much upon wing: is caught in snares, or by a baited hook. The flesh and eggs eaten; and the skin used, like that of most other Gulls, for garments. Is found in *Hudson's Bay* during summer: breeds there, and retires at approach of winter. It breeds likewise on the islands on the coast of *South Carolina*. Is frequent about the *Caspian* and *Black* seas, and their great rivers: also about the greatest lakes of *Sibiria*.

453. WAGEL.   *Br. Zool.* ii. N° 247.—Larus nævius, *Lin. Syst.* 225.
     Le Goeland varié, ou le Grisard, *De Buffon*, viii. 413.—*Pl. Enl.* 266.—
      Latham, iii.   —Lev. Mus.—Bl. Mus.

G. With a black bill: irides dusky: whole plumage, above and below, varied with brown, white, and cinereous: primaries dusky: tail mottled with dusky and white; near the end a black bar; tips whitish: legs of a dirty white. WEIGHT thirty-two ounces. LENGTH near two feet. EXTENT four feet eight.

PLACE. Inhabits the north of *Europe, Iceland, Hudson's Bay,* and *Newfoundland.* Frequent about the lakes of *Russia,* and the west of *Sibiria.*

454. LAUGHING.   *Catesby,* i. 89.—*Will. Orn.* 346. N° iv.—La Mouette rieuse, *De Buffon,* viii. 433.
     —*Pl. Enl.* 970.
     Larus Atricilla, *Lin. Syst.* 225.—*Latham,* iii. —Lev. Mus.

G. With a red bill: black head. the ends of the primaries black: back and coverts of the wings cinereous: all the rest of the plumage white: legs black and long. LENGTH about eighteen inches. EXTENT three feet.

PLACE. Inhabits the *Babama* islands. Their note resembles a coarse laugh.

Br. Zool. ii. N° 252.—Larus ridibundus, *Lin. Syſt.* 225.—La Mouette rieuſe, *De Buffon*, viii. 433 —*Latham*, iii. —LEV. MUS.

G. With a red bill and legs: head and throat black: neck, belly, and tail, white: back and wings aſh-colored: ends of the primaries marked with black. LENGTH about fifteen inches. EXTENT thirty-ſeven. WEIGHT ten ounces.

Inhabits *New England :* comes in *May*, leaves the country in *Auguſt.* In *Europe*, not farther north than *England.* In all parts of *Ruſſia* and *Sibiria*, and even *Kamtſchatka.* Has the laughing notes of the former, of which it ſeems a variety.

Br. Zool. ii. N°. 250.—*Phipps*, 187.—Larus Riſſa, *Lin. Syſt.* 224. Kutge-gehef, *Marten's Spitzbergen*, 82.—*Latham*, iii. —LEV. MUS ?

G. With a yellow bill: inſide of the mouth orange: head, neck, under ſide of the body, and tail, white: behind each ear is a black ſpot: the back and coverts pale grey: primaries duſky, with a white ſpot near the ends: legs duſky: no back toe. LENGTH fourteen inches. EXTENT three feet two.

Is found about *Newfoundland.* Inhabits *Spitzbergen, Greenland, Iceland,* and the north of *Europe*, the arctic coaſt of *Aſia*, and *Kamtſchatka.*

Larus eburneus, *Phipps's Voy.* 187.—Larus candidus, *Faun. Groenl.* N° 67. Rathſher, *Marten's Spitzb.* 77.—La Mouette blanche, *De Buffon*, viii. 422.— *Pl. Enl.* 994.—*Latham*, iii. —LEV. MUS.

G. With the bill and legs of a lead-color: whole plumage of a ſnowy whiteneſs. LENGTH, to the end of the tail, ſixteen inches. EXTENT thirty-ſeven.

3 Y

Inhabits

PLACE.

Inhabits *Spitzbergen* and *Greenland:* alſo very frequent in the frozen ſea between *Aſia* and *America*; and off cape *Denbigh,* a little to the ſouth of *Bering's Streights.* Keeps uſually far at ſea; but when it does alight, is very ſtupid, and eaſily killed. The young are ſpotted with black, and their bills are black.

458. COMMON.

Br. Zool. ii. N° 249.—Larus canus. Homaka. Mave. *Lappis* Straule, *Faun. Suec.* N° 153.—*Latham,* iii.   —LEV. MUS.

G. With a yellow bill: head, neck, tail, and all the under ſide of the body, white: back and coverts of wings light grey: primaries duſky; near their extremities a white ſpot: legs dull white, tinged with green. LENGTH ſeventeen inches. EXTENT three feet. WEIGHT twelve ounces and a half.

PLACE.

Inhabits as high as *Iceland*; and is common about the *Ruſſian* lakes. Is frequent on the coaſt of *Newfoundland.*

459. ARCTIC.

• *Br. Zool.* ii. N° 245.—*Phipps,* 187.—Le Labbe à longue queue, *De Buffon,* viii. 445.—*Pl. Enl.* 762.—*Ph. Tranſ.* lxii. 421.—*Latham,* iii.   —LEV. MUS.

G. With a duſky bill, much hooked at the end: upper part covered with a thin cere: crown black: back, wings, and tail, duſky: neck, breaſt, and belly, white: tail cuneiform; two middle feathers near four inches longer than the reſt. FEMALE wholly brown; under ſide lighteſt. LENGTH twenty-one inches.

PLACE.

Inhabits, in *America, Hudſon's Bay**; all the north of *Europe* to *Spitzbergen*; frequent in *Greenland.* Feeds almoſt entirely on fiſh caught by other birds, which it perſecutes till they drop their prey, or vomit for fear; when it catches their droppings before it falls into the water. The *Dutch* call it, from a now exploded notion, that it lives on the dung of fowl, the *Stront-jagger.* Extends along the *arctic* coaſt to *Kamtſchatka.* Aſcends the great rivers, the *Ob, Jeneſei,* and *Lena,* above a hundred leagues inland.

• *Ph. Tranſ.* lxii.

Br.

Br. Zool. ii. N° 244.—Catharacta cepphus. Strandhoeg, *Brunnich,* N° 126.—Le Labbe, ou Stercoraire, *De Buffon,* viii. 441. tab. 34.—*Pl. Enl.* 991.—*Latham* iii. —Lev. Mus.

460. Black-toed.

G. With a bill refembling the former: head and neck of a dirty white, marked with dufky fpots: back, fcapulars, coverts of wings, and tail, black, prettily edged with pale ruft: breaft and belly white, croffed with numerous dufky and yellowifh lines: the fides and vent barred croffways with black and white: tail black, tipt with white; the exterior webs of the outmoft, fpotted with ruft; the two middle feathers are near an inch longer than the others. Thefe birds vary into lighter and darker colors; but the color of the toes are fpecific marks: the legs are of a blueifh lead-color: the toes and webs have their lower parts of a deep black. Weight eleven ounces. Length fifteen inches. Extent thirty-nine.

Inhabits, in *America,* the coaft of *Newfoundland* and *Hudfon's Bay*: is hated by the natives, who have a notion that the birds are companions to the detefted *Efkimaux.* I cannot, in *Europe,* trace it higher than *Great Britain* and *Denmark*; yet it has been fhot, in the *Atlantic* ocean, as near to the line as north *lat.* 8, weft *long.* 22. 12.

Place.

Skua, *Br. Zool.* ii. N° 243.—Catharacta Skua, *Brunnich,* N° 125.—*Muller,* 167. Le Goeland brun, *De Buffon,* viii. 408.—*Latham,* iii. —Lev. Mus.

G. With a ftrong fharp black bill and cere: head, back, and coverts of the wings, brown and ruft-colored: primaries and fecondaries dufky; the fhafts of the primaries white: on the fecondaries a great

white

white fpot: breaft and belly of a rufty afh: tail brown, white at the bafe: legs black and fcaly: claws black, fharp, ftrong, and hooked like thofe of a KITE. LENGTH two feet. EXTENT four and a half. WEIGHT three pounds.

PLACE.

Inhabits *Europe* very locally; only from *Foula* and *Unft*, two of the *Schetland* ifles, to the *Feroe* ifles, *Norway*, and as far as *Iceland*. Its manners, fuch as its great courage, and fiercenefs in defending its young, in driving away the eagle from its haunts, and, as is firmly afferted by Mr. *Schroter*, a furgeon in the *Feroe* ifles, its preying on the leffer water fowl, like a rapacious land bird, are fully defcribed in the *Britifh Zoology*. They abound about *Port Egmont*, in the *Falkland* iflands, and are therefore ftiled by navigators, *Port Egmont Hens*. They have been obferved in many parts of the *Pacific* ocean, as low as *lat.* 36. 56 fouth, to the eaft * of *New Zeland*; and as high, in the fame hemifphere, as *lat.* 67. 15 †. The navigators found them in great plenty, in their breeding feafon, in the latter end of *December*, about *Chriftmas Sound*, in *Terra del Fuego*, making their nefts in the dry grafs. They have not been remarked in other parts of the globe, nearer than the *Schetlands*.

B. GLAUCOUS, Larus Glaucus, *Brunnich*, N° 148.—*Muller*, N° 169.—*Faun. Groenl.* N° 64.—*Latham*, iii.

G. With a yellow bill, and orange fpot near the end: head and lower part of the body white: back and wings of a fine hoary grey; primaries darkeft, and tipt with white: legs of a pale fulvous hue. In SIZE fuperior to the Herring Gull.

PLACE.

Inhabits *Norway*, *Lapmark*, *Iceland*, *Greenland*, and *Spitzbergen*. Is called by the *Dutch*, *Burgermeifter*, being the mafter of all other fea fowl. It builds its neft high on the cliffs: preys on dead whales: attends the Walrufes, in order to feed on their dung; and, as *Frederic*

---

* *Cook's Voy. Hawkfworth's Coll.* ii. 283.   † *Forfter's Voy.* i. 109.

Martens

Martens afferts, will even deftroy and eat the young of the Razor-bills. It alfo feeds on fifh; and does not defpife the berries of the *Empetrum Nigrum.* It is almoft continually on wing; and makes a hoarfe noife, like the Raven.

C. SILVERY. Larus argentatus, *Brunnich,* N° 149.—*Latham,* iii.

G. With a white head and neck, ftreaked downwards with cinereous lines: back and under part of the body like the former fpecies: lower part of the primaries greyifh; upper black; the tips white: bill yellow, with an orange fpot. SIZE of the Herring Gull. This and the former feem nearly.

Inhabits *Norway.* PLACE.

D. TARROCK, *Br. Zool.* ii. N° 251.—Larus tridactylus, *Faun. Suec.* N° 157.—La Mouette tachetée, *De Buffon,* viii. 424.—*Latham,* iii. —LEV. MUS.

G. With a ftrong, thick, black bill: with white head, neck, breaft, and belly: behind each ear a black fpot: on the hind part of the neck a black crefcent: back and fcapulars blueifh grey: ten middle feathers of the tail white, tipt with black; outmoft quite white: a protuberance inftead of the back toe. SIZE of the former.

Inhabits *Europe* quite to *Iceland* and *Spitzbergen;* the *Baltic* and *White* fea; and again in *Kamtfchatka.* PLACE.

E. RED-LEGGED.

G. With blood-red bill and legs: head and neck white, mottled about the former: back and coverts of wings fine grey: leffer coverts mottled: under fide of body and the tail white; the laft tipt with black. SIZE of the Black-cap Gull.

A bird of this fpecies was brought from *Kamtfchatka.* Another of the fame kind has been fhot in *Anglefey.* PLACE.

LV. P E-

LV. PETREL. *Gen. Birds*, LXXXIX.

461. FULMAR.

Br. Zool. ii. N° 257.—Procellaria glacialis, *Faun. Succ.* N° 144.—Petrel de l'ifle de St. Kilda, *Pl. Enl.* 59.—*Latham*, iii.   —LEV. MUS.

PLACE.

P. With a ftrong yellow bill: head, neck, tail, and under fide of the body, white: back and coverts of wings cinereous: primaries dufky: legs of a pale yellow. Rather larger than the Common Gull.

Abound in the feas of *Spitzbergen* and *Greenland*, and common in thofe between *Kamtfchatka* and *America*: the latter are darker colored than the former. They are equally abundant in the fouthern hemifphere. Captain *Cook* found them among the ice, in his voyage towards the fouth pole, in lat. 64. 55 *; in lat. 59, to the fouth of the ifle of *New Georgia* †; and even in the moderate climate of lat. 34. 45, not remote from the *Cape of Good Hope* ‡. They keep chiefly in the high feas, and feed on dead whales, or any thing that offers on the furface; but will, with their ftrong bills, pick the fat out of the backs of living whales, efpecially of the wounded; whofe bloody track they will follow by hundreds, to watch its rifing. Their flight refembles running on the top of the water; for which reafon the *Norwegians* call it *Hav-heft*, or Sea-horfe; and *Storm-fugl*, or Storm-fowl, as being fuppofed to be a prefage of tempefts. The *Dutch* call it *Mallmucke*, or the Foolifh Fly, from their multitudes, and their ftupidity. They very feldom come to land, unlefs they chance to lofe their way in the mifts, which are fo frequent on the coaft of *Greenland* during the month of *Auguft*. They breed on the broken rocks about *Difco*, and remote from the main land.

They are, by reafon of their food, exceffively fetid; yet the flefh is ufed as a food by the *Greenlanders*, both raw and dreffed. The fat

---

* *Cook's Voy. S. Pole*, i. 252.   † *Forfter's Voy.* ii. 534.   ‡ *Forfter*, i. 52.

is

is alfo eaten, and ferves to fupply their lamps with oil. The prey of thefe birds being chiefly the blubber of cetaceous fifh, it is quickly converted into oil, which ferves the *Fulmars* for a double end ; as a fuftenance for the young, and a defence againft their affailants; for they fpurt it, on being feized, out of their mouths and noftrils, into the faces of the perfons who lay hold of them. The *Greenlanders* take them by darting them in the water.

Br. Zool. ii. N° 258.—*Edw. Av.* 359.—Procellaria Puffinus, *Lin. Syft.* 213.— 462. SHEAR-
Latham, iii.   —LEV. MUS. WATER.

P. With a dufky bill, more flender than that of the former : head, wings, and whole upper part of the body, of a footy blacknefs: lower part, from chin to tail, and the inner coverts of the wings, white : legs weak, compreffed; whitifh before, dufky behind. LENGTH fifteen inches. EXTENT thirty-one. WEIGHT feventeen ounces.

Inhabits the northern parts of *Europe, Iceland,* and *Greenland.* Confort with the laft in *Greenland* : and, in mifty weather, quite cover the fea. It extends, in the *Atlantic* ocean, to *America,* and again almoft to the *Cape of Good Hope* * ; and is alfo found in the fouthern hemifphere, having been feen in fouth lat. 13. 13, in Captain *Cook's* paffage from *Eafter* ifland to *Otaheitè* † : and again, in numbers, as low as cape *Defeada,* in fouth latitude 53 ‡. PLACE.

P. With the whole upper and under parts of a cinerous grey : bill 463. FORK-TAIL. much hooked, and black : leffer coverts of wings dufky ; greater, deep grey : exterior webs of primaries dufky ; interior, light grey : tail forked, and of a light grey. LENGTH nine inches.

Taken among the ice between *Afia* and *America.* PLACE.

* *Cook's Voy. to S. Pole,* 12. 13. † *Ibid.* ‡ *Ibid.*

464. STORMY. *Br. Zool.* ii. N° 259.—Procellaria pelagica, Stormwaders Fogel, *Faun. Suec.* N° 143. Le Petrel, ou l'Oiſeau tempete, *Pl. Enl.* 993.—*Latham*, iii. —LEV. MUS.

P. With a black bill, much hooked at the end: rump and feathers of the vent, and each ſide of the tail, white; all the reſt black: ſecondaries tipt with white: tail ſhort: wings very long. LENGTH ſix inches. EXTENT thirteen.

PLACE. This ſpecies inhabits the north of *Europe*: is common about *Kamtſchatka*, where it is larger than in other places; but does not extend to the *Arctic* circle, at leſt is unmentioned by the Fauniſts of that region. Is, with the preceding, found at all diſtances from land, in all parts of the *Atlantic*, from *Great Britain* to the coaſt of *North America* *: flocks attend the ſhips the whole way, and uſually keep in the wake, where they pick up every thing that drops. They never are off wing; yet ſeem to ſettle. They are ſilent during day; clamorous in the dark. Are hated by the ſailors, who call them *Witches*, imagining they forebode a ſtorm. The *Norwegians* ſtile them *Sondenvinds Fugl*; the *Swedes*, *Stormwaders Fogel*; and the inhabitants of *Feroe*, *Strunkvit*.

A. KURIL. Black Petrel, *Edw.* 89.—*Latham*, iii.

P. With a ſtrong yellow bill: whole plumage of an unvaried ruſty black: legs the ſame, daſhed with red. SIZE of a Raven.

PLACE. Sent to Doctor *Pallas* from the *Kuril* iſles.

* *Kalm*, i. 22, 23.

LVI.

LVI. MERGANSER. *Gen. Birds,* XC.

Br. Zool. ii. Nº 260.—Mergus Merganſer, Wrakfagel, Kjorfagel, Skraka, *Faun.* 465. GOOSANDER.
Suec. Nº 135.—Le Harle, *De Buffon,* viii. 267.—*Pl. Enl.* 951, 953.—
Latham, iii.—   LEV. MUS.—BL. MUS.

M. With a red bill : head full of feathers, looſe behind, and of a mallard green : lower part of the neck and belly of a fine ſtraw-color : upper part of the back, and ſcapulars next to it, black : lower part of the back, and the tail, cinereous : primaries duſky ; ſecondaries white, edged with black : coverts on the ridge of the wing black ; the others white : legs a full orange. WEIGHT four pounds. LENGTH two feet four. EXTENT three feet two. Head and upper part of the neck of the FEMALE, or DUN DIVER, ferruginous : behind is a pendent creſt : throat white : back, coverts of wings, and the tail, cinereous : primaries duſky : breaſt and middle of the belly white.

Inhabits the province of *New York* in winter : retires in *April,* PLACE. probably to *Hudſon's Bay,* and other northern countries. It is alſo found as low as *South Carolina *.* Breeds in every latitude in the *Ruſſian* empire ; but moſtly in the north. Is common in *Kamtſchatka.* Extends through northern *Europe* to *Iceland* and *Greenland* †. Continues the whole year in the *Orknies* ; but viſits *South Britain* only in ſevere winters. Swims with its body very deep in the water : dives admirably ; and is a great devourer of fiſh.

Br. Zool. ii. Nº 261.—Mergus ferrator, Ptacka, *Faun. Suec.* Nº 136.—Le Harle 466. RED-
huppé, *De Buffon,* viii. 273.—*Pl. Enl.* 207.—*Faun Groenl.* Nº 48.—*La-* BREASTED.
tham, iii.   —LEV. MUS.

M. With a creſted head ; and part of the neck a mallard green : reſt of the neck, and whole belly, white : breaſt ferruginous, ſpotted with black : upper part of the back black : exterior ſcapulars

---

* The birds like a Duck, with a narrow bill, with ſets of teeth, called in *Carolina,* *Fiſhermen,* and deſcribed as having a fiſhy taſte, are of this ſpecies. See *Lawſon,* 150.
† *Olaffen Iceland*—and *Faun. Groenl.* Nº 49.

3 Z   black ;

black; interior white: coverts of the wings black and white: primaries dusky: lower part of the back, and sides under the wings, cinereous, barred with small lines of black: tail brown: legs orange. In the FEMALE the head and upper part of the neck are dull ferruginous: throat white: fore part of the neck, and the breast, marbled with deep ash color: back, scapulars, and tail, cinereous: primaries dusky. WEIGHT of the male two pounds. LENGTH one foot nine. EXTENT two feet seven.

PLACE.

Frequent *Newfoundland* and *Greenland* during summer; and appear, in the same season, in *Hudson's Bay* in great flocks. Is found in *Europe*, as high as *Iceland*, where it is called *Vatus-önd*. In the *Russian* dominions is gregarious, about the great rivers of *Sibiria* and lake *Baikal*.

467. HOODED.

Round-crested Duck, *Catesby*, i. 94.—*Edw.* 360.—*Latham*, iii. Mergus cucullatus, *Lin. Syst.* 207.—LEV. MUS.—BL. MUS.

M. With a large, upright, circular crest, beginning at the base of the bill, and ending at the hind part of the head; flabelliform, edged with black; the rest white; and on each side, above the eyes, streaked with a shorter set of black feathers: forehead, cheeks, neck, back, and tail, black: breast and belly white: sides yellowish rust, crossed by slender dusky lines. Head and neck of the FEMALE dark ash, mottled with black: crest short, and rust-colored: back, wings, and tail, dusky; the wings crossed with a white line: breast and belly white. In SIZE between a Wigeon and a Teal.

PLACE.

This species breeds in some unknown parts of the north. Appears in *New York*, and other parts of *North America*, as low as *Virginia* and *Carolina*, in *November*; and frequents fresh waters: retires in *March*.

Br. Zool. N° 262.—La Piette, *De Buffon,* viii. 275.—*Pl. Enl.* 449, 450. Le Harle couronné, *De Buffon,* viii. 280.—*Pl. Enl.* 935, 936.—*Latham,* iii. —Lev. Mus.

M. With a lead-colored bill: horizontal creft, white above, black beneath: eyes included in a large oval fpot, black, gloffed with green, which extends to the bafe of the bill: neck, and whole under fide of the body, pure white: wings and fcapulars particolored with black and white: tail deep afh-color: legs blueifh grey. Length eighteen inches. Extent twenty-fix. Weight thirty-four ounces. Head of the Female * ruft-colored, and flightly crefted: around the eyes a fpot of the fame color and form as in the male: neck grey, darkeft behind: in the other marks refembles the male, except the legs, which are grey.

This fpecies was fent to Mrs. *Blackburn* from *New York,* I think as a winter bird. In *Europe* it extends to *Iceland:* vifits *Britain* in the fevere feafon. In the *Ruffian* empire frequents the fame places with the Goosander. Each of thefe retire fouthward at approach of winter ; and are obferved returning up the *Volga* in *February,* tending towards the north. Migrates, during fummer, even as low as *Tinos* in the *Archipelago* †.

---

* Confiding in other writers, I made, in my *Britifh Zoology,* another fpecies of the female of the Smew, under the name of the *Red-headed,* N° 263. The bird I thought to be the female, and call the Lough Diver, is a diftinct kind. Mr. *Plymley* informs me that he diffected feveral, and found males and females without any diftinction of plumage in either fex.

† *Extracts,* ii. 146.—*Haffelquift,* 269.

A. Minute

A. MINUTE Lough Diver, *Br. Zool.* ii. p. 560.—Mergus minutus, *Faun. Suec.* Nº 138.
—*Latham,* iii.   —LEV. MUS.

M. With head and hind part of the neck ruft-colored; the head
 flightly crefted :. back, fcapulars, and tail, dufky : fore part of
the neck white : breaft clouded with grey : on the leffer coverts of
the wings a great bed of white ; on the primaries and greater coverts
two tranfverfe lines of white : legs dufky.

PLACE.   Inhabits the fhores of *Sweden :* found alfo, during winter, in *Great*
Britain ; at which feafon the whole genus quits *Sweden,* expelled by
the ice.

LVII.

LVII. DUCK. *Gen. Birds,* XCI.

Br. Zool. ii. N° 264.—Anas Cygnus ferus. Swan, *Faun. Succ.* N° 107.— 469. WHISTLING
Latham, iii. —LEV. MUS. SWAN.

D. With the lower part of the bill black; upper part, and space
between that and the eyes, covered with a naked yellow skin:
eye-lids naked and yellow: whole plumage pure white: legs black.
LENGTH, to the tip of the tail, four feet ten. EXTENT seven feet
three. WEIGHT from thirteen to sixteen pounds.

These birds inhabit the northern world, as high as *Iceland,* and as PLACE.
low as the soft climate of *Greece,* or of *Lydia,* the modern *Anatolia,* in
Asia Minor: it even descends as low as *Egypt* *. They swarm, during summer, in the great lakes and marshes of the *Tartarian* and
Sibirian desarts; and resort in great numbers to winter about the
Caspian and *Euxine* seas. Those of the eastern parts of *Sibiria* retire
beyond *Kamtschatka,* either to the coasts of *America,* or to the isles
north of *Japan.* In *Sibiria,* they spread far north, but not to the
Arctic circle. They arrive in *Hudson's Bay* about the end of *May:*
breed in great numbers on the shores, in the islands, and in the inland lakes; but all retire to the southern parts of *North America* in
autumn, even as low as *Carolina* and *Louisiana.* Mr. *Lawson,* who
was no inaccurate observer, says, that there were two sorts in *Carolina:*
the larger is called, from its note, the *Trumpeter.* These arrive in
great flocks to the fresh rivers in winter; and, in *February,* retire to
the great lakes to breed: the lesser are called *Hoopers,* and frequent
mostly the salt water. The Cygnets are esteemed a delicate dish.
The *Indians* of *Louisiana* make diadems for their chieftains with the
large feathers: the lesser are woven into garments for the women of
rank. The young of both sexes make tippets of the unplucked skin.

* *Catesby, App.* xxxvi.—*Lawson,* 146.—*Du Pratz,* ii. 78.

They

They breed in great multitudes in the lakes of *Lapland*; and refort towards the more fouthern parts of *Europe*, during the fevere feafon. Breed even in the *Orkney* ifles.

In *Iceland* they are an objeé of chace. In *Auguft* they lofe their feathers to fuch a degree as not to be able to fly. The natives ,at that feafon, refort in great numbers to the places where they moft abound; and come provided with dogs, and aétive and ftrong horfes, trained to the fport, and capable of paffing nimbly over the boggy foil and marfhes. The fwans will run as faft as a tolerable horfe. The greater numbers are taken by the dogs, which are taught to catch them by the neck, which caufes them to lofe their balance, and become an eafy prey. Great ufe is made of the plumage : the flefh is eaten; and the fkin of the legs and feet, taken off entire, looks like fhagreen, and is ufed for purfes. The eggs are colleéted in

the fpring for food *. In *Kamtfchatka*, where they abound both in winter and fummer, they are alfo taken with dogs, in the moulting feafon; or killed with clubs. During winter they are taken in the unfrozen rivers, and form a conftant difh at the tables of the natives †.

This fpecies has feveral diftinétions from the fpecies which we, in *England*, call the Tame Swan. In *Ruffia* this fpecies more fitly clames the name, it being the kind moft commonly tamed in that empire. The Whiftling Swan carries its neck quite ereét : the other fwims with it arched. This is far inferior in fize. This has twelve ribs on a fide; the Mute ‡ only eleven. But the moft remarkable is the ftrange figure of the windpipe, which falls into the cheft, then turns back like a trumpet, and afterwards makes a fecond bend to join the lungs. Thus it is enabled to utter a loud and fhrill note. The other *Swan*, on the contrary, is the moft filent of birds; it can do nothing more than hifs, which it does on receiving any provocation. The vocal kind emits its loud notes only when flying, or

---

* *Olaffen*, i. 118.   † *Defcr. Kamtfchatka*, 495.

‡ We change the name of the Tame Swan into Mute, as the former name is equivocal, and this fpecies emits no found.

calling:

calling : its found is, *whoogh, whoogh,* very loud and fhrill, but not dif-agreeable, when heard far above one's head, and modulated by the winds. The natives of *Iceland* compare it to the notes of a violin : in fact they hear it at the end of their long and gloomy winter, when the return of the Swans announces the return of fummer : every note muft be therefore melodious which prefages the fpeedy thaw, and the releafe from their tedious confinement.

It is from this fpecies alone that the antients have given the fable of the Swan being endued with the powers of melody : embracing the *Pythagorean* doctrine, they made the body of this bird the manfion of the fouls of departed poets : and after that, attributed to the birds the fame faculty of harmony which their inmates poffeffed in a pre-exiftent ftate. The vulgar, not diftinguifhing between fweetnefs of numbers and melody of voice, thought that real which was only in-tended figuratively. The MUTE Swan never frequents the *Padus*; and I am almoft equally certain that it never is feen on the *Cayfter,* in *Lydia*; each of them ftreams celebrated by the poets, for the great refort of Swans. The *Padus* was ftyled *Oloriferus,* from the numbers which frequented its waters ; and there are few of the poets, *Greek* or *Latin,* who do not truly make them its inhabitants. I fhall give one reference only, out of refpect to the extreme beauty of the imagery.

VOCAL SWAN OF THE POETS.

> Haud fecus *Eridani* ftagnis ripave *Cayftri*
> Innatat albus Olor, pronoque immobile corpus
> Dat fluvio : & pedibus tacitis emigrat in undas.
>
> *Silius Italicus,* lib. 14.

Tame Swan, *Br. Zool.* ii. N° 265.—Anfer Cygnus, N° 107. β.—*Latham.* iii. —LEV. MUS.

470. MUTE SWAN.

D. With a deep red bill, and black incurvated nail at the end : a triangular naked black fkin between the bill and the eyes : at the bafe of the upper mandible a large black rounded protube-

rance :

rance : legs black : whole plumage of a fnowy whitenefs. WEIGHT
fometimes twenty-five pounds.

PLACE.

The Mute Swan, or that which we call Tame, is found in a wild
ftate in fome parts of *Ruffia* ; but far more plentiful in *Sibiria*. It
arrives, in fummer, later from the fouth, and does not fpread fo far
north *. Thofe which frequent the provinces of *Ghilan* and *Mafen-
deran*, on the fouth of the *Cafpian* fea, grow to a vaft fize, and are
efteemed great delicacies. The *Mahometans* hold them in high veneration †.

471. CANADA
GOOSE.

Edw. 151.—*Catefby*, i. 91.—Anas Canadenfis, *Lin. Syft.* 198.—*Phil. Tranf.*
lxii. 412.—*Latham*, iii.　　—LEV. MUS.—BL. MUS.

D. With an elevated black bill : head, neck, primaries, and tail,
black : from the throat paffes, along the cheeks to the hind
part of each fide of the head, a triangular white fpot : bottom of
the neck, vent feathers, lower belly, and coverts of the tail, white :
breaft, upper belly, back, and wings (except primaries) of a dufky
brown : legs of a deep lead-color.

PLACE.

Inhabit the northern parts of *North America*. Immenfe flocks
appear annually in the fpring in *Hudfon's Bay*, and pafs far to
the north to breed; and return fouthward in the autumn. Numbers
alfo breed about *Hudfon's Bay*, and lay fix or feven eggs. The young
are eafily made tame. M. *Fabricius* fufpects that they are found,
during fummer, in *Greenland* ‡. They proceed, in their fouthern migration, as low as *South Carolina*, where they winter in the rice-
grounds. The *Englifh* of *Hudfon's Bay* depend greatly on Geefe, of
thefe and other kinds, for their fupport ; and, in favorable years, kill
three or four thoufand, which they falt and barrel. Their arrival
is impatiently attended ; it is the harbinger of the fpring, and the
month named by the *Indians* the *Goofe* moon. They appear ufually
at our fettlements in numbers, about *St. George's* day, O. S. and fly

* Doctor PALLAS.　　† *Extracts*, iii. 78.　　‡ *Faun. Groenl.* p. 66.

northward

northward to neſtle in ſecurity. They prefer iſlands to the continent, as further from the haunts of men. Thus *Marble Iſland*
was found, in *Auguſt*, to ſwarm with Swans, Geeſe, and Ducks ;
the old ones moulting, and the young at that time incapable of
flying *.

The *Engliſh* ſend out their ſervants, as well as *Indians*, to ſhoot
theſe birds on their paſſage. It is in vain to purſue them : they
therefore form a row of huts made of boughs, at muſquet-ſhot diſtance from each other, and place them in a line acroſs the vaſt
marſhes of the country. Each hovel, or, as they are called, *ſtand*, is
occupied by only a ſingle perſon. Theſe attend the flight of the birds,
and on their approach mimic their cackle ſo well, that the Geeſe will
anſwer, and wheel and come nearer the ſtand. The ſportſman keeps
motionleſs, and on his knees, with his gun cocked, the whole time ;
and never fires till he has ſeen the eyes of the Geeſe. He fires as they
are going from him, then picks up another gun that lies by him,
and diſcharges that. The Geeſe which he has killed, he ſets up on
ſticks as if alive, to decoy others ; he alſo makes artificial birds for
the ſame purpoſe. In a good day (for they fly in very uncertain
and unequal numbers) a ſingle *Indian* will kill two hundred. Notwithſtanding every ſpecies of Gooſe has a different call, yet the
Indians are admirable in their imitation of every one.

The vernal flight of the Geeſe laſts from the middle of *April* until the middle of *May*. Their firſt appearance coincides with the
thawing of the ſwamps, when they are very lean. The autumnal,
or the ſeaſon of their return with their young, is from the middle
of *Auguſt* to the middle of *Oƈtober* †. Thoſe which are taken in
this latter ſeaſon, when the froſts uſually begin, are preſerved in
their feathers, and left to be frozen for the freſh proviſions of the
winter ſtock. The feathers conſtitute an article of commerce, and
are ſent into *England*.

* *Drage,* i. 93.   † *Dobbs's Hudſon's Bay,* 52.

4 A   *B*.

472. BEAN GOOSE.   *Br. Zool.* ii. N° 267.—*Latham,* iii.   —LEV. MUS.

D. With a fmall bill, much compreffed near the end; bafe and nail black; middle of a pale red: head and neck cinereous brown, tinged with ruft: breaft and belly dirty white, clouded with afh-color: leffer coverts of the wings very light grey: back plain afh-color: fcapulars darker, edged with white: primaries and fe-condaries grey, edged with black: tail edged with white: legs faf-fron-color: claws white. LENGTH two feet feven. WEIGHT fix pounds and a half.

PLACE.   Obferved by Mr. *Hearne*, in *Hudfon's Bay*. Is in *Europe* a northern bird. Breeds in great numbers in *Lewis*, one of the *Hebrides*, and is moft deftructive to the green corn. Migrates at the latter end of *Auguft*, in flocks innumerable, into the wolds of *Yorkfhire*, and into *Lincolnfhire*; and among them are fome white *. They all difappear in the fpring. The appearance and difappearance of this kind in *Auftria* is fimilar †. Wild Geefe are feen flying over, but very rarely alight in the *Orknies*.

473. GREY LAG GOOSE.   *Br. Zool.* ii. N° 266.—Anas Anfer. Willgås, *Faun. Suec.* N° 114.—Wild Goofe of all authors.—*Latham*, iii.
L'Oye Sauvage, *Pl. Enl.* 995.—LEV. MUS.—BL. MUS.

D. With an elevated bill, flefh-colored, tinged with yellow, and with a white nail: head and neck cinereous, mixed with dirty yellow: neck ftriated downwards: back and primaries dufky; the laft tipt with black; fhafts white: fecondaries black, edged with white: leffer coverts dufky, edged with white: breaft and belly whitifh, clouded with

---

* *Lifter*, in *Ph. Tr. Abridg.* ii. 852. I cannot but fufpect, that fome of the SNOW GEESE, N° 477, may mix with them, as none of this genus vary in color in the wild ftate.
† *Kramer Anim. Auftr.* 339.

I   afh-color:

afh-color : rump and vent white : middle feathers of the tail dufky, tipt and edged with white ; the outmoft almoft entirely white : legs flefh-colored : claws black. LENGTH two feet nine. EXTENT five feet. WEIGHT fometimes ten pounds.

Inhabits the north of *Europe*, *Afia*, and *America*, and migrates into *Hudfon's Bay*. Frequents, during winter, *South Carolina*, and particularly the rice grounds, where it gleans the droppings of the harveft. This fpecies breeds in the fens of *Lincolnfhire*, and never migrates from that county. They are feen, early in the fpring, flying over *Sweden*, to the *Lapland* moors, and to the eaftern and fouthern parts of *Iceland* ; in which quarters of that ifland alone they breed *. Return in autumn : make a fhort ftay along the fhores ; but never winter in *Sweden* †. Abound in *Ruffia*, *Sibiria*, and *Kamtfchatka* ; but breed chiefly in the north.

PLACE.

Edw. 152.—Anfer Cærulefcens, *Lin. Syft.* 196.—*Latham, iii.* —LEV. MUS.

474. BLUE-WING-ED GOOSE.

D. With a red elevated bill : crown yellowifh ; reft of the head and neck white ; the hind part of the laft fpotted with black ; in fome the fpots are wanting : bafe of the neck, breaft, fides under the wings, and back, of a deep brown : coverts of the wings and tail of a light blueifh afh-color : belly and vent white : primaries dufky : fcapulars and tail white and grey, difpofed in ftripes : legs red. In SIZE rather leffer than the common Tame Goofe.

Migrates into *Hudfon's Bay*, and re-migrates like the former. The *Indians* have a notion, that to avoid the cold, it flies towards the fun, till it finges its pate againft that luminary. Few go very far north ; but are moft numerous about *Albany Fort* ; where, on the contrary, the SNOW GEESE are very fcarce.

PLACE.

---

* *Paulfon.* † *Amæn. Acad.* iv. 585.

D. With

475. BERING.

D. With a yellow excrefcence at the bafe of the bill, radiated in the middle with blueifh black feathers : round the ears a fpace of greenifh white : eyes black, encircled with yellow, and rayed with black : back, fore part of the neck, and belly, white : wings black : hind part of the neck blueifh. SIZE of a common Wild Goofe.—STELLER's *Defcr.* *

PLACE.

Obferved by Mr. *Steller,* in *July,* on the ifle of *Bering.* They probably came from *America.* It is the remark of that great natu-ralift, during his ftay on that ifland, that Geefe of various kinds migrated this way to and from *America* to *Afia,* in vaft flocks. In the fpring they came from the weft, in autumn from the eaft ; which proves, that the Water-Fowl of thefe latitudes prefer, for breeding-places, the *Afiatic* waftes to thofe of *America.*

476. WHITE-FRONTED GOOSE.

Br. Zool. ii. Nº 268.—Anas Erythropus Fiælgas, *Faun. Suec.* Nº 116.—*Latham,* iii. Laughing Goofe, *Edw.* 153.—LEV. MUS.—BL. MUS.

D. With a pale yellow elevated bill : forehead white : head and neck of a cinereous brown, darkeft on the crown : coverts of the wings grey, edged with brown : breaft of an afh-color, clouded with a deeper : belly white, marked with large black fpots : coverts of the tail and the vent white : tail dufky, edged with white : legs

SIZE.

orange. LENGTH two feet four. EXTENT four feet fix. WEIGHT five pounds and a half.

PLACE.

Inhabits, during fummer, *Hudfon's Bay,* and the north of *Europe.* Breeds alfo in the extreme north of *Afia* ; and in its migration is very frequently fcattered over *Sibiria.* Migrates over only the eaft of *Ruffia* ; and is very fcarce in the weft. Mr. *Fabricius* fufpects that they are found in *Greenland* †.

* See *Defcr. Kamtfchatka,* 496, 7.　　　　† *Faun. Groenl.* p. 66.

Anfer

Anſer Grandinis. Schnee Gans. *Schwenckfelt Sileſ.* 213.—*Phil. Tranſ.* lxii. 413.   477. Sɴᴏᴡ.
Anſer Hyperboreus, *Pallas Spicil. Zool.* faſc. vi. 26.—*Latham,* iii.
White Brant, *Lawſon,* 147.

D. With an elevated bill; upper mandible ſcarlet; lower whitiſh
forehead yellowiſh : head, neck, and body, of a ſnowy whiteneſs : primaries white at the bottoms, black to the tips : leſſer coverts uſually cinereous, with duſky tips : legs and feet deep red.
The young Geeſe are blue, and do not attain their proper colors in
leſs than a year. Lᴇɴɢᴛʜ two feet eight inches. Exᴛᴇɴᴛ three   Sɪᴢᴇ.
feet and a half. Wᴇɪɢʜᴛ between five and ſix pounds.

This ſpecies is common to the north of *Aſia,* and to *North*   Pʟᴀᴄᴇ.
America. They appear in flights about *Severn* river in *Hudſon's
Bay,* in the middle of *May,* on their way northward ; return in
the beginning of *September* with their young, and ſtay about the
ſettlement a fortnight; and proceed, about the tenth of *October,*
flying very high, ſouthward to paſs the winter. They come in flocks
of thouſands ; quite cover the country ; riſe in clouds, and with an
amazing noiſe. They viſit *Carolina* * in vaſt flocks ; and feed on   Fᴏᴏᴅ.
the roots of ſedge and graſs, which they tear up like hogs. It uſed
to be a common practice in that country, to burn a piece of a
marſh, which enticed the Geeſe to come there, as they could then
more readily get at the roots ; which gave the ſportſman opportunity
of killing as many as he pleaſed. In *Hudſon's Bay* thouſands are annually ſhot by the *Indians* for the uſe of the ſettlement; and are
eſteemed excellent meat.

They arrive in the earlieſt ſpring, before any other ſpecies of   Mɪɢʀᴀᴛɪᴏɴꜱ.
Water-fowl, in immenſe flights, firſt about the river *Kolyma.* Their

---

* *Lawſon,* 147.—*Quere,* The ſort of whitiſh fowl mentioned by Mr. *Lawſon,*
p. 150, which he calls *Bull-necks,* of the ſize of a Brant, which come to *Carolina*
after *Chriſtmas,* and frequent the rivers : are excellent meat ; but are very ſhy, and
ſuch good divers, as not to be ſhot without difficulty ‡

courſe

courfe is from the eaft, tending to the frozen ocean ; and fpreading to the eftuaries of the *Jana* and *Lena* before the ice is broken up. Finding the want of fubfiftence, they bend their journey a little fouthward, in fearch of the infects and plants which abound in the inland lakes and moors. In this manner they penetrate as low as *Jakut*, and very rarely farther, except in very fmall detachments, which ftray towards the *Olecma*, and fometimes by accident to the junction of the *Witim* with the *Lena*. They make very little ftay in thofe parts ; but again tend directly to the *Arctic* coafts of *Sibiria*, where they breed ; but they do not take the fame route, keeping more eafterly, towards the *Jana* and *Indigirka*. It is obfervable, that they never migrate weftward beyond long. 130, a little beyond the mouth of the *Lena*; neither is their migration by fo high a latitude as *Kamtfchatka*, where they are extremely rare * ; or their flight over that country may be fo lofty as to render their courfe imperceptible. In the beginning of winter they are feen flying at a great height over *Silefia*; but it does not appear that they continue there, being only on their paffage to fome other country †.

The general winter quarters of this fpecies feems to be the temperate and warm part of *North America*.

STUPIDITY.

They are the moft numerous and the moft ftupid of all the Goofe race. They feem to want the inftinct of others, by their arriving at the mouths of the *Arctic Afiatic* rivers before the feafon in which they can poffibly fubfift. They are annually guilty of the fame miftake, and annually compelled to make a new migration to the fouth in queft of food, where they pafs their time till the northern eftuaries are freed from the bonds of ice.

MANNER OF TAK-
ING.

They have fo little of the fhynefs of other Geefe, that they are taken in the moft ridiculous manner imaginable, about *Jakut*, and the other parts of *Sibiria* which they frequent. The inhabitants firft place, near the banks of the rivers, a great net, in a ftrait line,

---

* *Defcr. Kamtfch.* 496.   † *Schwenkfelt An. Silefia,* 215.

or elfe form a hovel of fkins fewed together. This done, one of the company dreffes himfelf in the fkin of a white rein-deer, advances towards the flock of Geefe, and then turns back towards the net or the hovel; and his companions go behind the flock, and, by making a noife, drive them forward. The fimple birds miftake the man in white for their leader, and follow him within reach of the net, which is fuddenly pulled down, and captivates the whole. When he chufes to conduct them to the hovel, they follow in the fame manner; he creeps in at a hole left for that purpofe, and out at another on the op-pofite fide, which he clofes up. The Geefe follow him through the firft; and as foon as they are got in, he paffes round, and fecures every one[*]. In that frozen clime, they afford great fubfiftence to the natives; and the feathers are an article of commerce. Each family will kill thoufands in a feafon. Thefe they pluck and gut; then fling them in heaps into holes dug for that purpofe, and cover them with no-thing more than the earth. This freezes, and forms over them an arch; and whenever the family has occafion to open one of thefe magazines, they find their provifion fweet and good.

Br. Zool. ii. N° 270.—Anas Hrota, *Muller*, N° 115.—Anas Bernicla. Belgis 478. BRENT. Rotgans. Calmariens Prutgas, *Faun. Suec.* N° 115.—*Latham*, iii. —LEV. MUS.

D. With a fhort, black, elevated bill: head, neck, and upper part of the breaft, black: a white fpot marks each fide of the neck near its junction with the head: primaries and tail black: belly, fcapulars, and coverts of the wings, cinereous, clouded with a deeper: coverts of tail and the vent white: legs black.

Is frequent in *Hudfon's Bay*. Breed in the iflands, and along the coafts; but never fly inland. Feed about high-water mark. Re-turn towards the fouth in vaft flocks in autumn. Probably they winter in *Carolina*; for *Lawfon* mentions a *Grey Brent* frequent in

P. ACT.

[*] The *Kamtfchatkans* ufe the fame method in taking Geefe. *Defcr. Kamtfchatka*, 496.

that feafon*. During winter, they fwarm in *Holland* and in *Ireland*: in the firft, every eating-houfe is full of them: in the laft, they are taken in flight-time, in nets placed acrofs the rivers; are fattened, and reckoned great delicacies. They appear in fmall flocks in *Hoy Sound*, in the *Orknies*; but do not continue there: on the contrary, they winter in *Horra Sound*, in *Schetland*, in flocks of two hundred, and are called *Horra Geefe* †. They retire from *Europe* to breed in the extreme north. A few, after flying over *Sweden*, ftop on the borders of *Lapland*; but the great bodies of them continue their flight even to the moft northern ifles of *Greenland* ‡, and to *Spitzbergen*. Fly in the fhape of a wedge, and with great clamor. Feed on grafs, water-plants, berries, and worms. Cannot dive. *Barentz* found multitudes fitting on their eggs, about the 21ft of *June* 1595, in the great bay called *Wibe Janz Water*; and, to his amazement, difcovered them to be the *Rotganfen*, which his countrymen, the *Dutch*, fuppofed to have been generated from fome trees in *Scotland*, the fruit of which, when ripe, fell into the fea, and were converted into Goflings ‖. Thefe birds arrive every year in the eaft part of *Sibiria*, in order to breed; but are not feen to the weft of the *Lena*, nor yet in *Ruffia*.

479. BERNACLE.	*Br. Zool.* ii. N° 269.—Anas Erythropus (maf.) *W. Botb.*—Fiælgâs, *Faun. Suec.* N° 116.—Anas Helfingen, *Olaffin Iceland*, ii. tab. 33.—*Latham*, iii. La Bernache, *Pl. Enl.* 855.—Lev. Mus.—Bl. Mus.

D. With white cheeks and forehead: from bill to the eyes runs a dufky line; the reft of the head, neck, and part of the breaft, black: belly, vent, and coverts of tail, white: back, fcapulars, and coverts of wings, barred with black, grey, and white:

---

* *Lawfon*, 147.   † Reverend Mr. *Low*.   ‡ *Faun. Groenl.* N° 41.
‖ *Navigation par la Nord, Amftelredam*, 1606, folio, p. 14.—The *Englifh* fabled the fame of the *Bernacle*. See *Gerard's Herbal*.

tail

tail and legs black. LENGTH two feet one inch. EXTENT four feet five. WEIGHT about five pounds.

These birds are seen, but extremely rarely, in *Hudson's Bay*. It is found, and I believe breeds, in the north of *Russia* and *Lapland*, in *Norway*, and in *Iceland* *; but not in *Sibiria*. They appear on the *British* shores and marshes, in vast flocks, during winter; but retire in *February*. *Linnæus* unaccountably makes the White-fronted Goose, N° 476, the female of this. PLACE.

Br. Zool. ii. N° 271.—Anas Mollissima, Ada, Eider, Gudunge, *Faun. Suec.* N° 117. 480. EIDER. —*Latham*, iii.

Great Black and White Duck, *Edw.* 98.—*Pl. Enl.* 208, 209.—LEV. MUS.— BL. MUS.

D. With a black bill, somewhat elevated : forehead of a velvet black : a broad black bar, glossed with purple, extends from thence beyond each eye : middle of the head, whole neck, upper part of the back, scapulars, and coverts of the wings, white : below the hind part of the head is a stain of pea-green : lower part of the back, tail, breast, and whole under side of the body, black : legs greenish. The FEMALE is almost entirely of a dull rust-color, barred with black : primaries and tail dusky. WEIGHT of the female is about three pounds and a half. The MALE is double the size of the common Tame Duck.

Inhabits the seas near *New York*, in the spring season; and breeds on the desert isles of *New England*, and from thence as far as the extreme coasts of the northern world, in *America*, *Europe*, and *Asia*; but never comes within land. Common in *Kamtschatka*. The most southern of its breeding-places are the *Fern* isles, on the coast of *Northumberland*. Lays seldom more than five eggs; those large, and of a pale green color. These birds afford the most luxurious of PLACE.

---

* Not in *Greenland* or *Spitzbergen*, as I once conjectured. See *Br. Zool.* ii. p. 578.

down,

down, which forms, in many of the regions, a confiderable article of commerce. Moft Ducks pluck off a certain down to form its neft : thefe have the greateft quantity, and the fineft and moft elaftic. It is cuftomary in fome places to take away the firft eggs, which occafions a fecond laying, and a fecond deplumation. In *Greenland* they lay among the grafs ; in *Sweden* among the juniper bufhes. Nature hath furnifhed them with fo warm a cloathing, that they brave the fevereft winter, even of the *Arctic* regions. In *Greenland*, they are feen in that feafon by hundreds, or even thoufands, in the fheltered fouthern bays : their breeding-places are in the moft northern. They take their young on their backs inftantly to fea, then dive, to fhake them off and teach them to fhift for themfelves. It is faid, that the males are five years old before they come to their full color ? that they live to a great age ; and will at length grow quite grey. They are conftant to their breeding-places : a pair has been obferved to occupy the fame neft twenty years. They dive to great depths for their food, which is fhells of all kinds. The *Greenlanders* kill them with darts ; purfue them in their little boats ; watch their courfe (when they dive) by the air-bubbles ; and ftrike them when they arife wearied. The flefh is valued as a food. The fkin of this and the next fpecies is the moft valuable of all, as a garment placed next to the fkin.

481. KING.   Grey-headed Duck, *Edw.* 154.—Anas Spectabilis, *Faun. Suec.* N° 112.—*Latham*, iii.   —LEV. MUS.

D. With a red bill, extending high up the forehead on each fide, in form of a broad bean-fhaped plate : head, and part of the hind part of the neck, light grey, bounded by a line of black dots : cheeks and neck, as low as the grey color, pea-green : a narrow black line from the bar of the bill bounds the lower part of the cheeks : throat, neck, and breaft, white : back, belly, and tail, black : leffer coverts of the wings, and primaries, dufky brown : fecondaries black, gloffed with rich purple ; coverts above them form

7   a great

a great bed of white : legs dirty red.   Size near double of the Mallard.   The Female differs greatly in color, being moftly black
and brown : the belly dufky : the plate on the bill flightly eminent.

This fpecies is found in *Hudfon's Bay*; and, in winter, as low as
New York.   Is as common in *Greenland* as the Eider.   Yields almoft as much down, and is as ufeful to the natives : has the fame
haunts, and is taken in the fame manner.   Inhabits the coaft of
Norway, and even has been killed in the *Orknies*.   Is frequent on
the *Arctic* fhores of *Sibiria*, and extends to *Kamtfchatka*.

Br. Zool. ii. N° 272.—Anas Fufca, Swârta, *Faun. Suec.* N° 109.
La grande Macreufe, *Pl. Enl.* 956.—*Latham*, iii.   —Lev. Mus.

D. With a broad bill, elevated near the bafe ; black in the middle ; yellow on the fides ; the nail red : behind each eye * is a
white fpot : a bar of the fame color croffes each wing : all the plumage befides is of a rich velvet black : legs red.   The Female is of
a deep brown ; but marked, like the male, with white.

Frequents the feas about *New York*.   Is very common in the great
lakes and rivers of the north and eaft of *Sibiria*, and on the fhores.
Extends to *Kamtfchatka*.   Is lefs common in *Ruffia*.   Lays from
eight to ten white eggs.   Notwithftanding they are Ducks which
at all other times frequent the fea, yet, in the laying feafon, go
far inland, and make their nefts : as foon as that tafk is over, the
males fly away ; but as foon as the young can fly, they are rejoined
by their mates ; followed by the brood †.

---

* Read *eye*, in the *Br. Zool.* inftead of *ear*.   † *Steller*, in *Nov. Com. Petrop.*
iv. 421.—*Strom.* p. 230.

Eaw.

483. BLACK.   Edw. 155.—Ph. Transf. lxii. 417.—Canard du Nord, ou le Marchand, Pl. Enl.
995.—Latham, iii.
Anas Perspicillata, Lin. Syst. 201.—Lev. Mus.—Bl. Mus.

D. With a compressed bill, rising into a knob at the base, each
side of which is marked with a patch of black; middle
white; sides of a deep orange; the edges black; nail red. fore
part of the head white: crown and cheeks black: just beneath the
hind part of the head, the neck is marked with a large white spot:
rest of the plumage of a dull black: legs and toes bright red; webs
black. WEIGHT two pounds two ounces. LENGTH twenty-one
inches. EXTENT thirty-five. The FEMALE is twenty inches long:
of a footy color: has no white on the hind part of the head, but
the cheeks are marked with two dull white spots.

PLACE.   Appears in *Hudson's Bay* as soon as the rivers are free from ice.
Breed along the shores: make their nests with grass, and line them
with feathers. Lay from four to six white eggs: hatch in the end
of *July*. Feed on grass. Extends to *New York*, and even to *South
Carolina* *.

484. SCOTER.   Br. Zool. ii. N° 273.—Anas Nigra, Faun. Suec. N° 110.—Latham, iii.
La Macreuse, Pl. Enl. 278.—Lev. Mus.

D. With a bill black; of a rich yellow in the middle; on the
base a green knob, divided longways with a furrow; no nail:
whole plumage black: head and neck glossed with purple: tail
cuneiform: legs black. WEIGHT two pounds two ounces. LENGTH
twenty-two inches. EXTENT thirty-four.

PLACE.   Sent to Mrs. *Blackburn*, from *New York*. Abounds on the great
lakes and rivers of the north and east of *Sibiria*, and on the shores;
but is less frequent in *Russia*. Inhabits *Sweden* and *Norway* †. Lives
much at sea. Is of a very fishy taste.

* *Catesby*, App.   † *Lawson*, 151.

Br.

Br. Zool. ii. N° 285.—Le Souchet, *Pl. Enl.* 971.—Anas Clypeata, *Faun.* Suec. N° 119.—*Latham,* iii.

Blue-wing Shoveler, *Catesby,* i. 96. (fem.)—LEV. MUS.—BL. MUS.

485. SHOVELER.

D. With a very large black bill, expanding greatly towards the end: head, and greatest part of the neck, of a mallard green; lower part of the neck, breast, and scapulars, white: belly bay: back brown: coverts of wings of a fine sky-blue: primaries dusky: speculum green: outmost feathers of the tail white; rest dusky, edged with white: legs red. Plumage of the FEMALE like that of the common Wild Duck; only the coverts of the wings are of the same colors with those of the Drake. LENGTH twenty-one inches. WEIGHT twenty-two ounces.

PLACE.

Found about *New York,* and even as low as *Carolina,* during winter. Is common in *Kamtschatka;* and breeds in every latitude of the *Russian* dominions; but chiefly in the north. Inhabits *Sweden* and *Norway.* We are to seek for the *Swaddle Bill,* an ash-colored Duck of *Carolina,* with an extraordinary broad bill, said not to be very common there, but to be very good food; we must therefore join it, for the present, to this species.

Br. Zool. ii. N° 276.—Anas Clangula, Knipa. Dopping, *Faun. Suec.* N° 722. Le Garrot, *Pl. Enl.* 802.—*Ph. Trans.* lxii. 417.—*Latham,* iii.   —LEV. MUS.—BL. MUS.

486. GOLDEN-EYE.

D. With a short broad black bill: large head, black, glossed with green: at each corner of the mouth a great white spot: breast and belly white: back, lesser coverts of the wings, and tail, black: scapulars black and white: greater coverts white: primaries dusky: legs orange. Head of the FEMALE rusty brown: neck grey: breast and belly white: coverts and scapulars dusky and cinereous: primaries and tail black: legs dusky. LENGTH nineteen inches. EXTENT thirty-one. WEIGHT two pounds.

Inhabits

Inhabits from *New York* to *Greenland:* in the laſt is very rare; and arrives in the bay on the breaking up of the ice : diſappears on the return of froſt. Frequents freſh-water lakes : makes a regular neſt of graſs, and feathers from its own breaſt. Lays from ſeven to ten white eggs. Is expelled *Sweden* by the froſt, except a few which haunt the unfrozen parts of rivers near the cataraġts : there they live, diving continually for ſhells. Extends to *Norway.*

487. SPIRIT.　　Little Black and White Duck, *Edw.* 100.—*Ph. Tranſ.* lxii. 416.
Anas Albeola, *Lin. Syſt.* 199.—*Latham,* iii.
FEM. Little Brown Duck, *Cateſby,* i. 98.
Sarcelle de la Louiſiane, dite la Religieuſe, *Pl. Enl.* 948.—LEV. MUS.

With a black bill : crown and fore part of the head of a gloſſy black, varying with green and purple : throat and upper part of the neck encircled with the ſame: cheeks and hind part of the head white : lower half of the neck, breaſt, belly, and ſcapulars, white : primaries, ſecondaries, and tertials, duſky ; upper ends of the ſecondaries white; coverts incumbent on them white; on the others duſky : back and tail duſky : legs orange. In the FEMALE the head and upper part of the neck duſky : a large white oblong ſpot marks the ſides of the head, beginning behind each eye : back, tail, primaries, and leſſer coverts, duſky : great coverts and ſecondaries white : breaſt and belly dirty white : legs orange. SIZE of a Wigeon.

Inhabits *North America,* from *Hudſon's Bay* to *Carolina.* Called ſometimes the *Spirit,* as is ſuppoſed, from its ſuddenly appearing again at a diſtance, after diving. Viſits *Severn* river, in *Hudſon's Bay,* in *June:* and makes its neſt in trees, among the woods near freſh waters.

D.　With

D. With the lower part of the bill black, the upper yellow: on the summit of the head is an oblong black spot: forehead, cheeks, rest of the head, and neck, white; the lower part encircled with black: scapulars and coverts of wings white: back, breast, belly, and primaries, black: tail cuneiform, and dusky: legs black. The bill of the supposed FEMALE? resembles that of the male: head and neck mottled with cinereous brown and dirty white: primaries dusky: speculum white: back, breast, and belly, clouded with different shades of ash-color: tail dusky and cuneiform: legs black. SIZE of a common Wild Duck.

Sent from *Connecticut*, to Mrs. *Blackburn*. Possibly the great flocks of pretty Pied Ducks, which whistled as they flew, or as they fed, seen by Mr. *Lawson* * in the western branch of *Cape Fear* inlet, were of this kind.

PLACE.

Buffel's-head Duck, *Catesby*, i. 95.—Anas bucephala, *Lin. Syst.* 200.—*Latham*, iii. —LEV. MUS.

489. BUFFEL.

D. With a short blue bill: head vastly increased in size by the fullness of the feathers; black, richly glossed with green and purple: neck white all round: upper part of the breast pure white; lower, and belly, clouded with pale brown: back, primaries, and secondaries, black: the coverts on the ridge of the wings mottled, bounding the others, which form a great bed of white: tail cinereous: legs orange.

Is found frequently in the fresh waters of *Carolina*, during winter.

PLACE.

* *Hist. Carolina*, 148.

Dusky

490. HARLEQUIN.

Dusky and Spotted Duck, *Edw.* 99; and the Female, *Edw.* 157 —*Catesby*, i. 98. Anas Histrionica, *Lin. Syst.* 204.—*Ph. Transf.* lxii. 419 —*Latham*, iii. Anas Brìnnond, *Olaffen Iceland.* ii. tab. xxxiv.—*Pl. Enl.* 798.—LEV. MUS.— BL. MUS.

D. With a small black bill : between the base and the eyes a great white patch : crown black, bounded by a light rusty line : cheeks, chin, and neck, black ; beneath each a white spot ; below that a short line of white, pointing down the neck : bottom of the neck, on each side, bounded by a transverse line of white ; beneath which is another of black : breast, back, scapulars, and part of the belly, of a pleasant slate-color : breast on each side marked with semilunar stripes of white, beginning at the shoulders, and bounded on each side with a stripe of black : wings and tail deep ash : rump, above and below, of a full black : legs black. The FEMALE is almost wholly dusky, and is marked at the base of the bill with a white spot, and another behind each ear. SIZE of a Wigeon.

PLACE.

Inhabits from *Carolina* to *Greenland :* in the last frequents, during summer, the rapid rivers, and the most shady parts. Nestles on the banks, among the low shrubs. Swims and dives admirably. In winter seeks the open sea. Flies high and swiftly, and is very clamorous. Feeds on shell-fish, spawn, and the larvæ of gnats. Is found in *Iceland*, and as low as *Sondmor* *. Is common from the lake *Baikal* to *Kamtschatka :* breeds there, as well as every where else, about the most rocky and rapid torrents.

491. POCHARD.

Br. Zool. ii. N° 284.—Anas Ferina, *Faun. S. ec.* N° 127.—*Latham*, iii. Le Millouin, *Pl. Enl.* 303.—LEV. MUS.—BL. MUS.

D. With a lead-colored bill : head and neck bright bay : breast and upper part of the back black : rest of the back, scapulars, and coverts of wings, pale grey, streaked transversely with lines of black :

* *Strom.* 243.

primaries

primaries dusky: belly grey and brown: tail deep grey: legs lead-colored. In the FEMALE the head rusty brown: breast rather darker: belly and coverts of wings cinereous: back like that of the male: legs lead-colored. LENGTH nineteen inches. EXTENT two feet and a half. WEIGHT one pound twelve ounces.

FEMALE.

Inhabits *North America*, in winter, as low as *Carolina* * ; and, I believe, is the Red-headed Duck of *Lawson*. Is found, in *Europe*, as high as *Drontheim*. Is met with in the great rivers and lakes in all latitudes of the *Russian* empire. A fresh-water Duck, and of excellent taste.

PLACE.

Black-billed Whistling Duck, *Edw.* 193 †.—*Latham*, iii.
Anas Arborea, *Lin. Syst.* 207.—Whistlers, *Catesby*, App. xxxvii.—*Lawson Carolina*, 149.—LEV. MUS.

492. WHISTLING.

D. With a black bill, and crown slightly crested: cheeks brown: hind part of the neck dusky; fore part white, spotted with black: back and wings brown; coverts spotted with black: tail and its coverts black: breast of a dark reddish color, spotted with black: belly white, mixed on the sides with black: legs long, and of a lead-color; hind claw placed high up the leg. Lesser than a Tame Duck. Described from Mr. *Edwards*.

Inhabits *South Carolina* and *Jamaica*. Is, from its voice, called the *Whistling Duck*: perches on trees. Placed here merely on the authority of the name given it by *Lawson* and *Catesby*. The last says, that it frequents the coasts of *Carolina* during winter; which makes me doubt, whether Mr. *Edwards*'s bird, a native of *Jamaica*, is the same: for it may be held as a rule, that the water-fowl of hot climates never retire in winter to colder; and that those of *Arctic* climates almost generally retire from them into warmer. Clouds of birds annually quit *Hudson's Bay*, and other severe climates, at approach of winter; stock the different latitudes of *North America*;

PLACE.

---

* *Catesby*, App.   † Probably not the female of *Edwards's* Duck, 194.

4 C

and

and return in fpring to encreafe and multiply.   To the conftitutions
of the SUMMER DUCK, a very few other water-fowl, and to many land-
birds, the warm temperature of the *Carolinas* is climate fufficiently
north.   They are driven, by the exceffive heat and arid foil of the
Antilles and *Guiana*, to the moift favannas and woods of thefe provinces, there to difcharge the firft great command.

493. SUMMER.   Summer Duck, *Catefby*, i. 97.—*Edw.* 101.—Anas Sponfa, *Lin. Syft.* 207.
Le beau Canard hupé de la Caroline, *Pl. Enl.* 980.—*Latham*, iii.   —LEV.
Mus.—BL. Mus.

D. With the ridge and nail of the upper mandible black; lower
part fcarlet: on the head a beautiful creft, hanging half down
the neck, and beginning at the bafe of the bill; upper part fhining
purple; beneath that a line of white; then fucceeds purple; and that
again is bounded by white: cheeks purplifh and green: throat, and
part of the neck, pure white: from the hind part of the neck a bead
of purple divides the white, and points towards the throat: reft of
the neck and breaft ferruginous, fpotted with white triangular fpots:
belly white: feathers of the fides, which hide part of the wings,
elegantly marked downwards with incurvated lines of black and
white: back deep brown, gloffed with copper and green: primaries
dufky: fecondaries refplendent blue: coverts of the tail, and tail itfelf, dufky, gloffed with green: legs dirty orange.   Head of the
FEMALE of a deep brown; crefted, but not fo much as the Drake:
back deep brown: cheeks brown: behind each eye a white fpot:
throat white: neck and breaft reddifh brown, with white fagittal
fpots: belly white.   LENGTH, from the bill to the tip of the tail,
near nineteen inches.   EXTENT about thirty.

PLACE.   This moft elegant fpecies is found from *New York* to the *Antilles*,
and alfo in *Mexico*.   It paffes the fummer in *Carolina*; and in a
fingular manner makes its neft in the holes made by Woodpeckers
in the loftieft trees, which grow near the water, efpecially the deci-
duous cyprefs.   When the young are hatched, they are conveyed

3   down

down on the backs of the old ones, to whom the Ducklings adhere closely with their bills. It often nestles on the bodies or boughs of trees which have fallen over the streams which run up the woods. It appears in *New York*, in the latter end of *February* or beginning of *March*, and retires towards the south at approach of winter. They are very delicate eating. The *Mexicans* call it *Yztactzonyayauhqui*, or the bird of the *various-colored head*. It is there migratory. The natives feign that, from the situation of its legs, it cannot stand.

Br. Zool. ii. N° 279.—Anas Boschas. Gräs-and, Blänacke, *Faun. Suec.* N° 131.— 494. **MALLARD.**
    Ph. Transf. lxii. 419.—*Pl. Enl.* 776, 777.—*Latham,* iii.   —**Lev. Mus.**
    —**Bl. Mus.**

D. With a bill of a yellowish green : head and neck of a shining changeable green : on the front of the lower part of the neck is a semicircle of white : breast of a purplish red : lower part of the back, and belly, grey, crossed with speckled lines of black : speculum purple : four middle feathers of the tail curled upwards : legs saffron-colored. **Female** is of a pale reddish brown, spotted with black. **Length** twenty-three inches. **Extent** thirty-five. **Weight** about two pounds and a half.

Inhabits the northern parts of *North America*, from *Hudson's Bay* to *Carolina* * : is frequent in *Greenland*, and continues there the whole year. Arrives in *Hudson's Bay* in *May* : retires in *October*. Is common in all latitudes of the *Russian* empire : and was observed by *Steller* in the *Aleutian* islands. In *Sweden* retires in winter to the shores of *Schonen*; but in severe seasons passes over to *Denmark* and *Germany*, possibly to *England*; for this island can hardly supply the vast wintery flocks.

Place.

* *Catesby,* App.

495. ILATHERA.   Ilathera Duck, *Catesby*, i. 93.—Anas Bahamenfis, *Lin. Syft.* 199.—*Latham*, iii.

D. With a large dufky blue bill; on the bafe of the upper mandible a great triangular orange-colored fpot: head, as far as the
eyes, hind part of the neck, and back, of a mixed grey, inclining to
yellow: fore part and fides of the neck white: belly of the fame color,
fpotted with darker: leffer coverts of the wings, and primaries,
dufky; great coverts green, tipt with black: fecondaries dull yellow:
legs lead-colored. In SIZE fomewhat lefs than the common Tame
Duck.

PLACE.   Inhabits the *Bahama* iflands; but is very rare: extends to the
Brafils, where the *Indians* call it *Marecu* *. This fpecies, the *Summer Duck*, and the *Whiftling Duck, Edw.* 193, perch and rooft on
trees; and are among the few of this clafs which do not migrate
northward to breed.

496. DUSKY.   **D.** With a long and narrow dufky bill, tinged with blue: crown
dufky: chin white: neck pale brown, ftreaked downwards with
dufky lines: back, and coverts of the wings, deep brown: breaft and
belly of the fame color, edged with dirty yellow: primaries dufky:
fpeculum of a fine blue, bounded above with a black bar: tail cuneiform; dufky, edged with white: legs in one fpecimen dufky, in another yellow. LENGTH near two feet.

PLACE.   From the province of *New York.*—BL. MUS.

497. WESTERN.   Anas *Stelleri*, PALLAS *Spicil. Zool.* fafc. v. p. 35. tab. v.—*Latham*, iii.
—LEV. MUS.

D. With the head, cheeks, and upper part of the neck, white: between the bill and the eyes a mallard-green fpot; another
acrofs the hind part of the head: chin and throat of a full black:

* *Marcgrave*, 214.

around

around the neck a black glossy color : back of the same color : coverts
of the wings white : primaries dusky : secondaries black, tipt with
white : breast and sides of a light yellowish brown : belly, vent, and
tail, black.   SIZE of a Wigeon.

Brought by the late navigators from the western side of *America*;
but had been before discovered by *Steller* to breed among the inac-
cessible rocks about *Kamtschatka*; to flyin flocks, and never to enter
the mouths of rivers.

Br. Zool. ii. N° 275.—Anas marila, *Faun. Suec.* N° 111.—*Ph. Transf.* lxii. 413.—   498. SCAUP.
Le Millouinan, *Pl. Enl.* 1002.—*Latham*, iii.   —LEV. MUS.—BL. MUS.

D. With a broad, flat, and blueish grey bill : irides yellow : head
and neck black, glossed with green : breast black : back, coverts
of the wings, and scapulars, marked with numbers of transverse lines
of black and grey : primaries dusky : secondaries white, tipt with
black : belly white : tail, coverts, and vent feathers, black : legs
dusky.   Male WEIGHS a pound and a half : female two ounces more.
LENGTH sixteen inches and a half.   EXTENT twenty inches.

Inhabits *America*, as high as *Hudson's Bay* : comes there in *May*;
retires in *October*.   Is found in *Iceland*, and most part of the north of
Europe.   Are common on the northern shores of *Russia* and *Sibiria*;
and are most frequent about the great river *Ob* : migrate southward :
dive much : and feed on shell-fish.

D. With a large blueish bill : head and neck of a very pale brown :   499. BROWN.
lower part of the last, and breast, of the same color, edged with
rust-color : wings cinereous grey : speculum blue, tipt with white :
tail and legs dusky.

Inhabits *Newfoundland*.

Br.

500. PINTAIL.

Br. Zool. ji. N° 282.—Anas acuta, Aler, Ahlfogel, *Faun. Suec.* N° 126.
Le Canard à longue queue, *Pl. Enl.* 959.—*Latham*, iii.　　　—Lev. Mus.
　—Bl. Mus.

D. With bill black on the middle ; blueifh on the fides : head and half the neck rufty brown : from the ears, half way of each fide of the neck, a white line, bounded by black, points downwards : lower hind part of the neck, back, and fides, marked with white and dufky waved lines : fore part of the neck, breaft, and belly, white : coverts of the wings cinereous ; loweft tipt with dull orange : fecondaries marked with green, black, and white : exterior feathers of the tail afh-colored ; middle black, and three inches longer than the reft : legs afh-colored. FEMALE brown, fpotted with black. WEIGHT twenty-four ounces. LENGTH two feet four. EXTENT three feet two inches.

PLACE. Appears about *New York* in winter : breeds in the north : in *Europe*, about the *White Sea*. Migrates fouthward at approach of the froft. Is feen in *Sweden* about fourteen days in the fpring, on its paffage northward : and in autumn repaffes the fame way to the fouth. Vifit the *Orknies* in great flocks in the winter. In the *Ruffian* empire, extends to *Kamtfchatka*.

501. LONG-
TAILED.

Br. Zool. ii. N° 283.—Anas hyemalis. Winter-and, *Faun. Suec.* N° 125.
Anas Glacialis, *Lin. Syft.* 203.—*Ph. Tranf.* lxii. 418.—Male, *Edw.* 280.
　Female, 156.—*Latham*, iii.
Le Canard de Miclon, *Pl. Enl.* 954.—Lev. Mus.—Bl. Mus.

D. With bill black, orange in the middle : forehead, fides of the head, and neck, pale brown, dafhed with rofe-color : beneath each ear a large dufky fpot points downwards : hind part of the head and neck, throat, and breaft, white : back and belly black : fides and vent feathers white : fcapulars long and white : coverts of the wings gloffy black : primaries dufky : fecondaries dark rufty brown : two middle feathers of the tail black, and four inches longer than the others, which are white : legs red. FEMALE ; crown dufky : cheeks white :

white: reft of the head, neck, back, and breaft, coverts of the wings, and primaries, deep brown: fcapulars and fecondaries rufty: belly white: tail and legs like thofe of the MALE.

Inhabits to the extreme north. Breeds in *Hudfon's Bay* and *Green-* PLACE. *land,* among the ftones and grafs: makes its neft, like the *Eider,* with the down of its own breaft; which is equal in value to that of the *Eider,* if it could be got in equal quantity; but the fpecies is fcarcer. It lays five eggs: fwims and dives admirably: and feeds on fhell-fifh, which it gets in very deep water. Flies irregularly, fome-times fhewing its back, fometimes its belly. Continues in *Greenland* the whole year, in unfrozen places *: but there are feafons fo very fevere, as at times to force them towards the fouth. Thofe which breed between *Lapland* and the polar circle, are often driven into *Sweden,* and the neighborhood of *Peterfburg*: thofe from the coaft of the *Icy* fea, as low as lat. 55; but on the fetting in of froft, retire ftill further fouth, unlefs where fome open fpots remain in the rivers. Vifit the frefh-water lakes in the *Orknies,* in *Oflober,* and continue there till *April.* At fun-fet they are feen, in great flocks, returning to and from the bays, where they frequently pafs the night, and make fuch a noife as to be heard fome miles in frofty weather. Their found is like *Aan-gitche,* and is faid not to be difagreeable. *Steller,* who obferved them in *Kamtfchatka* †, fays, that their larynx has three openings, covered with a thin (I fupofe valvular) mem-brane, which forms the fingularity of the voice. *La Sarcelle de Feroe,* or the *Feroe* Teal, of M. *Briffon* ‡, is probably conjeflured, by M. *Brunnich,* to be only a variety of this fpecies: feemingly a female.

Le Canard Jenfen de la Louifiane, *Pl. Enl.* 955.—*Latham,* iii.   —LEV. MUS. 502. AMERICAN WIGEON.

D. With a lead-colored bill, tipt with black: crown and forehead yellowifh white: hind part of the head, and whole neck, prettily fpeckled with black and white: behind each eye is a large black fpot,

* *Faun. Groenl.* p. 73.   † *Defcr. du Kamtfchatka,* 498.   ‡ vi. 466.
tab. xl.—*Pl. Enl.* 999.

gloffed

gloffed with green : back and fcapulars pale ruft and black, elegantly difpofed in narrow tranfverfe waving lines : coverts of the wings white : primaries, coverts of the tail, and vent, black : tail cuneiform; middle feathers black; the reft cinereous : legs dufky. In Size fuperior to the *Englifh* WIGEON ; with which it feems to agree in colors and marks, except thofe on the head.

PLACE.   Sent from *New York*, under the name of the *Pheafant Duck :* is a rare bird there. Found as low as *Louifiana.*—BL. MUS.

503. WHITE-FACED.   White-faced Teal, *Catefby,* i. 100. Male.—Blue-wing Teal, *Catefby,* i. 99. Fem.—Anas Difcors, *Lin. Syft.* 205.
Sarcelle mâle de Cayenne, dite le Soucrourou, *Pl. Enl.* 966.—*Latham,* iii. —LEV. MUS.—BL. MUS.

D. With bill and crown black; bafe of the bill bounded by black : between the laft and the eyes a white ftripe, ending on each fide of the chin : cheeks, hind part of head, and whole neck, purplifh green : breaft yellow, fpotted elegantly with black : back brown, waved with a lighter color; on the lower part feveral long, narrow, light brown feathers : coverts of the wings fine cærulean : primaries dufky : fpeculum green : vent black : tail brown : legs yellow. The FEMALE is almoft entirely brown; in parts marked with dufky fpots : the blue on the wings duller than that of the DRAKE. In SIZE a little larger than a Teal.

PLACE.   This fpecies is found as high as *New York.* Arrives in *Carolina* in great plenty, in *Auguft,* to feed on the rice ; and continues till *October,* when the rice is got in. In *Virginia,* where there is no rice, it feeds on wild oats. Is reckoned moft delicious meat. Extends as far fouth as *Guiana.*

A VARIETY ?   D. With crown and upper part of the neck dufky brown : cheeks, under part, and fides of the neck, whitifh brown, mottled with darker : back, breaft, and belly, marked with great dufky fpots, edged with dirty white : coverts of the wings pale fky-blue; lower order white : fpeculum rich purple, with a white edge : primaries and tail dufky

dufky. Size of the laft, with the female of which it has great affinity ; but in the purple *fpeculum* refembles the GADWALL.

Brought from *Newfoundland* by Sir *Jofeph Banks*.

Ph. Tranf. lxii. 419 —Anas circia? *Faun. Succ.* N° 130.—*Latham*, iii. Krik-and *Danis, Brunnich,* N° 130.

D. With head and upper part of the neck of a fine deep bay : from each eye to the hind part of the head is a broad bar of rich changeable green : wants the white line, which the *European* kind has above each eye, having only one below : lower part of the neck and breaft dirty white, beautifully fpotted with black : has over each fhoulder a lunated bar, another diftinction from our fpecies : coverts of wings brown : upper part of the back marked with waved lines of white and black ; lower part brown : tail dufky : *fpeculum* green : legs dufky.   Plumage of the FEMALE of a brownifh afh, tinged with red, and fpotted with black : wings refemble thofe of the male.

Inhabits *America*, as high as *Hudfon's Bay*, and as low as *Carolina*. Is found plentifully about *Severn* river, in the woods and plains near the frefh waters; and has from five to feven young at a time. Difappears in autumn : and is found, during winter, as low as *Carolina*, and perhaps *Jamaica*.   We feem here to have recovered the SUMMER TEAL of Mr. *Willughby* *, to which the *American* kind has great affinity. He calls it the left of Ducks : and muft be the fame with the fecond kind defcribed by *Lawfon* † as frequenting frefh waters ; being leffer than the common fort, and always nodding their heads.

* *Ornith.* 378.   † *Hift. Carol.* 149.

A. GREAT GOOSE. With a black bill, tawny at the bafe: a dufky body; white beneath: fcarlet legs. Of a vaft SIZE, weighing near twenty-five or thirty *Ruffian* pounds.

PLACE.

This fpecies is found in the eaft of *Sibiria*, from the *Lena* to *Kamt-fchatka*: and is taken in great numbers, together with the RED-NECK-

TAKEN IN GLADES.

ED GOOSE, in glades, as we do Woodcocks in *England*. The Geefe in the day-time repair to the corn-fields and meadows: in the evening refort to the lakes, to wafh themfelves and pafs the night. The *Sibirians* generally fix on a place where there are two or three lakes near each other, and cut between each an avenue through the thick birch woods of the country. If there is not the advantage of adjacent lakes, the avenue is made through the woods which border the fides of any which the birds frequent. At the entrance of the glades, on each fide, a tall birch-tree is left ftanding, and all their branches ftripped away: from the tops of thefe naked trees is placed a ftrong net, which fills the breadth of the avenue: this net is capable of being dropped or raifed at pleafure, by means of certain long cords which run along the top; and the ends of which are held by a man who conceals himfelf in the high grafs. The Geefe commonly leave the lakes an hour before fun-rife; and, as they do not chufe to fly high at that feafon, prefer going through the avenues; and with their long extended necks ftrike into the nets, which are fuddenly dropped; and twenty, and often more, of the Geefe are taken at a time. All forts of Ducks, and other water-fowl, are taken in the fame manner *.

* PALLAS's *Travels*, ii. 325, 326.

CHINESE

B. Chinese Goose. Anfer Cygnoides, *Lin. Syst.* 194. β.—Swan Goofe, *Wil Orn.* 360.
—*Raii. Syn. av.* 138.—*Briffon*, vi. 280.—*Latham*, iii.   —Lev. Mus.

D. With a black bill, and a large protuberance at the bafe, biggeft
in the males : on the chin is a naked, pendulous, black fkin : from
the crown to the back a black line runs down the hind. part of the
neck : the reft of the neck and breaft is of a cream-color, often dafhed
with tawny : belly white : between the bafe of the bill and the eyes
is a white line : the back and wings deep grey : tail of the fame
color, with whitifh tips : legs red : in fome the bill is of the fame
color.   In Length often reaches to three feet three.

This fpecies is found wild about lake *Baikal* ; in the eaft of *Sibiria* ;   Place.
and in *Kamtfchatka*.   They are very commonly kept tame in moft
parts of the *Ruffian* empire.   Will produce, with the Common Goofe,
a breed which preferves an exact medium between both fpecies.   As
an exception to the remark that a mulifh race will not breed, thefe
frequently couple with one another, and with the genuine kind *.
They are frequent in *China :* are very ftately birds, therefore are dig-
nified with the title of *Swan Goofe*.

C. Red-breasted Goose. Anfer ruficollis, *Pallas Spicil. Zool.* fafc. vi. 21. tab. iv.—
Lev. Mus.—Bl. Mus.

D. With a fhort black bill ; a great patch of white between the bafe
and the eyes, bounded by black : crown, chin, hind part of the
head and neck, back, wings, and tail, of an intenfe black : fides marked with a few white fpots : greater coverts tipt with the fame color :
coverts of the tail white : lower part and fides of the neck of a bright
bay, bounded by a narrow line of white : the breaft and lower part of
the neck divided from the belly and back by a circle of black and

---

* *Doctor* Pallas.

another

another of white : legs black. LENGTH one foot ten inches. Ex-
TENT three feet ten. WEIGHT three pounds *Troy*.

PLACE,

This moſt elegant of Geeſe is found to breed from the mouth of
the *Ob*, along the coaſts of the *Icy* ſea, to that of the *Lena*. The winter quarters of theſe birds is not certainly known. They are obſerved
in the ſpring, flying from the *Caſpian* ſea, along the *Volga*, northward,
in ſmall flocks ; and are ſeen about *Zarizyn*, between the ſixth and
tenth of *April*. They reſt a little time on the banks of the *Sarpa*,
but ſoon reſume their *Arctic* courſe*. Their winter retreat is probably in *Perſia*. They are highly eſteemed for the table, being quite
free from any fiſhy taſte.

D. SHIELDRAKE, *Br. Zool.* ii. Nᵒ 278.—Tadorne, *Pl. Enl.* 53.—*Latham*, iii.
Anas Tadorna Jugas *Gotlandis, Faun. Suec.* Nᵒ 113.—LEV. MUS.—BL. MUS.

D. With a ſcarlet bill : on the baſe of that of the male a large protuberance : head, and part of the neck, of a mallard-green :
reſt of the neck and belly white : the breaſt croſſed with a large band
of orange bay : coverts of wings, and the back, white : ſcapulars pied :
tail white ; tips of the outmoſt feathers black : legs fleſh-color.
WEIGHT of the male two pounds ten ounces. LENGTH two feet.
EXTENT three and a half.

PLACE.

Inhabits northern *Europe*, as high as *Iceland*. Viſits *Sweden* and the
Orknies in winter : returns in the ſpring. Continues in *England* the
whole year. Is found in *Aſia* about the *Caſpian* ſea, and all the ſalt
lakes of the *Tartarian* and *Sibirian* deſerts ; and extends even to
Kamtſchatka.

E. GULAUND.

D. With a narrowed bill : head of a mallard-green : breaſt and
belly white. SIZE between the Gooſe and Duck kind.

* *Extracts,* ii. 20.

Inhabits

Inhabits the moraffes of *Iceland.* Lays from feven to nine eggs. Is a fcarce fpecies. The account of it was communicated to me by M. *Brunnich,* from the catalogue of Doctor *Biorno Paulfen.* The *Icelanders* call it *Gulaund.*

F. MORILLON, *Br. Zool.* ii. N° 277.—Anas Glaucion. Brunnaeke, *Faun. Suec.* N° 123. —*Latham,* iii. —LEV. MUS.

D. With dufky ruft-colored head: irides gold-colored: neck with a white collar; and beneath that another, broader, of grey: back and coverts of wings dufky, marked with a few white ftripes: greater coverts dufky, with a few great white fpots: primaries and tail black: fecondaries white: breaft and belly white: above the thighs black: legs yellow. Rather lefs than the GOLDEN EYE.

Inhabits as high as *Sweden:* is found, but rarely, even in *Greenland*:* or may be fuppofed to be feen in the intermediate parts. Is frequent in every place in *Ruffia* and *Sibiria,* and even in *Kamtfchatka.*

G. TUFTED, *Br. Zool.* ii. N° 274.—Anas fuligula, Wigge, *Faun. Suec.* N° 132.— Le Morillon, *Pl. Enl.* 1001.—*Latham,* iii. —LEV. MUS.—BL. MUS.

D. With a thick, fhort, pendent creft: belly and under coverts of the wings pure white: primaries dufky; part of their inner webs white: fecondaries white, tipt with black: all the reft of the plumage black; about the head gloffed with violet: legs blueifh grey. WEIGHT two pounds. LENGTH fifteen inches.

Inhabits *Europe,* as high as *Norway.* Common in all latitudes of the *Ruffian* empire; but commonly travels northward to breed. Frequent in *Kamtfchatka.*

---

* *Faun. Groenl.* N°

H. Hrafn-ond, *Olaffen Iceland*, fect. 688.—*Muller*, N° 161.—*Latham*, iii.

D. With a crefted head, black above: under fide of the neck, breaft, and belly, white: legs faffron-colored.

PLACE.　　　Inhabits *Iceland*. Whether a variety of the former? for the *Icelanders* ftyle that fpecies *Hrafas-aund* *.

I. FALCATED. Anas Falcaria, *Pallas Itin.* iii. 701.—*Latham*, iii.

D. With a fmall dufky bill: feathers above the bafe of the upper mandible white: middle of the head pale ruft: reft of the head filky green, variable, and changing, on the fides of the neck, to refplendent copper: from the head to the hind part of the neck is a creft clofely compreffed, and ending in an angle: throat and half the fore part of the neck white; which color encircles the neck, and is bounded above by another of variable black and green; the reft of the neck, and the breaft, elegantly marked with femicircles of grey and black: the back and wings undulated with the fame colors: the *fpeculum* of the color of polifhed fteel, edged with white: five laft fecondaries long and falcated, of a violet-color edged with white: vent white, croffed with a black bar: legs dufky. SIZE of a Wigeon.

PLACE.　　　Found, but rarely, in *Kamtfchatka*. Frequent in the eaft of *Sibiria*, from the *Jenefei* to the *Lena*, and beyond lake *Baikal*. None in the weft. Probably winters in *China* and the *Mongalian* deferts.

K. WIGEON, *Br. Zool.* ii. N° 286.—Anas *Penelope*, Wriand, *Faun. Suec.* N° 124. —Le Canard fiffleur, *Pl. Enl.* 825.—*Latham*, iii.　　—LEV. MUS.—BL. MUS.

D. With forehead whitifh: head and upper part of the neck of a bright light bay: hind part of the head, and breaft, vinaceous: in other refpects like the AMERICAN kind. FEMALE colored like a

* *Biorne's Lift.*

Wild

XXIII

Falcated Duck I. p. 574.

Western Duck N.º 407.

P. Mazell sculp.

Wild Duck. Length twenty inches. Extent two feet three. Weight near twenty-three ounces.

Inhabits *Europe*, perhaps not higher than *Sweden*. Is not uncommon about the *Caspian* sea, and afcends its rivers, but not far up. Is fometimes feen in the great lakes on the eaft fide of the *Urallian* chain; but not in the reft of *Sibiria*. Is found in plenty about *Aleppo*, during winter *: and taken in great numbers in the *Nile*, in nets, juft before the waters have quite fubfided †. Thefe probably retire north to breed. The *Germans* call this fpecies *Pfeiff-ent*, or the *Fifing Duck*, from its acute note. The *French*, for the fame reafon, call it *le Canard fiffleur*: and the *Englifh*, the *Whewer*. My Bimaculated Duck, *Br. Zool.* ii. N° 287, nas been difcovered, by Doctor Pallas, along the *Lena*, and about lake *Baikal*; and a defcription fent by him to the Royal Academy at *Stockholm*, under the title of *Anas Glocitans*, or the *Clucking Duck*, from its fingular note.

Place.

Gadwall, *Br. Zool.* ii. N° 288.—Anas ftrepera, *Faun. Suec.* N° 121.—Le Chipeau, *Pl. Enl.* 958.—*Latham*, iii. —Lev. Mus.—Bl. Mus.

D. With a black flat bill: head and upper part of the neck reddifh, fpotted with black: breaft, upper part of the back, and fcapulars, elegantly marked with black and white lines: belly dirty white: coverts on the ridge of the wings reddifh brown; the next purplifh red, with a border of black: primaries dufky: fpeculum white: tail cinereous: legs orange. Breaft of the female reddifh brown, fpotted with black: other colors fimilar, but more dull. Rather lefs than a Wigeon.

This fpecies does not feem to advance higher in *Europe* than *Sweden*. In the *Ruffian* empire extends over moft of the latitudes of the *European* and *Sibirian* part, except the eaft of *Sibiria*, and *Kamtfchatka*.

Place.

* *Ruffell's Aleppo.* † *Haffelquift*, 288.

Lapmark,

M. Lapmark. Skoaara, *Leems Lapmark*, 266.—Anas latiroſtra, *Brunnich*, N° 91.—Le Canard brun ? *Pl. Enl.* 1007 —*Latham*, iii.

D. With a broad black bill and legs; the laſt reaching far beyond the tail: head, neck, and upper part of the body, duſky, thick ſet with ſmall ſpots: on each ſide of the baſe of the bill a great white ſpot. neck and breaſt clouded: on the wings an oblique white mark: belly duſky: feathers on the ſides ferruginous. Size of a Wild Duck.

Place.   Inhabits *Lapmark*, and frequents both ſea and freſh-water. Is alſo found in *Denmark*.

N. Red. Anas rutila, *Faun. Suec.* N° 134.—Ferruginous Duck, *Br. Zool.* ii. N° 285. —*Latham*, iii.

D. With a long pale blue bill, much flatted: head, neck, and upper part of the body, a fine reddiſh brown: throat, breaſt, and belly, paler: belly white: legs pale blue: webs black. Weight twenty ounces.

Place.   Found, but rarely, in the *Swediſh* rivers. Sent to me from *Denmark*, by the late Mr. *Fleiſcher*. Has been ſhot in *England*.

O. Garganey. *Br. Zool.* ii. N° 289.—Anas Querquedula, *Faun. Suec.* N° 128. La Sarcelle, *Pl. Enl.* 946.—*Latham*, iii.   —Lev. Mus.—Bl. Mus.

D. With a white line from the further corner of each eye, pointing to the nape: crown duſky, ſtreaked lengthways: cheeks and neck very pale purple, ſtreaked with white: chin black: breaſt light brown, marked with ſemicircular bars of black: ſpeculum green: ſcapulars long and narrow, hanging over the wings, and ſtriped with white, aſh, and black: tail duſky. Length ſeventeen inches. Extent twenty-eight. The Female has an obſcure whitiſh

7

mark

mark over the eyes: reſt of the plumage browniſh aſh, ſpotted. Wants the *ſpeculum*.

This elegant ſpecies ſeems not to inhabit *Europe* higher than *Sweden*; but is found in all latitudes of the *Ruſſian* empire, even to *Kamtſchatka*. PLACE.

P. EUROPEAN TEAL, *Br. Zool.* ii. N° 290.—Anas Crecia. Arta. Kræcka, *Faun. Suec.* N° 129.
La petite Sarcelle, *Pl. Enl.* 947.—*Latham*, iii.   —LEV. MUS.—BL. MUS. ‥

O U R ſpecies in all reſpects reſembles the *American*, except in having a white line above and beneath each eye, and in wanting the humeral ſtripe of white, which the latter has.

In *Europe* it is found as high as *Iceland*; and even in that ſevere climate lays from thirteen to nineteen eggs *. The *American* ſpecies appears to be far leſs prolific. Found in the *Ruſſian* empire, in the ſame places with the GARGANEY. PLACE.

Biorne's Liſt.

LVIII. PELECAN*   *Gen. Biras,* XCIII.

505. GREAT.   Pelecanus Onocratolus, *Lin. Syſt.* 215.—*Edw.* 92.—*Ph. Tranſ.* lxii. 419.
Le Pelican, *De Buffon,* viii. 282.—*Pl. Enl.* 87.—*Latham,* iii.   —LEV. MUS.

P. With a bill fifteen inches long, flat, dilated near the point, with a hook at the end, and a ridge from that to the baſe running along the middle ; on the midway of the ridge riſes a bony proceſs, an inch and ſeven tenths high, three inches broad at the baſe, and only two tenths of an inch thick. In ſome are ſeveral leſſer proceſſes between this and the point : a vaſt naked membranaceous pouch extends from the point of the lower mandible, widening gradually, and extending ten inches down the front of the neck : on the hind part of the head is a tuft of very narrow delicate feathers, not very diſcernible, as they uſually lie flat : the reſt of the head and neck is covered with moſt exquiſitely fine down, and very thick ſet : the reſt of the plumage white, except the primaries and baſtard wings, which are black : legs fleſh-color. The largeſt of web-footed Water-Fowl. Some are ſuperior in SIZE to a SWAN. One was killed off *Majorca,* which weighed twenty-five pounds. Their extent of wings from eleven to fifteen feet. Notwithſtanding their great bulk, they ſoar to a moſt ſurpriſing height. This is owing to the amazing lightneſs of the bones, which, all together, do not weigh a pound and a half. Add to this, the quantity of air with which its body is filled, which gives it a wonderful ſpecific lightneſs.

PLACE,   One of the birds from which this deſcription was taken, was ſhot at *Auguſta* in *South Carolina,* a hundred and fifty miles from the

---

* This genus, in the *Br. Zool.* is called by the more familiar name of *Corvorant,* there being none of the *Pelecan* ſpecies in *Britain.*

ſea.

fea. It agrees entirely with the Pelecan of the old continent, except in the bony proceffes on the bill. The other was fent, with other birds, from *Hudfon's Bay*, to the Royal Society. Inftead of the bony proceffes on the bill, was a tuft or fibrous fringe, fufficient to identify the fpecies. This fpecies extends over moft parts of the torrid zone, and many parts of the warmer temperate. Is found in *Europe* on the lower parts of the *Danube*, and in all parts of the *Mediterranean* fea, almoft all *Africa*, and *Afia Minor*. Are feen in incredible numbers about the *Black* and *Cafpian* feas; and come far up the rivers, and into the inland lakes of the *Afiatic Ruffian* empire; but grow fcarcer eaftward, and are feldom met with fo far north as the *Sibirian* lakes; yet are not unknown about that of *Baikal*. They are common on the coaft of *New Holland*, where they grow to an enormous fize *. They feed upon fifh; which they take fometimes by plunging from a great height in the air, and feizing, like the GANNET: at other times, they fifh in concert, fwimming in flocks, and forming a large circle in the great rivers, which they gradually contract, beating the water with their wings and feet, in order to drive the fifh into the center; which when they approach, they open their vaft mouths, and fill their pouches with their prey, then incline their bills, to empty the bag of the water; after which they fwim to fhore, and eat their booty in quiet. As the pouch is capable of holding a dozen quarts of water, a guefs may be made of the quantity of fifhes it can contain. The *French* very properly call them *Grand-gofiers*, or *Great-throats*. It is faid that when they make their nefts in the dry deferts, they carry the water to their young in their vaft pouches, and that the lions and beafts of prey come there to quench their thirft, fparing the young, the caufe of this falutary provifion. Poffibly, on this account, the *Egyptians* ftyle this bird the *Camel of the River*; the *Perfians*, *Tacab*, or the *Water-carrier*.

---

* *Cook's Firft Voy.* iii. 627.

Pelecanus

506. DUSKY.   Pelecanus Onocrotalus occidentalis, *Lin. Syst.* 215.—*Edw.* 93.
               Le Pelican Brun, *De Buffon,* viii. 306.—*Pl. Enl.* 957.—*Latham,* iii.
               —LEV. MUS.—BL. MUS.

P. With a red bill and black hook : the pouch extending half
way down the neck : between the bill and eyes naked and
red : head mottled with afh-color and white : the nape flightly
crefted : hind part of the neck covered with foft cinereous feathers :
back, fcapulars, primaries, and coverts, dufky, edged with dirty
white : tail deep afh : legs dufky green. In SIZE fcarcely equal to a
SWAN.

PLACE.   Inhabits, during fummer, *Hudfon's Bay.* One was fent to Mrs.
Blackburn, fhot near *New York,* I think in the winter. Extends
to *Louifiana* * ; to *Jamaica,* the bay of *Campechy* †, and as low
as *Carthagena.* They fit on rocks in the fea in a fluggifh manner,
with their bills refting on their breafts.

507. CHARLES-   P. Dufky above : white on the breaft and belly, with a pouch, be-
    TOWN.   ginning at the chin, and reaching to the breaft-bone, capable
of containing numbers of gallons of liquids. SIZE of a *Canada*
Goofe.

PLACE.   Abound in the bay of *Charles-town,* where they are continually
fifhing.—Doctor GARDEN ‡.

---

* *Du Pratz,* ii. 79.   † *Dampier's Voy.* Campechy, 70.
‡ The fame Gentleman inform d me, that the SNOWY OWL, N°   is frequent
near the fhores of *South Carolina,* among the *Palmetto* trees.

Wd

Wil. Orn. 330.—Pelecanus Graculus, *Faun. Suec.* N° 146.—*Latham*, iii.   508. SHAG.
   —Lev. Mus.

P. With head and neck black, gloffed like filk with green : the back and coverts of wings of the fame color, edged with purplifh black : belly dufky and dull ; the middle cinereous : tail confifts of twelve feathers, dufky, gloffed with green : legs black : middle claw ferrated. LENGTH two feet fix. EXTENT three feet eight. WEIGHT four pounds.

Frequent in many parts of *Great Britain.* Found in *Sweden, Norway,* and *Iceland.*   PLACE.

Br. Zool. N° 293.—Pelecanus Carvo, Haffs-tjader, *Faun. Suec.* N° 145.—   509. CORVORANT.
   Latham, iii.   —Lev. Mus.—Bl: Mus.

P. With a narrow bill, hooked at the end : a fmall dilatable pouch under the chin ; feathers at its bafe white, in the male : head and neck of a footy blacknefs, fometimes ftreaked with white : coverts of wings, back, and fcapulars, deep green, edged with black, gloffed with blue : breaft and belly black : on the thighs of the male a tuft of white : tail confifts of fourteen feathers, and is rounded. WEIGHT feven pounds. LENGTH three feet four. EXTENT four feet two.

Extends over all parts of the northern hemifphere, even to *Greenland,* where it continues all the year. The natives ufe the jugular pouch as a bladder to float their darts after they are flung The fkins are ufed in cloathing ; the flefh is eaten ; but the eggs are fo fetid as to be rejected, even by the very *Greenlanders.* Thefe birds are taken either by darts on the water ; by fnares dropt down the precipices, and placed before their haunts ; or, in winter, they are taken while afleep upon the ice. Are found in all the temperate latitudes of the *Ruffian* empire, and in immenfe numbers on the fhores of the RUSSIAN EMPIRE. *Cafpian* fea *. Reach even to *Kamtfchatka.* I believe this to be the   PLACE.   GREENLAND.

* *Extracts* i. 164.—ii. 405.

the

the kind which the *Chinese* train for fishing. They keep numbers, which fit on the edge of their boats; and, on a fignal given, plunge under water, and bring up their prey, which they are unable to fwallow, by reafon of a ring placed by their mafters round their necks *

Mrs. *Blackburn* received this fpecies from *New York*. There are great flocks in *Carolina*, efpecially in *March* and *April*, when the herrings run up the creeks; at which time they fit fifhing on the logs of wood which have fallen into the water †.

510. GANNET.	*Br. Zool.* ii. N° 293.—Pelecanus Baffanus. *Nautis,* Jaen Von Gent. *Faun. Suec.* N° 147.—*Latham,* iii. —LEV. MUS.—BL. MUS.

P With a ftrait dirty white bill, jagged at the edges: beneath the chin a naked black fkin, dilatable fo as to contain five or fix herrings: hind part of the head buff-colored: baftard wings and primaries dufky: all the reft of the plumage pure white: toes black, marked before with a pea-green ftripe: feathers of the tail fharp-pointed. WEIGHT feven pounds. LENGTH three feet one inch. EXTENT fix feet two.

PLACE.

Inhabits the coaft of *Newfoundland*; where it breeds, and migrates fouthward as far as *South Carolina*. The head of the bird which *Catefby* has engraven, and called the GREATER BOOBY, i. tab. lxxxvi. is of one in its young ftate. At that period it is deep afh-colored, fpotted with white. In *Europe* it is common on the coaft of *Norway* and *Iceland* ‡; but as it never voluntarily flies over land, is not feen in the *Baltic*. Wanders for food as far as the coaft of *Lifbon*, and *Gibraltar*, where it has been feen in *December*, plunging for *Sardinæ*. Straggles as high as *Greenland* ||. In northern *Afia*, it has been once feen by *Steller* off *Bering*'s ifle; but has been frequently met with in the fouthern hemifphere, in the *Pacific* ocean;

* *Du Halde,* i. 316.   † *Lawfon,* 150.   ‡ *Olaf. Iceland.* || *Faun. Groenl.* p. 92.

9   particularly,

particularly, in numbers about *New Zealand* and *New Holland* *
Captain *Cook* also saw them in his passage from *England* to the *Cape* of
Good Hope †, and remoter from land than they had been seen elsewhere. Among those observed in the South Sea, is the variety called
Sula ‡, with a few black feathers in the tail and among the secondaries. Found not only on the *Feroe* islands, but on our coasts, one
having been brought to me a few years ago, which had fallen down
wearied with its flight. A most ample account of the manners of the.
GANNET is given in the *Br. Zool.*

A. CRESTED CORVORANT. SHAG, *Br. Zool.* ii. N° 292.—*Latham*, iii.
Pelecanus Criſtatus. Top-ſkarv. *Brunnich*, N° 123.—*Faun. Groenl.* N° 58.—LEV. MUS.
—BL. MUS.

P. With a narrow dusky bill, hooked at the end : irides fine
green : on each side of the head is a long tuft of dusky feathers reaching beyond the crown : head, neck, and lower part of
the back, of a fine and glossy green : the upper part of the back, and
coverts of the wings, of the same color, edged with purplish black :
belly dusky : tail consists of twelve feathers, dusky tinged with
green. LENGTH two feet three. EXTENT three feet six. WEIGHT
three pounds three quarters.

Inhabits, in *Great Britain*, the vast precipices about *Holyhead* ; PLACE.
and is found in *Norway*, *Iceland* ||, and in the south of *Greenland* ¶ ;

---

* *Cook's First Voy.* ii. 382.—iii. 439, 627.   † *Cook's Voy. towards the South
Pole*, i. 10, 11.   ‡ *Wil. Orn.* 331. || *Olaffen.* ii. tab. xxxix.   ¶ *Faun.
Groenl.* N° 58.

but

but in the latter is scarce. The places which it inhabits are covered with its filthy excrements. The *Greenlanders* therefore call it *Ting-mingkpot*, or the bird *afflicted with a looseness*. It differs from the Shag in having a crest, and in being lesser. The *Norwegians* are well acquainted with both species, and distinguish them by different names[*]. I have seen several of the Shags shot among the *Hebrides*, but not one was crested. On the authority of the northern naturalists, I therefore separate them.

B. **Violet Corvorant.** Pelecanus Violaceus, **Pallas** MS. *List.*—*Latham*, iii.

PLACE.

P. With the body wholly black, glossed with violet color. Found about *Kamtschatka* and the isles.

C. **Red-faced Corvorant.** Ouril of the Kamtschatkans, *Descr. de la Kamtschatka,* 493.—*Latham*, iii.

P. With a slender bill ; upper mandible black ; lower red : from the bill to the eyes is a space covered with a blueish red naked skin : round each eye a white cutaneous circle : head crested : head, neck, and middle of the back, of a deep glossy green : on the fore part of the neck a few white slender feathers : sides of the back and scapulars glossed with purple : wings dusky : belly glossed with green : tail, consisting of twelve feathers only, is dusky : over each thigh is a tuft of white feathers : legs black. Length of one I measured thirty-one inches. *Steller* compares its size to that of a Goose.

PLACE.   Inhabits the high precipices on the coasts of *Kamtschatka*. Is very slow in rising ; but when on wing, flies most rapidly. Feeds on fish. During night they sit in rows on the cliffs, and often in their sleep fall off, and become the prey of *Arctic* Foxes ; who lie in

---

[*] *Brunnich*, N° 121, 123.

wait for thefe birds, which are a favorite food of thofe animals. They lay in *June*. Their eggs are green, and of the fize of thofe of a Hen. They are very bad tafted, and are not eafily dreffed ; yet are fo acceptable to the *Kamtfchatkans*, that, at the hazard of their necks, they will climb to the moft dangerous places in fearch of them, and often fall and lofe their lives. They catch thefe birds with nets, in which they are entangled in the places where they reft. They are alfo caught in fnares, with a running noofe hung to the end of a pole, with which the fowlers creep quietly to-wards the birds, and fling it round their necks, and draw them up the rock. The reft of the flock are fo ftupid, that, notwith-ftanding they fee the fate of their companions, they remain, fhak-ing their heads, on the fame fpot, till they are all taken. The flefh is exceffively hard and finewy. The *Kamtfchatkans* cook it af ter their fafhion, by putting the bird, without plucking or gutting into a hole filled with fire ; and when it is done enough, draw off the fkin, and make on it a favory repaft.

HAVING gone through the clafs of birds, let me remark, that there is the greateft probability, that numbers of thofe of *Kamtfchatka* are common to *North America* ; and that they pafs there the feafons of migration ; but not having actual proof of their being found on the new continent, I am obliged to place them in thefe appendages to each genus. The time may come, when it will be found neceffary to remove them into the *American* fections. It is alfo likely, that num-bers may feek a more fouthern retreat, and ftock *Japan* and *China* with their periodical flocks. I have done as much as the lights of my days have furnifhed me with. In fome remote age, when the *Britifh* offspring will have pervaded the whole of their vaft conti-nent, or the defcendants of the hardy *Ruffians* colonized the weftern parts from their diftant *Kamtfchatka*, the road in future time to new

4 F

conquefts :

conquefts : after, perhaps, bloody contefts between the progeny of *Britons* and *Ruffians*, about countries to which neither have any right ; after the deaths of thoufands of clamants, and the extirpation of the poor natives by the fword, and new-imported difeafes, a quiet fettlement may take place, civilization enfue, and the arts of peace be cultivated : learning, the luxury of the foul, diffufe itfelf through the nation, and fome naturalift arife, who, with fpirit and abilities, may explore each boundary of the ocean which feparates the *Afiatic* and *American* continents ; may render certain what I can only fufpect ; and, by his obfervations on the feathered tribe, their flights and migrations, give utility to mankind, in naval and œconomical operations, by auguries which the antients knew well to apply to the benefit of their fellow-creatures. He may, perhaps, fmile on the labors of the *Arctic* Zoologift (if by that time they are not quite obfolete) ; and, as the animate creation never changes her courfe, he may find much right ; and, if he is endowed with a good heart, will candidly attribute the errors to mifinformation, or the common infirmity of human nature.

I N D E X.

I N D E X.

Bunting,

I N D E X.

* A wrong repetition of name ; the Reader is therefore desired to distinguish by the addition of Second.

Dormouse,

INDEX.

* By inadvertency the word DUSKY is applied to this fpecies, a trivial before given to another Falcon: the Reader is therefore requefted to alter this with his pen.

black-headed

INDEX.

INDEX.

4 G   white-

INDEX.

I N D E X.

9

I N D E X.

Throftle

I N D E X.

I N D E X.

F I N I S.

ERRATA AND CORRECTIONS.

VOL. I.

Page IV, line 13, *for* but, *read* yet—P. XXVI, l. 31, fimiliarity, *read* fimilarity—
P. XXXII, l. 23, *Moura*, read *Mousa*—P. XXXVII, l. 2, *maen-hirion*, read *meini-hirion*—
P. XLII, l. 14, circumgirations, *read* circumgyrations—P. XLIII, laft line, *for* ‡ Same,
p. 7. § Same, p. 8. *Torfæus, &c.* ; read ‡ *Torfæus Hift. Norveg.* ii. p. 96.   § The
fame, p. 97—P. XLVI, l. 11, the laft to 1766, *read* the laft period it remained quief-
cent to 1766. l. 16, overflown, *read* overflowed—P. LVII, l. 16, *amata. Donec* ; read
amata donec. l. 19, *vidit*, read *vident*—P. LXII, l. 31, is, *read* are—P. LXIII, l. 18,
as low as that of 60, *read* and that of 60—P. LXXVI, l. 14, *Plearonectes*, read *Pleu-
ronectes*—P. LXXXII, l. 29, *infert, after the word* places, *the mark of reference* ‡, *and
blot it out of line* 31—P. LXXXVI, l. 13, 14, fmall and hard, *read* hard and fmall—
P. XCI, l. 26, *Lapes*, read *Lepas* ; l. 28, *carinotum*, read *carinatum.* l. 36, fee p. LV—
P. XCIX, l. 5, *dele* is—P. CIII, l. 10, *Salmon,* read *Salmo*—P. CVI, l. 6, yet is, *read*
which yet is—P. CVII, laft line, after *baccata*, add *Pallas Itin.* iii. 105   *Fl. Ross.*
23. tab. x—P. CVIII, note *, *read* COOK's *Voyage*—P. CXIV, l. 22, *Virg.* thofe, *read
Virg.* are diftinguifhed thofe—P. CXVI, l. 23, hieraciodes, *read* hieracioides—P. CXVIII,
l. 30, finally, of thofe, *read* finally, thofe — P. CXX. l. 10, is, *read* are — P. CXXIII,
laft line, 261, *read* 201—P. CXXXII, l. 28, *dele* either—P. CXLIV, l. 18, fhall, *read*
fhould—P. CLXVI, l. 24, had in the, *read* had been in the. l. 31, *dele* from—
P. CLXXIV, *after* N° 73, *add* 74 ; *after* N° 75, *add* 76 ; *after* N° 77, *add* 78—P. CXCI,
l. 1, œtus, *read* fœtus. l. 18, ovaria, *read* ova, l. 20, northernly, *read* northern—
P. CC, l. penult. for ; *read* ,

P. 3, l. 24, *Mivera*, read *Quivera*—P. 24, l. 9, Kungus, *read* Kungur—P. 33, l. 11,
is, *read* are—P. 34. note, *for* 9, 44 or 45, read 20, *read* lat. 60 to 20—P. 43, l. 23,
latter, *read* others—P. 50, l. 22, *add* The Lynx alfo inhabits the vaft forefts of the north
of *Europe* and *Asia* ; in the firft, as high as *Lapland*, in the laft, in moft parts of *Sibiria*, and even in the north of *India*, amidft the lofty mountains which bound that
country—P. 58, l. 26, carnivorous, *read* animal—P. 76, l. 16, *dele* in great plenty—
P. 89, l. 10, lat. 44, *read* 49—P. 90, l. 27, £. 25. *read* £. 20.—P. 98, l. 15, all
round, *read* in all parts of—P. 99, l. 3, Konyma, *read* Kowyma—P. 112, l. 23, *Hift.
Quad.* N° 265—P. 116, note *, *Hift. Quad.* 283. α.—P. 142, l. 16, *Sweden*, in the, *read
Sweden.* In the

VOL. II.

P. 220, l. 26, E DUSKY, *read* E GREENLAND—P. 223, l. 21, *Sea Eagle*, read
Ofprey—P. 244, l. 7, *for* north, *read* fouth—P. 368, l. 5, cychromi, *read* cychrami
—P. 407, l. 18, le, *read* la—P. 527, l. 18, *Non*, read *Nam :* and *dele* ?

OMITTED at p. 285, VOL. II.

L'Oifeau pourpre à bec de grimpereau, *De Buffon*, v. 526.—*Latham*, ii. 723.

175. A.
PURPLE CREEPER.

[C]R. wholly of a purple color. Length four inches and a half.
According to *Seba*, it inhabits *Virginia* ; and is faid to fing
ell.

Printed in the United States
By Bookmasters